Development of Plants

Development of Plants

Iqbal Hussain

Oxford Book Company
Jaipur ▪ New Delhi

Published By

Oxford Book Company

Regd. Off. "Bony Residency", Gate No. 2
Opp. Tilak Public School, Vishweswariya Nagar,
Gopalpura Road, Jaipur - 302018, India
Ph: 0141 - 2761280, Telefax: 2761381
email - oxfordbookjpr@gmail.com
website - oxfordbookcompany.com

Delhi Office: 102, 1st Floor, Satyam House,
4327/3, Ansari Road, Darya Ganj,
New Delhi - 110002, India
Ph: 011-45652440

Distributed by
KEYA INTERNATIONAL PRESS LLP
E-mail: keyainternationalpressllp@gmail.com

Published in India, 2025

ISBN : 9789380179292

Typeset by Shivangi Computers, Jaipur

Printed and bound in India by Replika Press Pvt. Ltd.

Preface

Plants that derive some or most of their nutrients (but not energy) from trapping and consuming animals or protozoans, typically insects and other Arthropods are known as carnivorous or insectivorous plants. These plants have made adaptations to grow in places where the soil is thin or poor in nutrients, especially nitrogen, such as acidic bogs and rock outcroppings. This trait of carnivory have evolved in at least ten separate lineages of plants, and these are now represented by more than a dozen genera in five families. Including about 625 species that attract and trap prey, produce digestive enzymes and absorb the resulting available nutrients.

Carnivorous plants use five basic types: pitfall traps, flypaper traps, shap traps, bladder traps and lobster-pot traps. These traps may be active or passive, depending upon whether movement aids the capture of prey. It has been studied that all trap types are modifications of a similar basic structure—the hairy leaf. Based on the radial symmetry of their flowers and their insect traps, these plants are placed in Droseraceae and Nepenthaceae in the order Nepenthales, but according to modern classification though the families are same but these are distributed in several diparate orders.

The present book makes an elaborate experimental study on various mechanisms of carnivorous plants: structure, movement of tentacles, aggregation of protoplasm, effects of heat, nitrogenous and non-nitrogenous fluids, salts, acids, alkaloids and other poisons, and vapours, sensitiveness and transmission of motor impulse. Basing the presentation on the common sun-dew, the book also describes various species such as Dionaea Muscipula, Aldrovanda Vesiculosa, etc. The book also makes an interesting presentation on various species of Drosophyllum, Pinguicula and Utricularia.

Iqbal Hussain

Contents

Contents

1

Drosera Rotundifolia, or The Common Sun-dew

Number of insects captured — Description of the leaves and their appendages or tentacles — Preliminary sketch of the action of the various parts, and of the manner in which insects are captured — Duration of the inflection of the tentacles — Nature of the secretion — Manner in which insects are carried to the centre of the leaf — Evidence that the glands have the power of absorption — Small size of the roots.

Number of Insects Captured

During the Summer we were surprised by finding how large a number of insects was caught by the leaves of the common sun-dew (*Drosera rotundifolia*) on a heath in Sussex. He had heard that insects were thus caught, but knew nothing further on the subject. He gathered by chance a dozen plants, bearing fifty-six expanded leaves and on thirty-one of these head insects or remnants of them adhered; and no doubt, many more would have been caught afterwards by these same leaves, and still more by those as yet not expanded. On one plant all six leaves, had caught their prey; and on several plants very many leaves had caught more than a single insect. On one large leaf were found the remains of thirteen distinct insects. Flies (Diptera) are captured much oftener than other insects. The largest kind which caught was a small butterfly (*Caenonympha pamphilus*); but the Rev. H.M. Wilkinson informed that he found a large living dragon-fly with its body firmly held by two leaves. As this plant is extremely common in some districts, the number of insect thus annually slaughtered must he prodigious. Many plants cause the death of insects, for instance, the sticky buds of the horse-chestnut *(Aesculus hippocastanum),* without thereby receiving, as far as we can perceive, any advantage; but it was soon evident that Drosera was excellently adapted for the special purpose of catching insects, so the subject seemed well worthy of investigation.

The results have proved highly remarkable; the more important ones being—firstly, the extraordinary sensitiveness of the glands to slight pressure and to minute doses of certain nitrogenous fluids, as shown by the movements of the so-called hairs or tentacles; second, the power possessed by the leaves or rendering soluble or digesting nitrogenous substances, and of afterwards absorbing them; thirdly, the changes which take places within the cells of the tentacles, when the glands are excited in various ways.

Leaves and Their Appendages/Tentacles

It is necessary, in the first place, to describe briefly the plants. It bears from two or three to five or six leaves generally extended more or less horizontally, but sometimes standing vertically upwards. The shape and general appearance of a leaf is shown, as seen in fig. 1, and as seen laterally, in fig. 2. The leaves are commonly a little broader than long, but this was not the case in the one here figured. The whole upper surface is covered with gland-bearing filaments, or tentacles, as we shall call them, from their manner of acting. The glands were counted on thirty-one leaves. But many of these were of unusually large size, and the average number was 192; the greatest number being 260, and the least 130. The glands are each surrounded by large drops of extremely viscid secretion, which, glittering in the sun, have given rise to the plant's poetical name of the sun-dew.

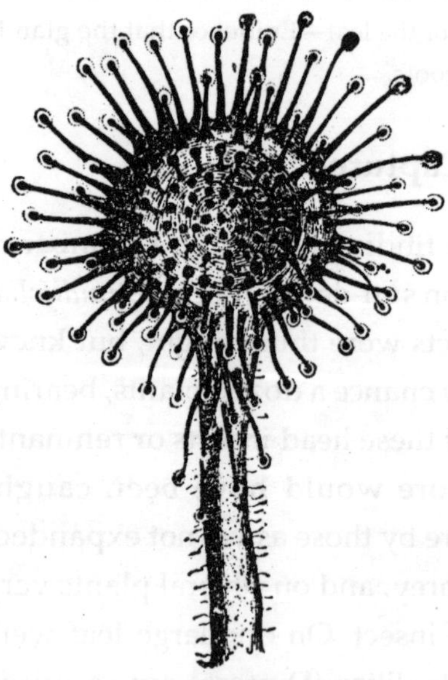

Fig. 1
(Drosera rotundifolia)
Leaf viewed from above; enlarged four times.

The tentacles on the central part of the leaf or disc are short and stand upright, and their pedicels are green. Towards the margin they become longer and longer and more inclined outwards, with their pedicels of a purple colour. Those on the extreme margin project in the same plane with the leaf, on more commonly (see fig. 2) are considerably reflexed. A few tentacles spring from the base of the footstalk or petiole, and these are the longest of all, being sometimes nearly ¼ of an inch in length. On a leaf bearing altogether 252 tentacles, the short ones on the disc, having green pedicels, were in number to the longer submarginal and marginal tentacles, having purple pedicels, as nine to sixteen.

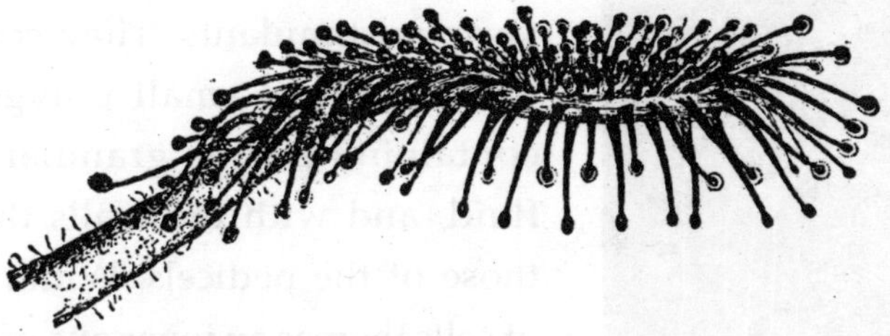

Fig. 2
(Drosera rotundifolia)
Old leaf viewed laterally; enlarged about five times.

A tentacle consists of a thin, straight, hair—like pedicel, carrying a gland on the summit. The pedicel is somewhat flattened, and is formed of several rows of elongated cells, filled with purple fluid of granular matter. There is, however, a narrow zone close beneath the glands of the longer tentacles, and a broader zone near their bases, of a green tint. Spiral vessels, accompanied by simple vascular tissue, branch off from the vascular bundles in the blade of the leaf, and run up all the tentacles into the glands.

Several eminent physiologists have discussed the homological nature of these appendages or tentacles, that is, whether they ought to be considered as hairs (trichomes) or prolongation of the leaf. Nitschke has shown that they include all the elements proper to the blade of a leaf; and the fact of their including vascular tissue was formerly thought to prove that they were prolongation of the leaf, but it is now known that vessels sometimes enter true hairs. The power of movement which they posses is a strong argument against their being viewed as hairs. The conclusion which seems the most probable will be given in chap. 15, namely, that they existed primordially as glandular hairs, or mere epidermic formation, and that their upper part should still be so considered; but that their lower part, which alone is capable of movement, consists of a prolongation of the leaf. We shall hereafter see that the terminal tentacles of the divided leaves of Roridula are still in an intermediate condition.

The glands, with the exception of those borne by the extreme marginal tentacles, are oval, and of nearly uniform size, *viz.*, about 4/500 of an inch in length. Their structure is remarkable, and their function complex, for they secrete, absorb, and are acted on by various stimulants. They consist of an outer layer of small polygonal cells, containing purple granular matter or fluid, and with the walls thicker than those of the pedicels. Within this layer of cells there is an inner one of differently shaped ones, likewise filled with purple fluid, but of a slightly different tint, and differently affected by chloride of gold. These two layers are sometimes well seen when a gland has been crushed or boiled in caustic potash. According to Dr. Warming, there it still another layer of much more elongated cells, as shown in the accompanying section (fig. 3) copied from his work; but these cells were not seen by Nitschke. In the centre there is a group of elongated, cylindrical cells of unequal lengths, bluntly pointed at their upper ends, truncated or rounded at their lower ends, closely pressed together, and remarkable from being surrounded by a spiral line, which can be separated as a distinct fibre.

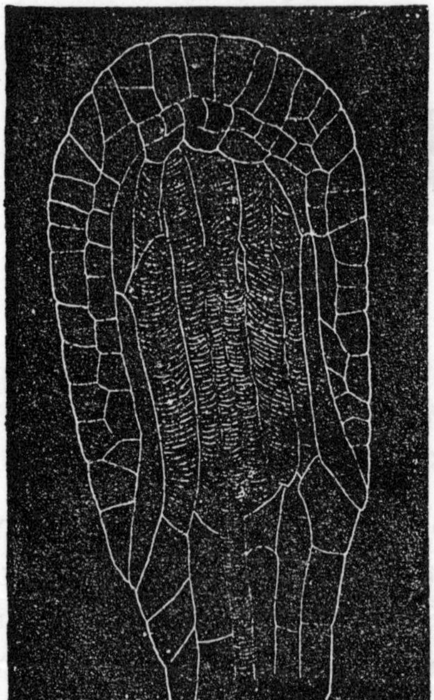

Fig. 3

(Drosera rotundifolia)

Longitudinal section of a gland; greatly magnified. From Dr. Warming.

These latter cells are filled with limpid fluid, which after long immersion in alcohol deposits much brown matter. It was presumed that they are actually connected with the spiral vessels which run up the tentacles, for on several occasions the latter were seen to divide into two or there excessively thin branches, which could be traced close up to the spiriferous cells. Their development has been described by Dr. Warming. Cells of the same kind have been observed in other plants, as reported by Dr. Hooker and were seen by in the margins of the leaves of Pinguicula. Whatever their function may be, they are not necessary for the secretion of the digestive fluid, or for absorption, or for the communication of a motor impulse to other parts of the leaf, as we may infer from the structure of the glands in some other genera of the Droseraceae.

The extreme marginal tentacles differ slightly from the others. Their bases are broader, and, besides their own vessels, they receive a fine branch from those which enter the tentacles on each side. Their glands are much elongated, and lie embedded on the upper surface of the pedicel, instead of standing at the apex. In other respects they do not differ essentially from the oval ones, and in one specimen was found every possible transition between the two states. In another specimen there were no long-headed gland. These marginal tentacles lost their irritability earlier than the others, and, when a stimulus is applied to the centre of the leaf, they are excited into action after the other. When cut-off leaves are immersed in water, they alone often become inflected.

The purple fluid, or granular matter which fills the cells of the glands, differs to a certain extent from that within the cells of the pedicels. For, when a leaf is placed in hot water or in certain acids, the glands become quite white and opaque, whereas the cells of the pedicels are rendered of bright red, with the exception of those close beneath the glands. These latter cells lose their pale red tint; and the green matter which they, as well as the basal cells, contain, becomes of a brighter green. The petioles bear many multicellular hairs, some of which near the blade are surmounted, according to Nitschke, by a few rounded cells, which appear to be rudimentary glands. Both surfaces of the leaf, the pedicels of the tentacles, especially the lower sides of the outer ones, and the petioles, are studded with minute papillae (hairs or trichomes), having a conical basis, and bearing on their summits two, and occasionally three, or even four, rounded cells, containing much protoplasm. These papillae are generally colourless, but sometimes include a little purple fluid. They vary in development, graduate, as Nitschke states, and as was repeatedly observed, into the long multicelluair hairs. The latter, as well as the papillae, are probably rudiments of formerly existing tentacles.

We may here add, in order not to recur to the papillae, that they do not secrete, but are easily permeated by various fluids; thus, when living or dead leaves are immersed in a solution of one part of chloride of gold, or of nitrate of silver, to 437 of water, they are quickly blackened, and the discoloration soon spreads to the surrounding tissue. The long multicellular hairs are not so quickly affected. After a leaf had been left in a weak infusion of raw meat for 10 hours, the cells of the papillae had evidently absorbed animal matter, for instead of limpid fluid they now contained small aggregated masses of protoplasm, which slowly and incessantly changed their forms. A similar result followed from an immersion of only 15 minutes in a solution of one part of carbonate of ammonia to 218 of water, and the adjoining cells of the tentacles, on which seated, not likewise contained aggregated masses of protoplasm. We may therefore conclude that, when a leaf has closely clasped a captured insect in the manner immediately to be described, the papillae, which project from the upper surface of the leaf and of the tentacles, probably absorb some of the animal matter dissolved in the secretion; but this cannot be the case with the papillae on the backs of the leaves or on the petioles.

Preliminary Sketch of the Action of the Several Parts and of the Manner in which Insects are Captured

If a small organic or inorganic object be placed on the glands in the centre of leaf, these transmit a motor impulse to the marginal tentacles. The nearer ones are first affected and slowly bend towards the centre, and then those farther off, until at last all become closely inflected over the object. This takes place in from one hour to four or five or more hours. The difference in the time required depends on many circumstances; namely, on the size of the object and on its nature, that is, whether it contains soluble matter of the proper kind; on the vigour and age of the leaf; whether it has lately been in action; and, according to Nitschke, on the temperature of the day, as likewise seemed to me to be the case. A living insect is a more efficient object than a dead one, as in struggling it presses against the glands of many tentacles. An insect such as a fly, with thin integuments, through which animal matter in solution can readily pass into the surrounding dense secretion, is more efficient in causing prologed inflection than an insect with a thick coat, such as a beetle. The inflection of the tentacles takes place indifferently in the light and darkness; and the plant is not subject to any nocturnal movement of so-called sleep.

If the glands on the disc are repeatedly touched or brushed, although no object is left on them, the marginal tentacles curve inwards. So again, if drops of various fluids, for instance of saliva or a solution of any salt of ammonia, are placed on the central glands, the same result quickly follows, sometimes in under half as hour.

Duration of Inflection of Tentacles

The tentacles in the act of inflection sweep through a wide space; thus a marginal tentacle, extended in the same plane with the blade, moves through an angle of 180°; and we have seen the much reflected tentacles of a leaf which stood upright move through an angle of not less than 270°. The bending part is almost confined to a short space near the base; but a rather larger portion of the elongated exterior tentacles becomes slightly incurved, the distal half in all cases remaining straight. The short tentacles in the centre of the disc, when

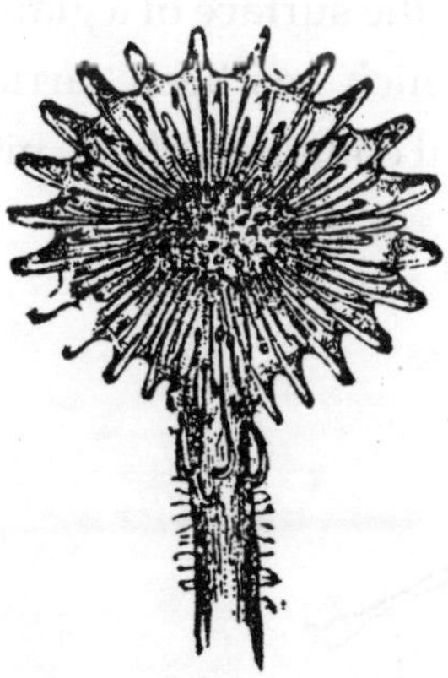

Fig. 4

(Drosera rotundifolia)

Leaf (enlarged) with all the tentacles closely inflected from immersion in a solution of phosphate of ammonia (one part to 87,500 of water).

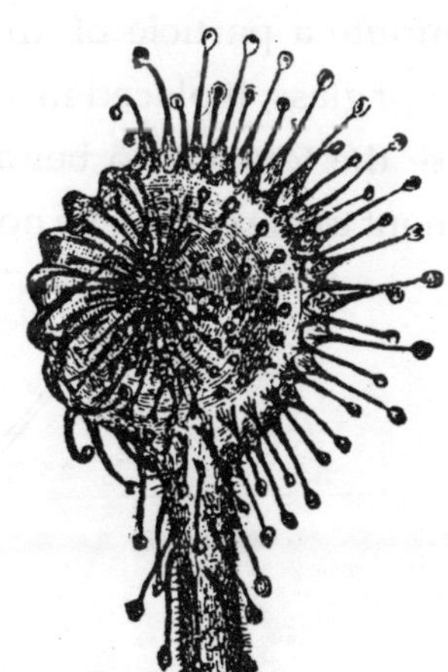

Fig. 5

(Drosera rotundifolia)

Leaf (enlarged) with the tentacles on one side inflected over a bit of meat placed on the disc.

directly excited, do not become inflected; if excited by a motor impluse received from other glands at a distance. Thus, if a leaf is immersed in an infusion of raw meat, or in a weak solution of ammonia (if the solution is at all strong, the leaf is paralysed), all the exterior tentacles bend inwards (see fig. 4) excepting those near the centre, which remain upright; but these bend towards any exciting object placed on one side of disc, as shown in fig. 5. The glands in fig. 4 may be seen to form a dark ring round the centre; and this follows from the exterior tentacles increasing in lengh in due proportion, they stand nearer to the circumference.

The kind of inflection which the tentacles undergo is best shown when the gland of one of the long exterior tentacles is in any way excited; for the surrounding ones remain unaffected. In the accompanying outline (fig.6) we see one tentacle, on which a particle of meat had been placed, thus bent towards the centre of the leaf, with two others retaining their original position. A gland may be excited by being simply touched three or four times, or by prolonged contact with organic or inorganic objects, and various fluids. I have distinctly seen, after an object had been placed on its gland; and we have often seen strongly pronounced inflection in under one minute. It is surprising how minute a particle of any substance, such as a bit of thread or hair or splinter of glass, if placed in actual contact with the surface of a gland, suffices to cause the tentacle to bend. If the object, which has been carried by this movement to the centre, be not very small, or if it contains soluble nitrogenous

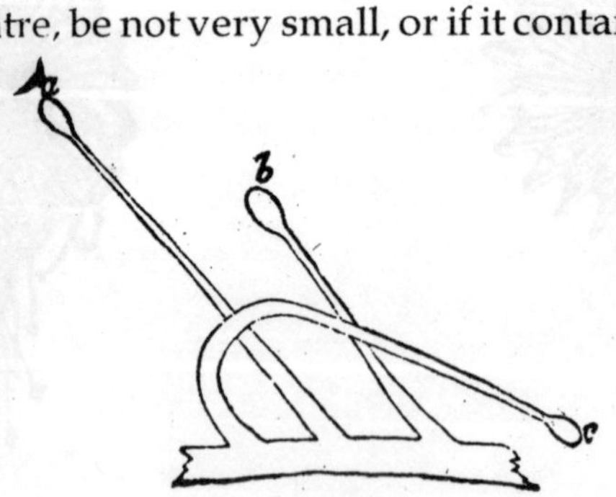

Fig. 6

(Drosera rotundifolia)

Diagram showing one of the exterior tentacles closely inflected; the two adjoining ones in their ordinary position.

matter, but acts on the central glands; and these transmit a motor impulse to the exterior tentacles, causing them to bend inwards.

Not only the tentacles but the blade of the leaf often but by no means always, becomes much incurved, when any strongly exciting substance or fluid is placed on the disc. Drops of milk and of a solution of nitrate of ammonia or soda are particularly apt to produce this effect. The blade is thus converted into a little cup. The manner in which it bends varies greatly. Sometimes the apex alone, sometimes one side, and sometimes both sides, becomes incurved. For instance we placed bits of hard-boiled egg on three leaves; one had the apex bent towards the base; the second had both distal margins much incurved, so that it became almost triangular in outline, and this perhaps is the commonest case; whilst the third blade also generally rises or bends upwards, and thus forms a smaller angle with the footstalk than it did before. This appears at first sight a distinct kind of movement, but it results from the incurvation of that part of the margin which is attached to the footstalk, causing the blade, as a whole, to curve or move upwards.

The length of time during which the tentacles as well as the blade remain inflected over an object placed on the disc, depends on various circumstances; namely, on the vigour and age of the leaf, and, according to Dr. Nitschke, on the temperature, for during cold weather, when the leaves are inactive, they re-expand at an earlier period than when the weather is warm. But the nature of the object is by far the most important circumstance; we have repeatedly found that the tentacles remain clasped for a much longer average time over objects which yield soluble nitrogenous matter than over those, whether organic or inorganic, which yield no such matter. After a period varying from one to seven days, the tentacles and blade re-expand, and are then ready to act again. We have seen the same leaf inflected three successive times over insects placed on the disc; and it would probably have acted a greater number of times.

Nature of Secretion

The secretion from the glands is extremely viscid, so that it can be drawn out into long threads. It appears colourless, but stains little balls of paper pale pink. An object, to secrete more freely; but the mere presence of the object renders this difficult to ascertain. In some cases, however, the effect was strongly marked, as when particles of sugar were added; but the result in this case is probably due merely to exosmose. Particles of carbonate and phosphate of ammonia and of some other salts, for instance sulphate of zinc, likewise increase the secretion. Immersion in a solution of one part of chloride of gold, or of some other salts, to 437 of water, excites the glands to largely increased secretion; on the other hand, tartrate of antimony produces no such effect. Immersion in many acids (of the strength of one part to 437 of water) likewise causes a wonderful amount of secretion, so that, when the leaves are lifted out, long ropes of extremely viscid fluid hang from them. Some acids, on the other hand, do not act in this manner. Increased secretion is not necessarily dependent on the inflection of the tentacle, for particles of sugar and of sulphate of zinc cause no movement.

It is a much more remarkable fact, that when an object, such as a bit of meat or an insect, is placed on the disc of a leaf, as soon as the surrounding tentacles become considerably inflected, their glands pour forth an increased amount of secretion. We ascertained this by placing bits of meat on one side of the disc; and as soon as the tentacles on this side became much inflected, but a record was kept of only thirteen cases, in nine of which increased secretion was plainly observed; the four failures being due either to the leaves being rather torpid, or to the bits of meat being too small to cause much inflection. We must therefore conclude that the central glands, when strongly excited, transmit some influence to the glands of the circumferential tentacles, causing them to secrete more copiously.

It is a still more important fact (as we shall see more fully when we treat of the digestive power of the secretion), that when the tentacles become inflected, owing to the central glands having been stimulated mechanically, or by contact with animal matter, the secretion not only increases in quantity, but changes its nature and becomes acid; and this occurs before the glands have touched the object on the centre of the leaf. This acid is of a different nature from that contained in the tissue of the leaves. As long as the tentacles remain closely inflected, the glands continue to secrete, and the secretion in acid; so that, if neutralised by carbonate of soda, it again becomes acid after a few hours. We have observed the same leaf with the tentacles closely inflected over rather indigestible substances, such as chemically prepared casein, pouring forth-acid secretion for eight successive days and over bits of bone for ten successive days.

The secretion seems to possess, like the gastric juice of the higher animals, some antiseptic power. During very warm weather we placed close together two equal-sized bits of raw meat, one on a leaf of the Drosera, and the other surrounded by wet moss. They were thus left for 48 hrs, and then examined. The bit on the moss swarmed with infusoria, and was so much decayed that the transverse striae on the muscular fibres could no longer be clearly distinguished; whilst the bit on the leaf, which was bathed by the secretion, was free from infusoria and its striae were perfectly distinct in the central and undissolved portion. In like manner small cubes of albumen and cheese placed on wet moss became threaded with filaments of mould, and had their surfaces slightly discoloured and disintegrated; whilst those on the leaves of Drosera remained clean, the albumen being changed into transparent fluid.

As soon as tentacles, which have remained closely inflected during several days over an object, begin to re-expand, their glands secrete less freely, or cease to secrete, and are left dry. In this state they are covered with a film of whitish, semi-fibrous matter, which was held in solution by the secretion. The drying of the glands during the act of re-expansion is of some little service to the plant; for it was often observed that objects adhering to the leaves could then be blown away by a breath of air; the leaves being thus left unencumbered and free for future action. Nevertheless, it often happens that all the glands do not become completely dry; and in this case delicate objects, such as fragile insects, are sometimes torn by the re-expansion of the tentacles into fragments, which remain scattered all over the leaf. After the re-expansion is complete, the glands quickly begin to re-secrete, and, as soon as full-size drops are formed, the tentacles are ready to clasp a new object.

Manner in which Insects are Carried to the Centre of the Leaf

When an insect alights on the central disc, it is instantly entangled by the viscid secretion and the surrounding tentacles after a time begin to bend and ultimately clasp it on all sides. Insects are generally killed, according to Dr. Nitschke, in about a quarter of an hour, owing to their tracheae being closed by the secretion. If an insect adheres to only a few of the glands of the exterior tentacles, these soon become inflected and carry their prey to the tentacles next succeeding them inward; these then bend inwards, and so onwards, until insect is ultimately carried by a curious sort of rolling movement to the centre of the leaf. Then, after an interval, the tentacles on all sides become inflected and bathe their prey with their secretion, in the same manner as if the insect had first alighted on the central disc. It is surprising how minute an insect suffices to cause this action: for instance, we have seen one of the smallest species of gnats (culex), which had just settled with its excessively delicate feet on the glands of the outermost tentacles, and these were already beginning to curve inwards, though not a single gland has as yet touched the body of the insect. Had we not interfered, this minute gnat would assuredly have been carried to the centre of the leaf and been securely clasped on all sides. We shall hereafter see what excessively small doses of certain organic fluids and saline solutions cause strongly marked inflection.

Whether insects alight on the leaves by mere chance, as a resting-place, or are attracted by the odour of the secretion is not clear. We suspect, from the number of insects caught by the English species of Drosera, and from what we have observed with some exotic species kept in greenhouse, that the odour is attractive. In this latter case the leaves may be compared with a baited trap; in the former case with a trap laid in a run frequented by game, but without any bait.

Evidence that Glands have the Power of Absorption

That the glands possess the power of absorption, is shown by their almost instantaneously becoming dark-coloured when given a minute quantity of carbonate of ammonia; the change of colour being chiefly or exclusively due to the rapid aggregation of their contents. When certain other fluids are added, they become by the widely different results which follow, from placing drops of various nitrogenous and non-nitrogenous fluids of the same density of the glands of the disc, or on a single marginal gland; and likewise by the very different lengths of time during which the tentacles remain inflected over objects, which yield or do not yield soluble nitrogenous matter. This same conclusion might indeed have been inferred from the structure and movements of the leaves, which are so admirably adapted for capturing insects.

The absorption of animal matter from captured insects explains how Drosera can flourish in extremely poor peaty soil—in some cases where nothing but sphagnum moss grows, and moss depend altogether on the atmosphere for their nourishment. Although the leaves at a hasty glance do not appear green, owing to the purple colour of the tentacles, yet the upper and lower surfaces of the blade, the pedicels of the central tentacles, yet the upper and lower surfaces of blade, the pedicels of the central tentacles, and the petioles contain chlorophyll, so that, no doubt, the plant obtains and assimilates carbonic acid from the air. Nevertheless, considering the nature of the soil where it grows, the supply of nitrogen would be extremely limited, or quite deficient, unless the plant had the power of obtaining this important element from captured insects.We can thus understand how it is that the roots are so poorly developed. These usually consist of only two or three slightly divided branches, from half to one inch in length, furnished with absorbent hairs. It appears therefore, that the roots serve only to imbibed water; though, no doubt, they would absorb nutritious matter if present in the soil; for as we shall hereafter see, they absord a weak solution of carbonate of ammonia. A plant of Drosera, with the edges of its leaves curled inwards, so as to from a temporary stomach, with the glands of the closely inflected tentacles pouring forth the acid secretion, which dissolves animal matter, afterwards to be absorbed, may be said to feed like an animal. But, differently from an animal, it drinks by means of its roots; and it must drink largely, so as to retain many drops of viscid fluid round the glands, sometimes as many as 260, exposed during the whole day to a glaring sun.

In one of our experiments plants were prevented from catching insects from themselves by means of covering of fine gauze, so that the only animal food which they obtained was supplied in very minute pieces of roast meat given to the "fed" plants but withheld from the "starved" ones. After only 10 days the difference between the fed and starved plants was clearly visible : the fed plants were of brighter green and the tentacles of a more lively red. At the end of August the plants were compared by number, weight, and measurement, with the following striking results:

	Starved	Fed
Weight (without flower-stems)	100	121.5
Number of flower-stems	100	164.9
weight of stems	100	231.9
Number of capsules	100	194.4
Total calculated weight of seed	100	379.7
Total calculated number of seeds	100	241.5

These results show clearly enough that insectivorous plants derive great advantage from animal food. It is of interest to note that the most striking difference between the two sets of plants is seen in what relates to reproduction—*i.e.,* in the flower-stems, the capsules, and the seeds.

After cutting off the flower-stems, their sets of plants were allowed to rest throughout the winter, in order to test (by a comparison of spring-growth) the amounts of reserve material accumulated during the summer. Both starved and fed plants were kept without food until April 3rd when it was found that the average weights per plants were 100 for the starved, 213.0 for the fed. This proves that the fed plants had laid by a far greater store of reserve material in spite of having produced nearly four times as much seed.

In Kellermann and Von Raumer's experiments aphides were used as food instead of meat—a method which adds greatly to the value of their results. Their conclusions are similar to my own, and they show that not only is the seed production of the fed plants greater, but they also form much heavier winter-buds than the starved plants.

Dr. M. Bülsgen has more recently published an interesting paper on the same subject. His experiments have the advantage of having been made on young Droseras grown from seed. The unfed plants are thus much more effectually starved than in experiments on full-frown plants possessing already a store of reserve matter. It is therefore not to be wondered at that Büsgen's results are more striking than Kellermann's and Von Raumer's—thus, for instance, he found that the "fed" plants, as compared with the starved ones, produced more than five times as many capsules, while my figures are 100:194. Büsgen given a good *resume* of the whole subject, and sums up by saying that the demonstrable superiority of fed over unfed plants is great enough to render comprehensible the organisation of the plants with reference to the capture of insects.

❑❑❑

2

The Movements of the Tentacles from the Contact of Solid Bodies

Inflection of the exterior tentacles owing to the glands of the disc being excited by repeated touches, or by objects left in contact with them – Difference in the action of bodies yielding and not yielding soluble nitrogenous matter – Inflection of exterior tentacles directly caused by objects left in contact with their glands – Periods of commencing inflection and of subsequent re-expansion – Extreme minuteness of the particles causing inflection – Action under water – Inflection of the exterior tentacles when their glands are excited by repeated touches – Falling drops of water do not cause inflection.

Introduction

We will give in this and the following chapters some of the many experiments made, which best illustrate the manner and rate of movement of the tenacles, when excited in various ways. The glands alone in all ordinary cases are susceptible to excitement. When excited they do not themselves move or change form, but transmit a motor impulse to the bending part of their own and adjoining tentacles, and are thus carried towards the centre of the leaf. Strictly speaking, the glands ought to be called irritable, as the term sensitive generally implies consciousness; but none supposes that the sensitive-plant is conscious, and, as we have found the term convenient, we shall use it without scruple. We will commence with the movements of the exterior tentacles, when indirectly excited by stimulants applied to the glands of the short tentacles on the disc. The exterior tentacles may be said in this case to be indirectly excited, because their own glands are not directly acted on. The stimulus proceeding from the glands of the disc acts on the bending part of the exterior tentacles, near their bases, and does not (as will hereafter be proved) first travel up the pedicels to the glands, to be then reflected back to the bending place. Neverthless, some influence does travel up to the glands, causing them to secrete more copiously, and the secretion to become acid. This latter fact is quite new in the physiology of plants; it has indeed only recently been established that in the animal kingdom an influence can be transmitted along the nerves to glands, modifying their power of secretion, independently of the state of the blood-vessels.

The Inflection of the Exterior Tentacles from Glands of the Disc Being Excited by Repeated Touches, or by Objects Left in Contact with Them

The central glands of a leaf were irritated with a small stiff camel-hair brush, and in 70 m. (minutes) several of the outer tentacles were inflected; in 5hrs. (hours) all the sub-marginal tentacles were inflected; next morning after an interval of about 22 hrs., they were fully re-expanded. In all the following cases the period is reckoned from the time of first irritation. Another leaf treated in the same manner had a few tentacles inflected in 20 m.; in 4 hrs. all the sub-marginal and some of the extreme marginal tentacles, as well as the edge of the leaf itself, were inflected; in 17 hrs. They had recovered their proper, expanded position. We then put a dead fly in the centre of the last-mentioned leaf, and next morning it was closely clasped; five days afterwards the leaf re-expanded, and the tentacles, with their glands surrounded by secretion, were ready to act again.

Particles of meat, dead flies bits of paper, wood dried moss, sponge, cinders, glass, &c., were repeatedly placed on leaves, and these objects were well embraced in various periods from 1 hr. to as long as 24 hrs., and set free again, with the leaf fully re-expanded, in from one or two, to seven or even ten days, according to the nature of the object. On a leaf which had naturally caught two flies, and therefore had already closed and reopened either once, or more probably twice, we put a fresh fly: in 7 hrs. it was moderately, and in 21 hrs. thoroughly well, clasped, with the edges of the leaf inflect. In two days and a half the leaf had nearly re-expanded; as the exciting object was an insect, this unusually short period of inflection was, no doubt, due to the leaf having recently been in action. Allowing this same leaf to rest for only a single day, we put on another fly, and it again closed, but now very slowly; nevertheless, in less than two days it succeeded in thoroughly clasping the fly.

When a small object is placed on the glands of the disc, on one side of a leaf, as near as possible to its circumference, the tentacles on this side are first affected, those on the opposite side much later, or, as after occurred, not at all. This was repeatedly proved by trials with bits of meat; but we will here give only the case of a minute fly, naturally caught and still alive, which was found adhering by its delicate feet to the glands on the extreme left side of the central disc. The marginal tentacles this side closed inwards and killed the fly, and after a time the edge of the leaf of this side also became inflected, and thus remained for several days. Whilst neither the tentacles nor the edge on the opposite side were in the least affected.

If young and active leaves are selected, inorganic particles not larger than the head of a small pin, placed on the central glands, sometimes cause the outer tentacles to bend inwards. But this follows much more surely and quickly, if the objects contains nitrogenous matter which can be dissolved by the secretion. On one occasion we observed the following unusual circumstance. Small bits of raw meat (which acts more energetically than any other substance), of paper, driedmoss, and of the quill of a pen were placed on several leaves, and they were all embraced equally well in about 2 hrs. On other occasion the above-named substances, or more commonly particles of glass, coal-cinder (taken from the fire), stone, gold-leaf, dried grass, cork, blotting-paper, cotton-wool, and hair rolled up into little balls, were used, and these substances, though they were sometimes well embraced, often caused no movement whatever in the outer tentacles, or an extremely slight and slow movement. Yet these same leaves w ere proved to be in an active condition, as they were excited to move by substances yielding soluble nitrogenous matter, such as bits of raw or roast meat, the yolk or white of boiled eggs, fragments of insects of all orders, spiders, and c. We will give only two instances. Minute flies were placed on the discs of several leaves, and on others ball of paper, bits of moss and quill of about the same size as the flies, and latter were well embraced in a few hours; whereas after 25 hrs. only a very few tentacles were inflected over the other objects. The bits of paper, moss, and quill were then removed from these leaves, and bits of raw meat placed on them; and now all the tentacles were soon energetically inflected.

Again, particles of coal cinder (weighing rather more than the flies used in the last experiment) were placed on the centers of three leaves: after an interval of 19 hrs. one of the particles was tolerably well embraced; a second by a very few tentacles; and a third by none. We then removed the particles from the two latter leaves. And put them on recently killed flies. These were fairly well embraced in 7 ½ hrs. and thoroughly after 20 ½ hrs. the tentacles remaining inflected for many subsequent days. On the other hand, the one leaf which had in the course of 29 hrs. embraced the bit of cinder moderately well, and to which no fly was given, after an additional 33 hrs. (*i.e.*, in 52 hrs. from the time when the cinder was put on) was completely re-expanded and ready to act again.

From these and numerous other experiments not worth giving, it is certain that inorganic substances, or such organic substances as are not attacked by the secretion, act much less quickly and efficiently than organic substances yielding soluble matter which is absorbed. Moreover, we have met with very few exceptions to the rule, and these exception apparently depended on the leaf having been too recently in action, that the tentacles remain clasped for a much longer time over organic bodies of the nature just specified than over those which are not acted on by the secretion, or over inorganic objects.

The Inflection of the Exterior Tentacles as Caused by Objects Left in Contact with Their Glands

A vast number of trials were made by placing, by means of a fine needle moistened with distilled water, and with the aid of a lens, particles of various substances on the viscid secretion surrounding the glands of the outer tentacles. We experimented on both the oval and long-headed glands. When a particle is thus placed on a single gland, the movement of the tentacle is particularly well seen in contrast with the stationary condition of the surrounding tentacles. (See previous fig 6.) In four cases small particles of raw meat caused the tentacles to be greatly inflected in between 5 and 6 m. Another tentacle similarly treated and observed with special care, distinctly, though slightly, changed its position in 10 s. (seconds); and this is the quickest movement seen. In 2 m. 30 s. it had moved through an angle of about 45°. The movement as seen through a lens resembled that of the hand of a large clock. In 5 m. it had moved through 90°and when we looked again after 10 m., the particle had reached the centre of the leaf; so that the whole movement was completed in less than 17 m. 30 s. In the course of some hours this minute bit of meat, from having been brought into contact with some of the glands of the central disc, acted centrifugally on the outer tentacles, which all became closely inflected. Fragments of flies were placed on the glands of four of the outer tentacles, extended in the same plane with that of the blade, and three of these fragments were carried in 35 m. through an angle of 180° to the centre. The fragment on the fourth tentacle was very minute, and it was not carried to the centre until 3 hrs. had elapsed. In three other cases minute flies or portions of large ones were carried to the centre in 1 hr. 30 s. In these seven cases, the fragments or small tentacle to the central glands, were well embraced by the other tentacles after an interval of from 4 to 10 hrs.

We also placed in the manner just described six small balls of writing paper (rolled up by the aid of pincors, so that they were not touched by fingers) on the glands of six exterior tentacles on distinct leaves; three of these were carried to the centre in about 1 hr., and the other three in rather more than 4 hrs; but after 24 hrs. only two of the six balls were well embraced by the other tentacles. It is possible that the secretion may have dissolved a trace of glue or animalised matter from the balls of paper. Four particles of coal-cinder were then placed on the glands of four exterior tentacles; one of these reached the centre in 3 hrs. 40 m.; the second in 9 hrs; whilst the fourth moved only a very short distance in 24 hrs., and never moved any farther. Of the above three bits of cinder which were ultimately carried to the centre, one alone was well embraced by many of the other tentacles. We here see clearly that such bodies as particles of cinder or little balls of paper, after being carried by the tentacles to the central glands, act very differently from fragments of flies, in causing the movement of the surrounding tentacles.

We made, without carefully recording the times of movement, many similar trials with other substances, such as splinters of white and blue glass, particles of cork, minute bits of gold-leaf, &c.; and the proportional number of cases varied much in which the tentacles reached the centre, or moved only slightly, or not at all. One evening, particles of glass and cork, rather larger than those usually employed, were placed on about a dozen glands, and next morning, after 13hrs, every single tentacle had carried its little load to the centre; but the unusually large size of the particles will account for this result. In another case, 6/7 of the particles of cinder, glass and thread, placed on separate glands, were carried towards, or actually, the centre; in another case 7/9 and in the last case only 7/26 were thus carried inwards, the small proportion being here due, at least in part, to the leaves being rather old and inactive. Occasionally a gland, with its light load, could be seen through a strong lens to move an extremely short distance and then stop; this was especially apt to occur when excessively minute particles, much less than those of which the measurements will be immediately given, were placed on glands; so that we here have nearly the limit of any action.

We were so much surprised at the smallness of the particles which caused the tentacles to become greatly inflected that it seemed worth while carefully to ascertain how minute a particle would plainly act. Accordingly, measured lengths of a narrow strip of blotting-paper, of fine cotton-thread, and of a women's hair, were carefully weighed by Mr. Trenhan Reeks, in an excellent balance, in the laboratory in Germany street. Short bits of the paper, thread, and hair were then cut off and measured by a micrometer, so that their weights could be easily calculated. The bits were placed on the viscid secretion surrounding the gland of the exterior tentacles, with the precautions already stated, and we are certain that the gland itself was never touched; nor indeed would a single touch have produced any effect. A bit of the blotting-paper, weighing 1/465 of a grain, was placed so as to rest on three glands together, and all three tentacles slowly curved inwards; each gland, therefore, supposing the weight to be distributed equally, could have been pressed on by only 1/1395 of a grain, or .0464 of a tried, and all acted. The shortest of these was 1/50 of an inch in length, and weighed 1/8197 of grain. The tentacle in this case was considerably inflected in 1 hr. 30 m., and the bit of thread was carried to the centre of the leaf in 1 hr. 40 m. Again, two particles of the thinner end of woman's hair, one of these being 18/1000 of an inch in length, and weighing 1/35714 of a grain, the other 19/1000 of an inch in length, and weighing of course a little more, were placed on two glands of opposite sides of the same leaf, and these two tentacles were inflected halfway towards the centre in 1 hr. 10 m; all the many other tentacles round the same leaf remaining motionless. The appearance of this one leaf showed in an unequivocal manner that these minute particles sufficed to cause the tentacles to bend. Altogether, ten such particles of hair were placed on ten glands on several leaves,.and seven of them caused the tentacles to move in a conspicuous manner. The smallest particle which was tried, and which acted plainly, was only 8/1000 of an inch (.203 millimeter) in length, and weighed the 1/78740 of a grain, or .000822 milligram. In these several cases, not only was the inflection of the tentacles conspicuous, but the purple fluid within their cells became greeted into little masses of protoplasm, in the manner to be described in the next chapter; and the aggregation was so plain that we could, by this clue alone, have readily picked out under the microscope all the tentacles which had carried their light loads towards the centre, from the hundreds of other tentacles on the same leaves which had not thus acted.

Our surprise was greatly excited, not only by the minuteness of the particles which caused movements, but how they could possibly act on the glands; for it must be remembered that they were laid with greatest care on the convex surface of the secretion. At first we thought—but, as we now know, erroneously—that particles of such low specify gravity as those cork, thread and paper, would never come into contact with the surfaces of the glands. The particles cannot act simply by their weight being added to that of the secretion, for small drops of water, many times heavier than the particles, were repeatedly added, and never produced any effect. Nor does the disturbance of the secretion produce any effect, for long threads were drawn out by a needle, and affixed to some adjoining object, and thus left for hours; but the tentacles remained motionless.

We also carefully removed the secretion from four glands with a sharply pointed piece of blotting-paper, so that they were exposed for a time naked to the air, but this caused no movement; yet glands were in an efficient state, for, after 24 hours had elapsed, they were tried with bits of meat, and all became quickly inflected. It then occurred that particles floating on the secretion would cast shadows on the glands, which might be sensitive to the interception of the light. Although this seemed highly improbable, as minute and thus splinters of colourless glass acted powerfully, nevertheless, after it was dark, we put on, by the aid of a single tallow candle, as quickly as possible, particles of cork and glass on the glands of a dozen tentacles, as well as some of meat on other glands, and covered them up so that not a ray of light could enter; but by next morning, after an interval of 13 hrs., all the particles were carried to the centres of the leaves.

These negative results led us to try many more experiments, by placing particles on the surface of the drops of secretion, observing, as carefully as we could, whether they penetrated it and touched the surface of the glands. The secretion, from its weight, generally forms a thicker layer on the under than on the upper sides of the glands, whatever may be the position of the tentacles. Minute bits of dry cork, thread, blotting-paper, and coal-cinders were tried, such as those previously employed; and we now observed that they observed much more of the secretion in the course of a few minutes, than we should have thought possible; and as they had been laid on the upper surface of the secretion, where it in thinnest, they were often drawn down, after a time, into contact with at least some one point of the gland. With respect to the minute splinters of glass and particles of hair, it was observed that the secretion slowly spread itself a little over their surfaces, by which means they were likewise drawn downwards or sideways, and thus one end, or some minute prominence, often came to touch, sooner or later, the gland.

In the foregoing and following cases, it is probable that the vibrations, to which the furniture in every room is continually liable, aids in bringing the particles into contact with glands. But as it was sometimes difficult, owing to the refraction of the secretion, to feel sure whether the particles were in contact, we tried the following experiment. Unusually minute particles of glass, hair, and cork were gently placed on the drops round several glands, and very few of the tentacles moved. Those which were not affected were left for about half an hour, and the particles were then disturbed or tilted up several times with a fine needle under the microscope, the glands not being touched. And now in the course of few minutes almost all the hitherto motion less tentacles began to move; and this, no double, was caused by one end or some prominence of the particles having come into contact with the surface of the glands. But, as the particles were unusually minute, the movement was small.

Lastly, some dark blue glass pounded into fine splinters was used, in order that the points of the particles might be better distinguished when immersed in the secretion; and thirteen such particles were placed in contact with the depending and therefore thicker part of the drops round so many glands. Five of the tentacles began moving after an interval of few minutes, and in these cases we clearly saw that the particles touched the lower surface of the glands. A sixth tentacle moved after 1 hr., 45m., and the particle was now in contact with the gland, which was not the case at first. So it was with the seventh tentacle, but its movement did not begin until 3 hrs. 45 m. had elapsed. The remaining six tentacles never moved as long as they were observed; and the particles apparently never came into contact with the surfaces of the glands.

From these experiments we learn that particles not containting soluble matter; when placed on glands, often cause the tentacles to begin bending in the course of from one to five minutes; and that in such cases the particles have been from the first in contact with the surfaces of the glands. When the tentacles do not begin moving for a much longer time, namely, from half an hour to three or four hours, the particles have been slowly brought into contact with the glands either by the secretion being absorbed by the particles or by its gradual spreading over them, together with its consequent quicker evaporation. When the tentacles do not move at all, the particles have never come into contant with the glands, or in some cases the tentacles may not have been in an active condition. In order to excite movement, it is indispensable that the particles should actually rest on the glands; for a touch once, twice or even thrice repeated by any hard body, is not sufficient to excite movement.

Another experiment, showing that extremely minute particles act on the glands when immersed in water, many here be given. A grain of sulphate of quinine was added to an ounce of water, which was not afterwards filtered; and, on placing three leaves in ninety minims of this fluid, we were much surprised to find that all three leaves were greatly inflected in 15 m.; for we knew from previous trials that the solution does not act so quickly as this. It immediately occurred that the particles of the undissolved salt, which were so light as to float about, might have come into contact with glands, and caused this rapid movement. Accordingly, we added to some distilled water a pinch of quite innocent substance, namely, precipitated carbonate of lime, which consists of an impalpable power; shook the mixture, and thus got a fluid like thin milk. Two leaves were immersed in it and in 6 m. almost every tentacle was much inflected. We placed one of these leaves under the microscope, and saw innumerable atoms of lime adhering to the external suffice to the secretion. Some, however, had penetrated it, and were lying on the surfaces of the glands; and no doubt it was these particles which caused the tentacles to bend. When leaf is immersed in water, the secretion instantly swells much; and we presume that it is ruptured here and there, so that little eddies of water rush in. If so, we can understand how the atoms of chalk, which rested on the surface of the glands, had penetrated the secretion. Any one who had rubbed precipitated chalk between his fingers will have perceived how excessively fine the power is. No doubt there must be a limit, beyond which a particle would be too small to act on a gland. We have often seen fibres and dust, which had failed from the air, on the glands of plants kept in my room, and these never induced any movement; but then such particles lay on the surface of the secretion and never reached the gland itself.

Finally, it is an extraordinary fact that little bit of soft thread, $\frac{1}{50}$ of an inch in length and weighing $\frac{8}{8197}$ of grain, or of a human hair, $\frac{8}{1000}$ of an inch in length and weighing only $\frac{1}{78740}$ of grain (.000822 milligram), or particles of precipitated chalk, after resting for short time on a gland, should induce some change in its cells, exciting them to transmit a motor impulse throughout the whole length of the pedicel, consisting of about twenty cells, to near its base, causing this part to bend, and the tentacle to sweep through an angle of above 180°. That the contents of the cells of the glands, and afterwards those of the pedicels, are affected in a plainly visible manner by the pressure of minute particles, we shall have abundant evidence when we treat of the aggregation of protoplasm. But the case in much more marketable than as yet stated; for the particles are supported by the viscid and dense secretion; nevertheless, even smaller ones than those of which the measurements have been given, when brought by an insensibly slow movement, through the means above specified, into contact with the surface of gland, act on it, and the tentacle bends. The pressure excerpted by the particle of hair, weighing only $\frac{1}{78740}$ of a grain and supported by a dense fluid, must have been inconceivably slight. We may conjecture that it could hardly have equaled the millionth of grain; and we shall hereafter see that far less than the millionth of grain of phosphate of ammonia in solution, when absorbed by a gland, acts on it and induces movement. A bit of hair, $\frac{1}{50}$ of an inch in length, and therefore much larger than those used in the above experiments, was not perceived when placed on my tongue; and it is extremely doubtful whether any nerve in the human body, even if in an inflamed condition, would be in any way affected by such a particle supported in dense fluid, and slowly brought into contact with the nerve. Yet the cells of the glands of Drosera are thus excited to transmit a motor to be that hardly any more remarkable fact than this has been observed in the vegetable kingdom.

The Inflection of the Exterior Tentacles, When Their Glands are Excited by Repeated Touches

We have already seen that, if the central glands are excited by being gently brushed, they transmit a motor impulse to the exterior tentacles, causing them to bend; and we have now to consider the effects which follow from the glands of the exterior tentacles begging themselves touched. On several occasion, a large number of glands were touched only once with a needle or fine brush, hard enough to bend the whole flexible tentacle; and though this must have caused a thousands-fold greater pressure than the weight of the above described particles, not a tentacle moved. On another occasion forty-five glands on eleven leaves were touched once, twice, or even thrice, with a needle or stiff bristle. This was done as quickly as possible, but with force sufficient to bent the tentacles; yet only six of the them became inflected – three plainly, and three in a slight degree. In order to ascertain whether these tentacles which where not affected were in an efficient state, bits of meat were placed on ten of them, and they all soon became greatly incurved. On the other hand, when a large number of glands were struck four, five, or six times with the same force as before, a needle or sharp splinter of glass being used, a much larger proportion of tentacles became inflected; but the result was so uncertain as to seem capricious. For instance, we struck in the above manner three glands, which happened to be extremely sensitive, and all three were inflected almost as quickly as if bits of meat had been placed upon them. On another occasion, we gave a single forcible touch to a considerable number of glands, and not one moved; but these same glands, after an interval of some hours, being touched four or five times with a needle, several of the tentacles soon became inflected.

The fact of a single touch or even of two or three touches not causing inflection must be of some service to the plant; as, during stormy weather, the glands cannot fail to be occasionally touched by the tall blades of grass, or by other plants growing near; and it would be great evil if the tentacles were thus brought into action, for the act of re-expansion takes a considerable time, and until the tentacles are re-expanded they cannot catch prey. On the other hand, extreme sensitiveness to slight pressure is of the highest service to the plant; for, as we have seen, if the delicate feet of a minute struggling insect press ever so lightly on the surfaces of two or three glands, the tentacles bearing these glands soon curl inwards and carry the insect with them to the centre, causing, after a time all the circumferential tentacles to embrace it. Nevertheless, the movements of the plant are not perfectly adapted to its requirements; for if a bit of dry moss, peat, or other rubbish, blown on to the disc, as often happens, the tentacles clasp it in a useless manner. They soon, however, discover their mistake and release such in nutritious objects.

It is also a remarkable fact, that drops of water falling from height, whether under the from of natural or artificial rain, do not cause the tentacles to move; yet the drops must strike the glands with considerable force, more especially after the secretion has been all washed away by heavy rain; and this often occurs, though the secretion is so viscid that it can be removed with difficulty merely by waving the leaves in water. If the falling drops of water are small, they adhere to the secretion, the weight of which must be increased in much greater degree, as before remarked, than by the addition of minute particles of solid matter; yet the drops never cause the tentacles to become inflected. It would obviously have been a great evil to the plant (as in the case of occasional touches) if the tentacles were excited to bend by every shower of rain; but this evil has been avoided by the glands either having become through habit insensible to the blows and prolonged pressure of drops of water, or to their having been originally rendered sensitive solely to the contact of solid bodies. We shall hereafter see that the impact of fluids, though exquisitely sensitive to momentary touches from any solid body.

When the pedicel of a tentacle is cut off by a sharp pair of scissors quite close beneath the gland, the tentacle generally becomes inflected. We tried this experiment repeatedly, as we were much surprised at the fact, for all other parts of the pedicels are insensible to any stimulus. These headless tentacles after a time re-expand. On the other hand, we occasionally succeeded in crushing a gland between a pair of pincers, but this caused no inflection. In this latter case the tentacles seem paralysed as likewise follows from the action of too strong solution of certain salts, and of too great heat, whilst weaker solution of the same salts and a more gently heat cause movement. We shall also see in future chapters that various other fluids, some vapour, and oxygen (after the plant has been for some time excluded from its action), all induce inflection, and this likewise results from an induced galvanic current.

□□□

3

Aggregation of the Protoplasm within the Cells of the Tentacles

Nature of the contents of the cells before aggregation – Various causes which excite aggregation – The process commences within the glands and travels down the tentacles – Description of the aggregated masses and of their spontaneous movements – Currents of protoplasm along the walls of cells – Action of carbonate of ammonia – The granules in the protoplasm which flows along the walls coalesce with the central masses – Minuteness of the quantity of carbonates of ammonia causing aggregation – Action of other saltes of ammonia – Of other substances, organic fluids, and c. – Of water – Of heat – Redissolution of the aggregated masses – Proximate causes of the aggregation of the protoplasm – Summary and concluding remarks – Supplementary observations on aggregation in the roots of plants.

Nature of the Contents of the Cells before Aggregation

If the tentacles of a young, yet fully matured leaf, that has never been excited or become inflected, be examined, the cells forming the pedicels are seen to be filled with homogeneous, purple fluid. The walls are lined by layer of colourless, circulating protoplasm; but this can be seen with much greater distinctness after the process of aggregation has been partly effected then before. The purple fluid which exudes from a crushed tentacle is somewhat coherent, and does not mingle with the surrounding water; it contains much flocculent of granular matter. But this matter may have been generated by the cells having been crushed; some degree of aggregation having been thus almost instantly caused.

If a tentacle is examined some hours after the gland has been excited by repeated touches, or by an inorganic or organic particle placed on it, or by the absorption of certain fluids, it presents a wholly changed appearance. The cells, instead of being filled with homogeneous purple fluid, now contain variously shaped masses of purple matter, suspended in a colouless or almost colourless fluid. The change is so conspicuous that it is visible through a weak lens, and even sometimes with the naked eye; the tentacles now have a mottled appearance, so that one thus affected can be picked out with case from all the other. The same result follows if the glands on the disc are irritated in any manner, so that the exterior tentacles become inflected; for their contents will then be found in an aggregated condition, although their glands have not as yet touched any object. But aggregation may occur independently of inflection, as we shall presently see. By whatever cause the process may have been excited, it commences within the glands, and then travels down the tentacles. It can be observed much more distinctly in the upper cells of the pedicels than within the glands, as these are somewhat opaque. Shortly after the tentacles have re-expanded, the little masses of protoplasm are all redissolved, and the purple fluid within the cells becomes as homogeneous and transparent as it was at first. The process of redissolution travels upwards from the bases of the tentacles to the glands, and therefore in reversed direction to that of aggregation. Tentacles in an aggregated condition were also seen by Prof. Huxley, Dr. Hooker, and Dr. Burden Sanders, who observed the changes under the microscope, and were much struck with the whole phenomenon.

The little masses of aggregated matter are of the most diversified shapes, often spherical or oval, sometimes much elongated, or quite irregular with thread- or necklace-like or club-formed projection. They consist of thick, apparently viscid matter, which in the exterior tentacles is of a purplish and in the short discal tentacles of greenish colour. These little masses incessantly change their forms and positions, being never at rest. A single mass will often separate into two, which afterwards reunite. Their movements are rather slow, and resemble those of Amoebae or of the white corpuscles of the blood. We may therefore conclude that they consist of protoplasm. If their shapes are sketched at intervals of few minutes, they are invariably seen to have undergone great changes of form; and the same cell has been observed for several hours. Eight rude, though accurate sketches of the same cell, made at intervals of between 2 m. or 3 m., are here given (fig. 7), and illustrate some of the simpler and commonest changes. The cell A, when first sketched, included two oval masses of purple protoplasm touching each other. These became separate, as shown at B, and then reunited, as

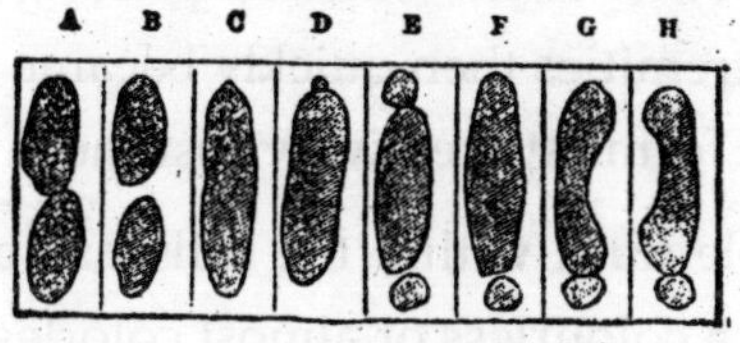

Fig. 7
(Drosera rotundifolia)
Diagram of the same cell of a tentacle, showing the various forms successively assumed by the aggregated masses of protoplasm.

at C. After the next interval a very common appearance was presented–D, namely, the formation of an extremely minute sphere at one end of an elongated mass. This rapidly increased in size, as shown in E, and was then reabsorbed, as at F, by which time another sphere had been formed at the opposite end.

The cell drawn in fig. 7 was from a tentacle of a dark red leaf, which had caught a small moth, and was examined under water. As we at first thought that the movements of the masses might be due to the absorption of water, we placed a fly on leaf, and, when after 18 hrs, all the tentacles were well inflected, these were examined without being immersed in water. The cell here represented (fig. 8) was from this leaf, being sketched eight times in the course of 15m. These sketches exhibit some of the more remarkable changes which the protoplasm

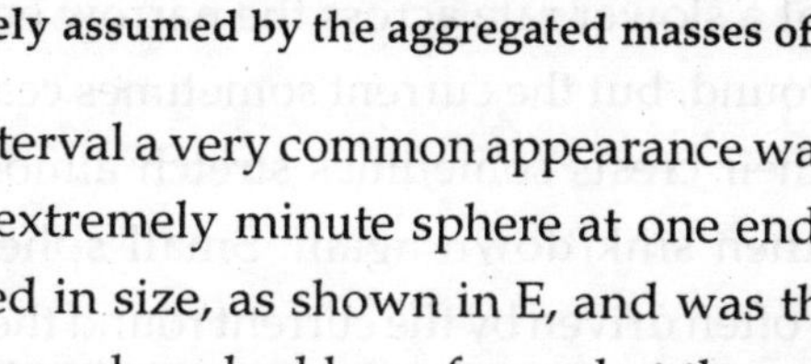

Fig. 8
(Drosera rotundifolia)
Diagram of the same cell of a tentacle, showing the various forms successively assumed by the aggregated masses of protoplasm.

undergoes. At first, there was at the larger mass near the upper end, and these seemed quite separate. Nevertheless, they may have been connected by a fine and invisible thread of protoplasm, for on two other occasion, whilst one mass was rapidly increasing, and another in the same cell rapidly decreasing, we were able, by varying the light and using a high power, to detect a connecting thread of extreme tenuity, which evidently served as the channel of communication between the two. On the other hand, such connecting threads are sometimes seen in break and their extremities then quickly became club-headed. The other sketches in fig. 8 show the forms successively assumed.

Shortly after the purple fluid within the cells has become aggregated, the little masses float about in a colourless or almost colorless fluid; and the layer of white granular protoplasm which flows along the walls can now be seen much more distinctly. The stream flows at an irregular rate, up one wall and down the opposite one, generally at a slower rate across the narrow ends of the elongated cells, and so round and round, but the current sometimes ceases. The movement is often in waves, and their crests sometimes stretch almost across the whole width of the cell, and then sink down again. Small spheres of protoplasm, apparently quite free, are often driven by the current round the cells; and filaments attached to the central masses are swayed to and fro, as if struggling to escape. Altogether, one of these cells with the ever-changing central masses, and with the layer of protoplasm flowing round the walls, presents a wonderful scene of vital activity.

Many observations were made on the contents of the cells whilst under going the process of aggregation, but we shall detail only a few cases under different heads. A small portion of a leaf was cut off, placed under high power, and the glands very gently pressed under a compressor. In 15 m. we distinctly saw extremely minute spheres of protoplasm aggregating themselves in the purple fluid; these rapidly increased in size, both within the cells of the glands and of the upper ends of the pedicels. Particles of glass, cork, and cinders were also placed on the glands of many tentacles; in 1 hr. several of them were inflected, but after 1 hr. 35m. there were no aggregate. Other tentacles with these particles were examined after 8 hrs., and now all their cells had undergone aggregation; so had the cells of the exterior tentacles which had become inflected through the irritation transmitted from the glands of the disc, which had not as yet become inflected. This latter fact shows that the process of aggregation is independent of the inflection of the tentacles, of which indeed we have other and abundant evidence. Again, the exterior tentacles on there leaves were carefully examined, and found to contain only homogeneous purple fluid; little bits of thread were then placed on the glands of three of them, and after 22 hrs. The purple fluid in their cells almost down to their bases was aggregated into innumerable spherical, elongated, or filamentous masses of protoplasm. The bits of thread had been carried some time previously to the central disc, and this had caused all the other tentacles to become somewhat inflected; and their cells had likewise undergone aggregation, which, however, it should be observed, had not as yet extended down to their bases, but was confined to the cells close beneath the glands.

Not only do repeated touches on the glands and the contact of minute particles cause aggregation, but if glands, without being themselves injured, are cut off from the summits of the pedicels, this induces a moderate amount in aggregation in the headless tentacles, after they have become inflected. On the other hand, if glands are suddenly crushed between pincers, as was tried in six cases, the tentacles seem paralysed by so great a shock, for they neither becomes inflected nor exhibit any sins of aggregation.

Carbonate of Ammonia

Of all the cause inducing aggregation, that which, as far as we have seen, acts the quickest, and is the most powerful, is a solution of carbonate of ammonia. Whatever its strength may be, the glands are always affected first, and soon become quite opaque, so as to appear black. For instance, we placed a leaf in a few drops on a strong solution, namely, of one part to 146 of water (or 3 grs. to 1 oz.) and observed it under a high power. All the glands began to darken in 10 s. (seconds); and in 13 s. were conspicuously darker. In 1 m. extremely small spherical masses of protoplasm could be seen arising in the cells of the pedicels close beneath the glands, as well as in the cushions on which the long-headed marginal glands rest. In several cases the process travelled down the pedicels of a length twice or thrice as great as that of the glands, in about 10 m. It was interesting to observe the process momentarily arrested at each transverse partition between two cells, and then to see the transparent contains of the cell next below almost flashing into a cloudy mass. In the lower part of the pedicels, the action proceeded slower, so that it took about 20 m. before the cells halfway down the long marginal and submerging tentacles aggregated.

We may infer that the carbonate of ammonia is absorbed by the glands, not only from its action being so rapid, but from its effect being somewhat different from that of other salts. As the glands, when excited, secrete an acid belonging to the acetic series, the carbonate is probably at once converted into a salt of this series; and we shall presently see that the acetate of ammonia causes aggregation almost or quite as energetically as does the carbonate. If a few drops of a solution of one part of the carbonate to 437 of water (or 1 gr. to 1 oz.) be added to the purple fluid which exuded from crushed tentacles, or to paper stained by being rubbed by them the fluid and the paper are changed into a pale dirty green. Nevertheless, some purple colour could still be detected after 1 hr., 30 m. within the glands of leaf left in a solution of twice the above strength (*viz.*, 2 grs. to 1 oz.); and after 24 hrs. The cells of the pedicels close beneath the glands still contained spheres of protoplasm of a fine purple tint. These facts show that the ammonia has not entered as carbonate, for otherwise the colour would have been discharged. It was, however, sometimes observed, especially with the long-headed tentacles on the margins of very pale leaves immersed in a solution, that the glands as well as the upper cells of the pedicels were discoloured; and in these cases we presume that the unchanged carbonate had been observed. The appearance above described, of the aggregating process being arrested for a short time a each transverse partition, impresses the mind with the ideal of matter passing downwards from cell to cell. But as the cells one beneath other undergo aggregation when inorganic and insoluble particles are placed on the gland, the process must be, at least in these cases, one of molecular change, transmitted from the glands., independently of the absorption any matter. So it may possibly be in the case of the carbonic of ammonia. As however the aggregation caused by this salt travels down the tentacles at a quicker rate then when insoluble particles are placed on the glands, it is probable that ammonia in some form is absorbed not only by the glands, but passes down the tentacles.

Having examined a leaf in water, and found the contents of the cells homogeneous, we placed it in a few drops of solution of one part of the carbonate of 437 of water, and attended to the cells immediately beneath the glands, but did not use a very high power. No aggregation was visible in 3 m.; but after 15 m. small spheres of protoplasm were formed, more especially beneath the long-headed marginal glands; the process, however, in case took place with unusual slowness. In 25 m. conspicuous spherical masses were present in the cells of the pedicels for a length about equal to that of the glands; and in 3 hrs. to that of a third or half of the whole tentacle.

If tentacles with cells containing only very pale pink fluid, and apparently but little protoplasm, are placed in a few drops of weak solution of one part of the carbonate to 4375 of water (1 gr. to 10 oz.), and the highly transparent cells beneath the glands are carefully observed under a high power, these may be seen first to become slightly cloudy from the formation of numberless, only just perceptible granules, which rapidly grow larger either from coalescence or from attracting more protoplasm from the surrounding fluid. On one occasion we chose a singularly pale leaf, and gave it, whilst under the microscope, single drop of a stronger solution of part of 437 of water; in this case the contents of the cells did not become cloudy, but after 10 m. minute irregular granules of protoplasm could be detected, which soon increased into irregular masses and globules of greenish or very pale purple tint; but these never formed perfect spheres, though incessantly changing their shapes and positions.

With moderately red leaves the first effect o a solution of the carbonate generally is the formation of two or three, or of several, extremely minute purple spheres which rapidly increase in size. To give an idea of the rate at which such spheres increase in size, we may mention that a rather pale purple leaf placed under a slip of grass was given a drop of a solution of one part to 292 of water, and in 13 m. a few minute sphere of protoplasm were formed; one of these, after 2 hrs., 30m., was about two-thirds of the diameter of the cell. After 4 hrs. 25m. it nearly equalled the cell in diameter; and a second sphere about half as large as the first, together with a few other minute ones, were formed. After 6 hrs. the fluid in which these spheres floated was almost colourless. After 8 hrs. 35 m. (always reckoning from the time when the solution was first added) four new minute spheres had appeared. Next morning, after 22 hrs., there were, besides the two large spheres, seen smaller ones, floating in absolutely colourless fluid, in which some flocculent greenish matter was suspended.

At the commencement of the process of aggregation, more especially in dark red leaves, the contents of the cells often present a different appearance, as if the layer of protoplasm (primordial utricle) which lines the cells had separated itself and shrunk from the walls; an irregularly shaped purple bag being thus formed. Other fluids, besides a solution of the carbonate, for instance an infusion of raw meat, produce this same effect. But the appearance of the primordial utricle shrinking from the walls is certainly false; for before giving the solution, I saw on several occasion that the walls were lined with colourless flowing protoplasm, and after the bag-like masses were formed, the protoplasm was still flowing along the walls in a conspicuous manner, even more so than before. It appeared indeed as if the stream of protoplasm was strengthened by the action of the carbonate. But it was impossible to ascertain whether this was really the case. The bag-like masses, when once formed, soon begin to glide slowly round the cells, sometimes sending our projections which separate into little spheres; other spheres appear in the fluid surrounding the bags, as these travel much more quickly. That the small spheres are separate, and sometimes they revolve round each other. We have occasionally seen spheres of this kind proceeding up and down the same side of cell, instead of round it. The bag-like masses after a time generally divide into two rounded or oval masses, and these undergo the changes shown in figs. 7 and 8. At other times spheres appear within the bags; and these coalesce and separate in an endless cycle of change.

After leaves have been left for several hours in a solution of the carbonate, and complete aggregation has been effected, the stream of protoplasm on the walls of the cells ceases to be visible; we observed this fact repeatedly, but will give only one instance. A pale purple leaf was placed in a few drops of solution of one part to 292 of water, and in 2 hrs., some fine purple spheres were formed in the upper cells of the pedicels, the stream of protoplasm round their walls being still quite distinct; but after an *additional* 4 hrs., during which time many more spheres were formed, the stream was no longer distinguishable on the most careful examination; and this no doubt was due to the contained granules having become united with the spheres, so that nothing was left by which the movement of the limpid protoplasm could be perceived. But minute free spheres still travelled up and down the cells, showing that there was still a current. So it was next morning, after 22 hrs., by which time some new minute spheres had been formed; these oscillated from side to side and changed their position, proving that the current had not ceased, though no stream of protoplasm was visible. On another occasion, however, a stream was seen flowing round the cell-walls of vigorous, dark-coloured leaf, after it had been left for 24 hrs., in rather stronger solution, namely, of one part of the carbonate to 218 of water. This leaf, therefore, was not much or at all injured by an immersion for this length of time in the above solution of two grains to the ounce; and on being afterwards left for 24 hrs. in water, the aggregated masses in many of the cells were redissolved, in the same manner as occurs with leaves in a state of nature when they re-expand after having caught insects.

In a leaf which had been left for 22 hrs. in a solution of one part of the carbonate to 292 of water, some spheres of protoplasm (formed by the self-division of a bag-like mass) were gently pressed beneath a covering glass, and then examined under a high power. They were now distinctly divided by well-defined radiating fissures, or were broken up into separate fragments with sharp edges, and they were solid to the centre. In the larger broken spheres the central part was more opaque, darker coloured, and less brittle then the exterior; the latter alone being in some cases penetrated by the fissures. In many of the spheres the line of separation between the outer and inner parts was tolerably well defined. The outer parts were of exactly the same very pale purple tint, as that of the last-formed smaller spheres; and these latter did not include any darker central core.

From these several facts we may conclude that, when vigorous dark- coloured leaves are subjected to the action of carbonate of ammonia, the fluid within the cells of the tentacles often aggregates exteriorly into coherent viscid matter, forming a kind of bag. Small spheres sometimes appear within this bag, and the whole generally soon divides into two or more spheres, which repeatedly coalesce and redivide. After a longer or shorter time the granules in the colorless layer of protoplasm, which flows round the walls, are drawn to and unite with the larger spheres, or from small independent spheres; these latter being of a much paler colour, and more brittle than the first aggregated masses. After the granules of protoplasm have been thus attracted, the layer of flowing protoplasm can no longer be distinguished, though a current of limpid still flows the walls.

If a leaf is immersed in a very strong, almost concentrated, solution of carbonate of ammonia, the glands are instantly blackened, and they secrete copiously; but no movement of the tentacles ensues. Two leaves thus treated become after 1 hrs. Flaccid, and seem killed; all the cells in their tentacles contained spheres of protoplasm, but these were small and discoloured. Two other leaves were placed in a solution not quite so strong, and there was well-marked aggregation in 30m. After 24 hrs. the spherical or more commonly oblong masses of protoplasm became opaque and granular, instead of being as usual translucent: and in the lower cells there were only innumerable minute spherical granules. It was evident that the strength of the solution had interfered with the completion of the process, as we shall see likewise follows from too great heat.

All the foregoing observations relate to the exterior tentacles, which are of a purple colour; but the green pedicels of the short central tentacles are acted on the carbonate, and by an infusion of raw meat, in exactly the same manner, with the sole difference that the aggregate masses are of a greenish colour; so that the process is in no way dependent on the colour of the fluid within the cells.

Finally, the most remarkable fact with respect to this salt is the extraordinary small amount, which suffices to cause aggregation. Full details will be given in the seventh chapter, and here it will be enough to say that with a sensitive leaf the absorption by gland of 1/134400 of a grain (000482 mgr.) is enough to cause in the course of one hour well-marked aggregation in the cells immediately beneath the gland.

The Effects of Certain other Salts and Fluids

Two leaves were placed in a solution of one part of acetate of ammonia to about 146 of water, and were acted on quite as energetically, but perhaps not quite so quickly as by the carbonate. After 10 m. the glands were black, and in the cells beneath them there were traces of aggregation, which after 15 m., mass well marked, extending down the tentacles for a length equal to that for the glands. After 2 hrs. the contents of almost all the cells in all the tentacles were broken up into masses of protoplasm. A leaf was immersed in a solution of one part of oxalate of ammonia to 146 of water; and after 24 m. some, but not a conspicuous, change could be seen within the cells beneath the glands. After 17 m. plenty of spherical masses of protoplasm were formed, and these extended down the tentacles for about the length of the glands. This salt, therefore, does not act so quickly as the carbonate. With respect to the citrate of ammonia, a leaf was placed in little solution of the above strength, and there was not even a trace of aggregation in the cells beneath the glands, until 56 m. had elapsed; but it was well marked after 2 hrs. 20 m. On another occasion a leaf was placed in a stronger solution, of one part of the citrate to 109 of water (4 grs. to 1 oz.), and at the same time another leaf in a solution of the carbonate of the same strength. The glands of the latter were blackened in less then 2 m., and after 1 hr. 45 m. the aggregated masses which were spherical and very dark- coloured, extended down all the tentacles, for between half and two-thirds of their lengths; whereas in the leaf immersed in the citrate the glands, after 30 m., were of dark red, and the aggregated messes in the cells beneath them pink and elongated. After 1 hr. 45 m. these masses extended down for only about one–fifth or one-fourth of the length of the tentacles.

Two leaves were placed, each in ten minims of a solution of one part of nitrate of ammonia to 5250 of water (1 gr. to 12 oz.). So that each leaf received 1/576 of grain (.1124 mgr). This quantity caused all the tentacles to be inflected, but after

24 hrs. There was only a trance of aggregation. One of these same leaves was then placed in a weak solution of the carbonate, and after 1 hr. 45 m. the tentacles for half their lengths showed an astonishing degree of aggregation. Two other leaves were then placed in a much stronger solution of one part of the nitrate to 146 of water (3 grs. to 1 oz.); one of these there was no marked change after 3 hrs; but in the other there a trace of aggregation after 52 m., and this was plainly marked after 1 hr. 22 m., but even after 2 m. 12 m. There was certainly not more aggregation than would have followed from an immersion of from 5 m. to 10 m. in an equally strong solution of the carbonate.

Lastly, a leaf was placed in thirty minims of a solution of one part of phosphate to 43,750 of water (1 gr. to 100 oz.), so that it received $\frac{1}{1600}$ of a grain (.04079 mgr.); this soon caused the tentacles to be strongly inflected; and after 24 hrs. the contents of the cells were aggregated into oval and irregularly globular masses, with a conspicuous current of protoplasm flowing round the walls. But after so long interval aggregation would have ensued, whatever had caused inflection.

Only a few other salts, besides those of ammonia, were tried in relation to the process of aggregation. A leaf was placed in a solution of one part of chloride of sodium to 218 of water, and after 1 hr. the contents of the cells were aggregated into small, irregularly globular, brownish masses; these after 2 hrs., were almost disintegrated and pulpy. It was evident that the protoplasm had been injuriously affected; and soon afterwards some of the cells appeared quite empty. These effects differ altogether from those produced by the several salts of ammonia, as well as by various organic fluids, and by inorganic particles placed on the glands. A solution of the same strength of carbonate of soda and carbonate of potash acted in nearly the same manner as the chloride, and here again, after 2 hrs. 30 m., the outer cells of some of the glands had emptied themselves of their brown pulpy contents. We shall see in the eight chapter that solutions of several salts of soda of half the above strength cause inflection, but do not injure the leaves. Weak solution of sulphate of quinine, of nicotine, camphor, poison of the cobra, & c., soon induce well-marked aggregation; camphor certain other substance (for instance, a solution of curare) have no such tendency.

Many acids, though much diluted, are poisonous; and though, as will be shown in the eight chapter, they cause the tentacles to bend, they do not excite true aggregation. Thus leaves were placed in a solution of one part of benzoic acid to 437 of water, and in 15 m. the purple fluid within the cells had shrunk a little from the walls; yet, when carefully examined after 1 hr. 20 m., there was no true aggregation; and after 24 hrs. the leaf was evidently dead. Other leaves in iodic acid, diluted to the same degree, showed after 2 hrs. 15 m. the same shrunken appearance of the purple fluid within the cells; and these, after 6 hrs. 15 m., were seen under a high power to be filled with excessively minute spheres of dull reddish protoplasm, which by the next morning, after 24 hrs., had almost disappeared, the leaf being evidently dead. Nor was there any true aggregation in leaves immersed in propionic acid of the same strength; but in this case the protoplasm was collected in irregular masses towards the bases of the lower cells of the tentacles.

A filtered infusion of raw meat induces strong aggregation, but not very quickly. In one leaf thus immersed there was a little aggregation after 1 hr. 20 m., and in another after 1 hr. 50 m. With other leaves a considerably longer time was required: for instance, one immersed for 5 hrs. showed no aggregation, but was plainly acted on in 5 m., when placed in a few drops of solution of one part of carbonate of ammonia to 146 of water. Some leaves were left in the infusion for 24 hrs., and these became aggregated to a wonderful degree, so that the inflected tentacles presented to the naked eye a plainly mottled appearance. The little masses of purple protoplasm were generally oval or beaded, and not nearly so often spherical as in the case of leaves subjected to carbonate of ammonia. They under went incessant changes of form; and the current of colorless protoplasm round the walls was conspicuously plain after and immersion of 25 hrs. Raw meat is too powerful a stimulant, and even small bits generally injure, and sometimes kill the leaves to which they are given: the aggregated masses of protoplasm become dingy or almost colourless, and present an unusual granular appearance, as is likewise the case with leaves which have been immersed in a very strong solution of carbonate of ammonia. A leaf placed in milk had the contents of its cells somewhat aggregated in 1 hr. Two other leaves, one immersed in human saliva for 2 hrs. 30m., and another in unboiled white of egg for 1 hr. 30 m., were not acted on in this manner though they undoubtedly would have been so, had more time been allowed. These same two leaves, one being afterwards placed in a solution of carbonate of ammonia (3 grs. to 1 oz.), had their cells aggregated, the one in 10 m. and other in 5 m.

Several leaves were left for 4 hrs. 30 m. in a solution of one part of white sugar to 146 of water, and no aggregation ensued; on being placed in a solution of this same strength of carbonate of ammonia, they were acted on in 5 m., as was likewise a leaf which had been left for 1 hr. 45 m. in a moderately thick solution of gum Arabic. Several other leaves were immersed for some hours in denser solutions of sugar, gum, and starch, and they had the contents of their cells greatly aggregated. This effect may be attributed to exosmose; for the leaves in the syrup became quite flaccid, and those in the gum and starch somewhat flaccid, with their tentacles twisted about in the most irregular manner, the longer ones like corkscrews. We shall hereafter see that solutions of these substance, when placed on the discs of leaves, do not incite inflection. Particles of soft sugar were added to the secretion round several glands and were soon dissolved causing a great increase of the secretion, no doubt by exosmose; and after 24 hrs. the cells showed a certain amount of aggregation, though the tentacles were not inflected. Glycerine causes in a few minutes well-pronounced aggregation, commencing as usual within the glands and then travelling down the tentacles; and this we presume may be attributed to the strong attraction of this substance for water. Immersion for several hours in water causes some degree of aggregation. Twenty leaves were first carefully examined and re-examined after having been left immersed in distilled water for various periods, with the following results. It is rare to find even trace of aggregation until 4 or 5 and generally not until several more hours have elapsed. When, however, leaf becomes quickly inflected in water, as sometimes happens, especially during very warm weather, aggregation may occur in little over 1 hr. In all cases leaves left in water for more than 24 hrs. have their glands blackened, which shows that their contents are aggregated; and in the specimens, which were carefully examined, there was fairly well-marked aggregation in the upper cells of the pedicels. These trials were made with cut-off leaves and it occurred to us that this circumstance might influence the result, as the footstalks would not perhaps absorb water quickly enough to supply the glands as they continued to secrete. But this view was proved erroneous, for a plant with uninjured roots, bearing four leaves, was submerged in distilled water for 47 hrs., and the glands were blackened, though the tentacles were very little inflected. In one of those leaves there was only a slight degree of aggregation in the tentacles; in the second rather more, the purple contents of the cells being a little separated from the walls; in the third and fourth, which were pale leaves, the aggregation in the upper parts of the pedicels was well-marked. In these leaves the little masses of protoplasm, many of which were oval, slowly changed their forms and positions; so that a submergence for 47 hrs. had not killed the protoplasm. In a previous trial with a submerged plant the tentacles were not in the least inflected.

Heat induces aggregation. A leaf with the cells of the tentacles containing only homogeneous fluid, was waved about for 1 m. in water at 130°F (54°.4 C), and was then examined under the microscope as quickly as possible, that is in 2 m., or 3m.; and by this time the contents of the cells had undergone some degree of aggregation. A second leaf was waved for 2 m., in water at 125°F (51°.6 C) and quickly examined as before; the tentacles were well inflected; the purple fluid in all the cells had shrunk a little from the walls, and contained many oval and elongated masses of protoplasm, with a few minute spheres. A third leaf was left in water at 125°F, until it coded, and, when examined after 1 hr. 45 m., the inflected tentacles showed some aggregation, which became after 3 hrs. more strongly marked, but did not subsequently increase. Lastly, a leaf was waved for 1 m. in water at 120°F (48°.8 C) and then left for 1 hr. 26 m. in cold water; the tentacles were but little inflected, and there was only here and there a trance of aggregation. In all these and other trials with warm water the protoplasm showed much less tendency to aggregate into spherical masses than when excited by carbonate of ammonia.

Redissolution of the Aggregated Masses of Protoplasm

As soon as tentacles which have clasped an insect or any inorganic object or have been in any way excited, have fully re-expanded, the aggregated masses of protoplasm are redissolved and disappear; the cells being now refilled with homogeneous purple fluid as they were before the tentacles were inflected. The process of redissolution in all cases commences at the bases of the tentacles, and proceeds up then towards the glands. In old leaves, however, especially in those which have been several times in action, the protoplasm in the uppermost cells of the pedicels remains in a permanently more or less aggregated condition. In order to observe the process of redissolution, the following observations were made; a leaf was left for 24 hrs., in a little solution of one part of carbonate of ammonia to 218 of water, and the protoplasm was as usual aggregated into numberless purple spheres, which were incessantly changing their forms. The leaf was then washed and placed in distilled water, and after 3 hrs. 15 m. some few of the spheres began to show by their less clearly defined edges signs of redissolution. After 9 hrs. many of them had become elongated, and the surrounding fluid in the cells was slightly more coloured, showing plainly that redissolution had commenced. After 24 hrs., though many cells still contained spheres, here and there one could be seen filled with purple fluid, without a vesting of aggregated protoplasm; to whole having been redissolved. A leaf with aggregated masses, caused by its having been waved for 2 m., in water at the temperature of 125° F, was left in cold water, and after 11 hrs. the protoplasm showed traces of incipient redissolution. When again examined three days after its immersion in the warm water, there was a conspicuous difference, though the protoplasm was still somewhat aggregated. Another leaf, with the contents of all the cells strongly aggregated from the action of a weak solution of phosphate of ammonia, was left for between three and four days in a mixture (known to be innocuous) of one drachm of alcohol to eight drachms of water, and when re-examined every trace of aggregation had disappeared, the cells being now filled with homogeneous fluid.

We have seen that leaves immersed for some hours in dense solution of sugar, gum, and starch have the contents of their cells greatly aggregated, and are rendered more or less flaccid, with their tentacles irregularly contorted. These leaves, after being left for four days in distilled water, became less flaccid, with their tentacles partially re-expanded, and the aggregated masses of protoplasm were partially redissolved. A leaf with its tentacles closely clasped over a fly, with the contents of the cells strongly aggregated, was placed in a little sherry wine; after 2 hrs. several of the tentacles had re-expanded, and the others could by a mere touch be pushed back into their properly expanded positions, and now all traces of aggregation had disappeared, the cells being filled with perfectly homogeneous pink fluid. The redissolution in these cases may be attributed to endosmose.

Proximate Causes of the Process of Aggregation

As most of the stimulants which cause the inflection of the tentacles likewise induce aggregation in the contents of their cells, this latter process might be thought to be the direct result of inflection; but this is not the case. If leaves are placed in rather strong solution of carbonate of ammonia, for instance, of three or four, and even sometimes of only two grains to the ounce of water (*i.e.*, one part to 109, or 146, or 218, of water), the tentacles are paralysed, and do not become inflected, yet they soon exhibit strongly marked aggregation. Moreover, the short central tentacles of leaf which has been immersed in a weak solution of any salt of ammonia, or in any nitrogenous organic fluid, do not become in the least inflected; nevertheless, they exhibit all the phenomena of aggregation. On the other hand, several acids cause strongly pronounced inflection, but no aggregation.

It is an important fact that when an organic or inorganic object is placed on the glands of the disc, and the exterior tentacles are thus caused to bend inwards, not only is the secretion from the glands of the latter increased in quantity are rendered acid, but the contents of the cells of their pedicels become aggregated. The process always commences in the glands, although these have not as yet touched any object. Some force or influence must, therefore, be transmitted for the central glands to the exterior tentacles, first to near their bases causing this part to bend, and next to the glands causing them to secrete more copiously. After a short time the glands, thus indirectly excited, transmit or reflect some influence down their own pedicels, inducing aggregation in cell beneath cell to their bases.

It seems at first sight a probable view that aggregation is due to the glands being excited to secrete more copiously, so the sufficient fluid is not left in their cells, and in the cells of the pedicels, to hold the protoplasm in solution. In favour of this view is the fact that aggregation follows the inflection of the tentacles, and during the movement the glands generally, or, as I believe, always, secrete more copiously than they did before. Again, during the re-expansion of the tentacles, the glands secrete less freely, or quite cease to secrete, and the aggregated masses of protoplasm are then redissolved. Moreover, when leaves are immersed in dense vegetable solution, or in glycerine, the fluid within the gland-cells passes outwards, and there is aggregation; and when the leaves are afterwards immersed in water, or in an innocuous fluid of less specific gravity than water, the protoplasm is redissolved, and this, no doubt, is due to endosmose.

Opposed to this view, that aggregation is caused by the outward passage of fluid from the cells, are the following facts. There seems no close relation between the degree of increased secretion and that of aggregation. Thus a particle of sugar added to the secretion round a gland causes a much greater increase of secretion, and much leas aggregation, than does a particle of carbonate of ammonia given in the same manner. It does not appear probable that pure water would cause much exosmose, and yet aggregation often follows from an immersion in water of between 16 hrs. and 24 hrs., and always after from 24 hrs. to 48 hrs. Still less probable is it that water at a temperature of from 125° to 130° F (51°.6 to 54°.4 C) should cause fluid to pass, not only from the glands, but from all the cells of the tentacles down to their bases, so quickly that aggregation is induced within 2 m. or 3 m. Another strong argument against this view is that, after complete aggregation, the sphere and oval masses of protoplasm float about in an aduandant supply of thin, colorless fluid; so that at least the latter stages of the process cannot be due to the want of fluid to hold the protoplasm in solution. There is still stronger evidence that aggregation independent of secretion; for the papillae described in the first chapter, with which the leaves are studded are not glandular, and do not secrete, yet they rapidly absorb carbonate of ammonia or an infusion of raw meat, and their contents then quickly undergo aggregation, which afterwards spreads into the cells of the surrounding tissues. We shall hereafter see that the purple fluid within the sensitive filaments of Dionaea, which do not secrete, likewise undergoes aggregation from the action of weak solution of carbonate of ammonia.

The process of aggregation is a vital one; by which we mean that the contents of the cells must be alive and uninjured to be thus affected, and they must be in an oxygenated condition for the transmission of the process at the proper rate. Some tentacles in a drop of water were strongly pressed beneath a slip of glass; many of the cells were ruptured, and pulpy matter of a purple colour, with granules of all sizes and shapes, exuded, but hardly any of the cells were completely emptied. We then added a minute drop of a solution of one part of carbonate of ammonia to 109 of water, and after 1 hr. examined the specimens. Here and there a few cells, both in the glands and in the pedicels, had escaped being rupture, and their contents were well aggregated into spheres which were constantly changing their forms and positions, and a current could still be seen flowing along the walls; so that the protoplasm was alive. On the other hand, the exuded matter, which was nor almost colourless instead of being purple, did not exhibit a trace of aggregation. Nor was there a trace in many cells which were ruptured, but which had not been completely emptied of their contents. Though we looked carefully, no signs of a current could be seen within these ruptured cells. They had evidently been killed by the pressure; and the matter which they still continued did not undergo aggregation any more than that which had exuded. In these specimens, the individuality of the life of each cell was well illustrated.

A full account will be given in the next chapter of the effects of heat on the leaves and it need here only state that leaves immersed for a short time in water at temperature of 120° F (48°.8 C) which, as we have seen, does not immediately induce aggregation, were then placed in a few drops of strong solutions of one part of carbonate of ammonia to 109 of water, and became finely aggregated. On the other hand, leaves after an immersion in water at 150° F (65°.5 C), on being placed in the same strong solution, did not undergo aggregation, the cells becoming filled with brownish, pulpy, or muddy matter. With leaves subjected to temperatures between these two extremes of 120° and 150° F (48°.8 and 65°.5 C), there were gradations in the completeness of the process; the former temperature not preventing aggregation from the subsequent action of carbonate of ammonia, the latter quite stopping it. Thus, leaves immersed in water, heated to 130° F (54°.4 C), and then in the solution, formed perfectly defined spheres, but these were decidedly smaller than in ordinary cases. With other leaves heated to 140° F (60° C) the spheres were extremely small, yet well defined, but many of the cells contained, in addition, some brownish pulpy matter.

The fluid within the cells of the tentacles must be in an oxygenated condition, in order that the force or influence which induces aggregation should be transmitted at the proper rate from cell to cell. A plant, with its roots in water, was left for 45 m., in a vessel containing 122 fluid oz. of carbonic acid. A leaf from this plant, and for comparison, one from a fresh plant, were both immersed. They were then compared, and certainly there was much less aggregation in the leaf which had been subjected to the carbonic acid than in the other. Another plant was exposed in the same vessel for 2 hrs. to carbonic acid, and one of its leaves was then placed in a solution of one part of the carbonante to 437 of water; the glands were instantly blackened, showing that they had absorbed, and that their contents were aggregated; but in the cells close beneath the glands there was no aggregation even after an interval of 3 hrs. After 4 hrs. 15 m. a few minute spheres of protoplasm were formed in these cells, but even after 5 hrs. 30 m. the aggregation did not extend down the pedicels for a length equal to that of the glands. After numberless trials with fresh leaves immersed in a solution of this strength, we have never seen the aggregating action transmitted at nearly so slow a rate. Another plant was left for 2 hrs. in carbonic acid, but was then exposed for 20 m. to the open air, during which time the leaves, being of a red colour would have absorbed some oxygen. One of them, as well as a fresh leaf for comparison, were now immersed in the same solution as before. The former were looked at repeatedly, and after an interval 65 m. a few spheres of protoplasm were first observed in the cells close beneath the glands, but only in two or three of the longer tentacles. After 3 hrs. the aggregation had travelled down the pedicels of few of the tentacles for a length equal to that of the glands. On the other hand, in the fresh leaf similarly treated, aggression was plain in many of the tentacles after 15 m; after 65 m. it had extended down the pedicels for five, or more times the length of the glands; and after 3 hrs. the cells of all the tentacles were affected for one-third or one half of their entire length. Hence there can be no doubt that the exposure of leaves to carbonic acid either stops for a time the process of aggregation, of checks the transmission of the proper influence when the glands are subsequently excited by carbonate of ammonia; and this substance acts more promptly and energetically than any other. It is known that the protoplasm of plants exhibits its spontaneous movements only as long as it is in an oxygenated condition; and so it is with the while corpuscles of the blood, only as long as they receive oxygen from the red corpuscles; but the cases above given are somewhat different, as they relate to the delay in the generation or aggregation of the masses of the protoplasm by the exclusion of oxygen.

Summary and Concluding Remarks

The process of aggregation is independent of the inflection of the tentacles and apparently of increases secretion from the glands. It commences within the glands, whether these have been directly excited, or indirectly by a stimulus received from other glands. In both cases the process is transmitted from cell to cell down the whole length of the tentacles, being arrested for a short time at each transverse partition. With pale-coloured leaves to first change which in perceptible, but only under a high power, is the appearance of the finest granules in the fluid within the cells, making it slightly cloudy. These granules soon aggregate into small globular masses. We have seen a cloud of this kind appear in 10 s. after a drop of solution of carbonate of ammonia had been given to a gland. With dark red leaves the first visible change often is the conversion of the outer layer of the fluid within the cells into bag-like masses. The aggregated masses, however they may have been developed, incessantly change their forms and position. They are not filled with fluid, but are solid to their centres. Ultimately the colourless granules in the protoplasm which flows round the walls coalesce with the central spheres of masses; but there is still a current of limpid fluid flowing within the cells. As soon as the tentacles fully re-expand, the aggregated masses are redissolved, and the cells become filled with homogeneous purple fluid, as they were at first. The process of redissolution commences at the bases of the tentacles, thence proceeding upwards to the glands; and, therefore, in reversed direction to that of aggregation.

Aggregation is excited by the most diversified causes—by the glands being several times touched, by the pressure of particles of any kind, and as these are supported by the dense secretion, they can hardly press on the glands with the weight of millionth of a grain, by the tentacles being cut off close beneath the gland, by the glands absorbing various fluids or matter dissolved out of certain bodies, by exosmose, and by certain degree of heat. On the other hand, a temperature of about 150° F (65°.5 C) does not excite aggregation; nor does the sudden crushing of a glands. If a cell is ruptured, neither the exceeded matter nor that which still remains within the cell undergoes aggregation when carbonate of ammonia is added. A very strong solution of these salt and rather large bits of raw meat prevent the aggregated masses being well developed. From these facts we may conclude that the protoplasmic fluid within a cell does not become aggregated unless it be in a living state, and only perfectly if the cell has been injured. We have also seen that the fluid must be in an oxygenated state, in order that the process of aggregation should travel from cell to cell at the proper rate.

Various nitrogenous organic fluids and salts of ammonia induce aggregation, but in different degrees and at very different rates. Carbonate of ammonia is the most powerful of all known substances; the absorption of $\frac{1}{134400}$ of a grain (.000482 mg.) by a gland suffices to cause all the cells of the same tentacle to become aggregated. The first effect of the carbonate and of certain other salts of ammonia, as well as of some other fluids, is the darkening or blackening of the glands. This follows even from long immersion in cold distilled water. It apparently depends in chief part on the strong aggregation of their cell-contents, which thus become opaque and do not reflect light. Some other fluids render the glands of a brighter red; whilst certain acids, though much diluted, the poison of the cobra-snake, &c., make the glands perfectly white and opaque; and this seems to depend on the coagulation of their contents without any aggregation. Nevertheless, before being thus affected, they are able, at least in some cases, to excite aggregation in their own tentacles.

That the central glands, if irritated, send centrifugally some influence to the exterior gland, causing them to sent back a centripetal influence inducing aggregation, is perhaps the most interesting fact given in this chaper. But the whole process of aggregation is in itself a striking phenomenon. Whenever the peripheral extremity of a nerve is touched or pressed, and a sensation is felt, it is believed that an invisible molecular change is sent from one end of the nerve to the other; but when a gland of Drosera is repeatedly touched or gently pressed, we can actually see a molecular change proceeding from the gland down the tentacle; though this change is probably of a very different nature from that in a nerve. Finally, as so many and such widely different causes excite aggregation, it would appear that the living matter within the gland-cells is in so unstable condition that almost any disturbance suffices to change its molecular nature, as in the case of certain chemical compounds and this change in the glands, whether excited directly, or indirectly by a stimulus received from other glands, is transmitted from cell to cell, causing granules of protoplasm either to be actually generated in the previously limpid fluid or to coalesce and thus to become visible.

Supplementary Observations on the Process of Aggregation in the Roots of Plants

It will hereafter be seen that a weak solution of the carbonate of ammonia induces aggregation in the cells of the roots of Drosera; and this led to make a few trials on the roots of other plants. We dug up in the latter part of October the first weed which we met with, *viz., Exphorbia peplus*, being careful not to injure the roots; these were washed and placed in a little solution of one part of carbonate of ammonia to 146 of water. In less than one minute we saw a cloud travelling from cell to cell up the roots, with wonderful rapidity. After from 8 m. to 9 m. the fine granules, which caused this cloudy appearance became aggregated towards the extremities of the roots into quadrangular masses of brown matter; and some of these soon changed their forms and became spherical. Some of the cells, however, remained unaffected. We repeated the experiment with another plant of the same species, but before we could get the specimen into focus under the microscope, clouds of granules and quadrangular masses of Reddish and brown matter were formed, and had run far up all the roots. A fresh root was now left for 18 hrs. in a drachm of a solution of one part of the carbonate to 437 of water, so that it received 1/81 of a grain or 2.024 mg. When examined, the cells of all the roots throughout their whole length contained aggregated masses of reddish and brown matter. Before making these experiments, several roots were closely examined, and not a trace of the cloudy appearance or of the granular masses could be seen in any of them. Roots were also immersed for 35 m. in solution of a part of carbonate of potash to 218 of water; but this salt produced no effect.

Thin slices of the stem of the euphoria were placed in the same solution, and the cells which were green instantly became cloudy, whilst others which were before colourless were clouded with brown, owing to the formation of numerous granules of this tint. It was also seen with various kinds of leaves, left for some time in a solution of carbonate of ammonia, that the grains of chlorophyll ran together and partially coalesced; and this seems to be a form of aggregation.

Plants of duck-weed (Lemna) were left for between 30 m. and 45 m. in a solution of one part of this same salt to 146 of water, and three of their roots were then examined. In two of them, all the cells which had previously contained only limpid fluid now included little green sphere. After from 1½ hr. to 2 hrs. similar spheres appeared in the cells on the borders of the leaves; but whether the ammonia had travelled up the roots or had been directly absorbed by the leaves, is not clear. As one species, *Lemna arrhiza*, produces no roots, the latter alternative is perhaps the most probable. After about 2½ hrs. some of the little green spheres in the roots were broken up into small granules which exhibited Brownian movement. Some duck-weed was also left for 1 hr. 30 m. in a solution of one part of carbonate of potash to 218 of water, and no decided change could be perceived in the cells of the roots; but when these same roots were placed for 25 m. in a solution of carbonate of ammonia of the same strength, little green spheres were formed.

A green marine alga was left for some time in this same solution but was very doubtfully effected. On the other hand, a red marine alga, with finely pinnated fronds, was strongly affected. The contents of the cells aggregated themselves into broken rings, still of a red colour, which very slowly and slightly changed their shapes and the central spaces within these rings became cloudy with red granular matter. The facts here given indicate that interesting results would perhaps be gained by observing the action of various saline solutions and other fluids on the roots of plants.

❑❑❑

4

The Effects of Heat on the Leaves

Nature of the experiments — Effects of boiling water — Warm water cause rapid inflection — Water at a higher temperature does not causes mediate inflection, but does not kill the leaves, as shown by their subsequent re-expansion and by the aggregation of the protoplasm — A still higher temperature kills the leaves and coagulates the albumin's contents of the glands.

Nature of Experiments

In *Drosera rotundifolia,* the leaves seemed to be more quickly inflected over animal substances and to remain inflected for a longer period during very warm then during cold weather. We wished, therefore, to ascertain whether heat almost would, induce inflection, and what temperature was the most efficient. Another interesting point presented itself, namely, at what degree life was extinguished; for drosera offers unusual facilities in this respect, not in the loss of the power of inflection, but in that of subsequent re-expansion, and more especially in the failure of the protoplasm to become aggregated, when the leaves after being heated are immersed in a solution of carbonate of ammonia.

Experiments were tried in the following manner. Leaves were cut-off, and does not in the least interfere with their powers; for instance, three cut-off leaves, with bits of meat placed on them, were kept in a damp atmosphere, and after 23 hrs. closely embraced the meat both with their tentacles and blades; and the protoplasm within their cells was well aggregated. Three ounces of doubly distilled water was heated in a porcelain vessel, with a delicate thermometer having a long bulb obliquely suspended in it. The water was gradually raised to the required temperature by a spirit-lamp moved about under the vessel; and in all cases the leaves were continually waved for some minutes close to the bulb. They were then placed in cold water, or in a solution of carbonate of ammonia. In other cases they were left in the water, which had been raised to a certain temperature, until it cooled. Again in other cases the leaves were suddenly plugged into water of certain temperature, and kept there for a specified time. Considering that the tentacles are extremely delicate, and that their coats are very thin, it seems scarcely possible that the fluid contents of their cells should not have been heated to within a degree or two of the temperature of the surrounding water. Any further precautions would have been superfluous, as the leaves from age or constitutional cause differ slightly in their sensitiveness to heat.

Effects of Boiling Water

It will be convenient first briefly to describe the effects of immersion for thirty seconds in boiling water. The leaves are rendered flaccid with their tentacles bowed backwards, which as we shall see in a elasticity for a longer period than their inner surfaces retain the power of contraction. The purple fluid within the cells of the pedicels is rendered finely granular. But the most remarkable change is that the glands become opaque and uniformly white; and this may be attributed to the coagulation of their albuminous contents.

Our first and preliminary experiment consisted in putting seven leaves in the same vessel of water, and warming it slowly up to the temperature of 110° F (43°.3 C); a leaf being taken out as soon as the temperature rose to 80° F (26°.6 C), another at 85° F, another at 90° F, and so on. Each leaf when taken out was placed in water at the temperature of my room, and the tentacles of all soon became slightly, though irregularly, inflected. They were now removed from the cold water and kept in damp air, with bits of meat placed on their discs. The leaf which had been exposed to the temperature of 110° F became in 15 m. greatly inflected; and in 2 hrs. Every single tentacle closely embraced the meat. So it was, but after rather longer intervals, with the six other leaves. It appears, therefore, that the warm bath had increased their sensitiveness when excited by meat.

Next, it was observed the degree of inflection which leaves underwent within stated periods, whilst still immersed in warm water, kept as nearly as possible at the same temperature; but we will here and elsewhere give only a few of the many trials made. A leaf was left for 10 m. in water at 100° F (37°.7 C), but no inflection occurred. A second leaf, however, treated in the same manner, had a few of its exterior tentacles very slightly inflected in 6 m., and several irregularly but not closely inflected in 10 m. A third leaf, kept in water at 105° to 106° F (40°.5 to 40°.1 C) was very moderately inflected in 6 m. A fourth leaf, in water at 110° F (43°.3 C), was somewhat inflected in 4 m., and considerably so in from 6 m. to 7 m.

Three leaves were placed in water which was heated rather quickly, and by the time the temperature rose to 115°-116° F (46°1 to 46°.06 C), all three were inflected. We then removed the lamp, and in a few minutes every single tentacle was closely inflected. The protoplasm within the cells was not killed, for it was seen to be in distinct movement; and the leaves, having been left in cold water for 20 hrs., re-expanded. Another leaf was immersed in water at 100° F (37°.7 C), which was raised to 120° F (48°.8 C); and all the tentacles leaf was now placed in cold water, and in 7 hrs. 30 m. it had partly, and in 10 hrs. fully, re-expanded. On the following morning it was immersed in a weak solution of carbonate of ammonia, and the glands quickly became black, with strongly marked aggregation in the tentacles, showing that the protoplasm was alive, and that the glands had not lost their power of absorption. Another leaf was placed in water at 110° F (43°.3 C) which was raised to 120° F (48°.8 C); and every tentacle, excepting one, was quickly and closely inflected. This leaf was now immersed in a few drops of strong solution of carbonate of ammonia (one part to 109 of water); in 10m. all the glands became intensely black, and in 2 hrs. the protoplasm in the cells of the pedicels was well aggregated. Another leaf was suddenly plugged, and as usual waved about, in water at 120° F and the tentacles became inflected in from 2 m. to 3 m., but only so as to stand at right angles to the disc. The leaf was now placed in the same solution (*viz.*, one part of carbonate of ammonia to 10-9 of water, of 4 grs. to 1 oz., which we will for the future disunite as the strong solution), and when we looked at it again after the interval of an hour, the glands were blackened. And there was well-marked aggregation. After an additional interval of 4 hrs. the tentacles had become much more inflected. It deserves notice that a solution as strong as this never causes inflection in ordinarily cases. Lastly, a leaf was suddenly placed in water at 125° F (51°.6 C), and was left in it until the water cooled; the tenacles were rendered of a bright red and soon became inflected. The contents of the cells underwent some degree of aggregation, which in the course of three hours increase; but the masses of protoplasm did not become spherical, as almost always occurs with leaves immersed in a solution of carbonate of ammonia.

Experiments

We shall now see that a temperature of 130° F (54°.4 C) is too high to cause immediate inflection, yet does not kill the leaves.

Experiment 1: A leaf was plunged, and as in all cases waved about for a few minutes, in water at 130° F (51°.4 C), but there was no trace of inflection; it was then placed in cold water, and after an interval of 15 m. very slow movement was distinctly seen in a small mass of protoplasm in one of the cells of a tentacle. After a few hours all the tentacles and the blade became inflected.

Experiment 2: Another leaf was plugged into water at 130° to 131° F, and as before, there was no inflection. After being kept in cold water for an hour, it was placed in the strong solution of ammonia, and in the course of 55 m. the tentacles were considerably inflected. The glands, which before had been rendered of a brighter red, were now blackened. The protoplasm in the cells of the tentacles was distinctly aggregated; but the spheres were much smaller than those usually generated in unheated leaves when subjected to ammonia. After an additional 2 hrs. all the tentacles, excepting six or seven, were closely inflected.

Experiment 3: A similar experiment to the last, with exactly the same results.

Experiment 4: A fine leaf was placed in water at 100° F (37°.7 C), which was then raised to 145° F (62°.7 C). Soon after immersion, there was, as might have been expected, strong inflection. The leaf was now removed and left in cold water; but from having been exposed to so high a temperature. It never re-expanded.

Experiment 5: Leaf immersed at 130° F (54°.4 C), and the water raised to 145° F (62°.7 C), there was no immediate inflection; it was then placed in cold water, and after 1 hr. 20m. some of the tentacles on one side became inflected. This leaf was now placed in the strong solution, and in 40 m. all the submarginal tentacles were well inflected, and the glands blackened. After an additional interval of 2 hrs. 45 m. all the tentacles, except eight or ten, were closely inflected, with their cells exhibiting a slight degree of aggregation; but the spheres of protoplasm were very small, and the cells of the exterior tentacles continued some pulpy or disintegrated brownish matter.

Experiment 6 and 7: Two leaves were plunged in water at 135° F (57°.2 C) which was raised to 145° F (62°.7 C); neither became inflected. One of these, however been left for 31 m. in cold water, exhibited some slight inflection, which increased after an additional interval of 1 hr. 45m., until all the tentacles, except sixteen or seventeen, were more or less inflected; but the leaf was so much injured that it never re-expanded. The other leaf, after having been left for half an hour in cold water, was put into the storing solution, but so inflection ensued; the glands, however, were blackened, and in some cells there was a little aggregation, the spheres of protoplasm being extremely small; in other cells, especially in the exterior tentacles, there was much greenish- brown pulpy matter.

Experiment 8: A leaf was plunged and waved about for a few minutes in water at 140° F (60° C), and was then left for half an hour in cold water, but there was no inflection. It was now placed in the strong solution, and after 2 hrs. 30 m. the inner submarginal tentacles were well inflected, with their glands blackened, and some imperfect aggregation in the cells of the pedicels. Three or four of the glands were spotted with the white porcelain-like structure, like that produced by boiling water. We have seen this result in no other instance after an immersion of only a few minutes in water at so low a temperature as 140° F, and in only one leaf out of four, after a similar immersion at a temperature of 145° F (62°.7 C), and the other in water at 140° F (60° C), both being left therein until the water cooled, the glands of both became white and porcelain- like, so that the duration of the immersion is an important element in the result.

Experiment 9: A leaf was placed in water at 140° F (60° C) which was raised to 150° F (65°.5 C); there was no inflection; on the contrary, the outer tentacles were somewhat bowed backwards. The glands become like porcelain, but some of them were a little mottled with purple. The bases of the glands were often more affected then their summits. This leaf having been left in the strong solution did not undergo any inflection or aggregation.

Experiment 10: A leaf was plunged in water at 150° to 150½° F (65°.5 C); it became somewhat flaccid, with the outer tentacles slightly reflexed, and the inner ones a little bent inwards, but only towards their tips; and this latter fact shows that the movements was not one of true inflection, as the basal part alone normally bends. The tentacles were as usual rendered of a very bright red, with the glands almost white like porcelain, yet tinged with pink. The leaf having been placed in the strong solution, the cell-contents of the tentacles became of a muddy brown, with no trace of aggregation.

Experiment 11: A leaf was immersed in water at 145° F (62°.7 C), which was raised to 156° F (68°.8 C). The tentacles became bright red and somewhat reflexed, with almost all the glands like porcelain; those on the disc being still pinkish, those near the margin quite white. The leaf being placed as usual first in cold water and then in the strong solution, the cells in the tentacles became of a muddy greenish brown, with the protoplasm not aggregated. Nevertheless, four of the glands escaped being rendered like porcelain and the pedicels of this glands were spirally curled, like a French horn, towards their upper ends; but this can by no means be considered as a case of true inflection. The protoplasm within the cells of the twisted portions was aggregate into distinct though excessively minute purple spheres. This case shown clearly that the protoplasm, after having been exposed to a high temperature for a few minutes, is capable of aggregation when afterwards subjects to the action of carbonate of ammonia, unless the heat has been sufficient to cause coagulation.

Concluding Remarks

As the hair-like tentacles are extremely thin and have delicate walls, and as the leaves were waved about for some minutes close to the bulb of the thermometer, it seems scarcely possible that they should not have been raised very nearly to the temperature which the instrument indicated. From the eleven last observation we see that a temperature of 130° F (54°.4 C) never causes the immediate inflection of the tentacles, though a temperature from 120° to 125° F (48°.8 to 51°.6 C) quickly produces this effect. But the leaves are paralysed only for a time by a temperature of 130° F as afterwards, whether left in simple water or in a solution carbonate of ammonia, they become inflected and their protoplasm undergoes aggregation. This great difference in the effects of higher and lower temperature may be compared with that from immersion in strong and weak solution of the salts of ammonia; for the former do not excite movement, whereas the latter act energetically. A temporary suspension of the power of movement due to heat is called by Sachs heat rigidity; and this in the case of the sensitive plant (mimosa) is induced by its exposure for a few minutes to humid air, raised to 120°-122° F, or 49 to 50° C. It deserves notice that the leaves of Drosera, after being immersed in water at 130° F, are excited into movement by a solution of the carbonate so strong that it would paralyse ordinary leaves and cause no inflection.

The exposure of the leaves for a few minutes even to a temperature of 145° F (62°.7 C) does not always kill them; as when afterwards left in cold water, or in a strong solution of carbonate of ammonia, they generally, though not always, become inflected; and the protoplasm within their cells undergoes aggregation, though the spheres thus formed are extremely small, with many of the cells partly filled with brownish muddy matter. In two instances, when leaves were immersed in water, at a lower temperature than 130° F (54°.4 C) which was then raised to 145° F (62°.7 C), they became during the earlier period of immersion inflected, but on being afterwards left in cold water were incapable of re-expansion. Exposure for a few minutes to a temperature of 145° F sometimes causes some few of the more sensitive glands to be speckled with the porcelain—like appearance; and on one occasion this occurred at a temperature of 140° F (60° C). On another occasion, when a leaf was placed in water at this temperature of only 140° F and left therein till the water cooled, every gland became like porcelain. Exposure for a few minutes to a temperature of 150° F (65°.5 C) generally produces this effect, yet many glands retain a pinkish colour, and many present a specked appearance. This high temperature never causes true inflection; on the contrary, the tentacles commonly become reflexed, though to a less degree than when immersed in boiling water; and this apparently is due to their passive power of elasticity. After exposure to temperature of 150° F, the protoplasm, if subsequently subjected to carbonated of ammonia, instead of undergoing aggregation, is converted into disintegrated or pulpy discoloured matter. In short, the leaves are generally killed by this degree of heat, but owing to differences of age or constitution, they vary somewhat in this respect. In one anomalous case, four out of the many glands on a leaf, which had been immersed in water raised to 156° F (68°.8 C), escaped being rendered porcellanous; and the protoplasm in the cells close beneath these glands underwent some slight, though imperfect, degree of aggregation.

Finally, it is a remarkable fact that the leaves of *Drosera rotundifolia*, which flourished on bleak upland moors throughout Great Britain, and exists (Hooker) within the Arctic Circle, should be able to withstand for even a short time immersion in water heated to a temperature of 145° F.

It may be worth adding that immersion in cold water does not cause any inflection: We suddenly placed four leaves, taken from plants which had been kept for several days at a high temperature, generally about 75° F (23°.8 C), in water at 45° F (7°.2 C), but they were hardly at all affected; not so much as some other leaves from the same plants, which were at the same time immersed in water at 75° F; for these became in a slight degree inflected.

❑❑❑

5

The Effects of Non-nitrogenous and Nitrogenous Organic Fluids on the Leaves

Non-nitrogenous fluids – Solutions of gum arabic – Sugar – Starch – Diluted alcohol – Olive oil – Infusion and decoction of tea – Nitrogenous fluids – Milk – Urine – Liquid albumen – Infusion of raw meat – Impure mucus – Saliva – Solution of isinglass – Difference in the action of these two sets of fluids – Decoction of green peas – Decoction and infusion of cabbage – Decoction of grass leaves.

Effects of Non-nitrogenous Fluids

In all the following cases a drop was allowed to fall from the same pointed instrument on the centre of the leaf; and by repeated trials one of these drops was ascertained to be on an average very nearly half a minim, or $\frac{1}{960}$ of fluid ounce, or .0295 c.c. But these measurements obviously do not pretend to any strict accuracy; moreover, the drops of the viscid fluids were plainly larger than those of water. Only one leaf on the same plant was tried, and the plants were collected from two distant localities. The experiments were made during August and September. In judging of the effects, one caution is necessary; if a drop of any adhesive fluid is placed on an old or feeble leaf, the gland of which have ceased to secrete copiously, the drop sometimes dries up, especially if the plant is kept in a room, and some of the central and submarginal tentacles are thus drawn together, giving to them the false appearance of having become inflected. This sometimes occurs with water, as it is rendered adhesive by months with the viscid secretion. Hence the only safe criterion, and to this alone we have trusted, is the bending inwards of the exterior tentacles, which have not been touched by the fluid, or at most only at their bases. In this case the movement is wholly due to the central glands having been stimulated by the fluid and transmitting a motor impulse to the exterior tentacles. The blade of the leaf likewise often curves inwards, in the same manner as when an insect or bit of meat is placed on the disc. This latter movement is never caused, as far as we have seen by the mere drying up of an adhesive fluid and the consequent drawing together of the tentacles.

First for the non-nitrogenous fluids. As a preliminary trial, drops of distilled water were placed on between thirty and forty leaves, and no effect whatever was produced; nevertheless, in some other and rare cases, a few tentacles became for a short time inflected; but this may have been caused by the glands having been accidentally touched in getting the leafs into a proper position. That water should produce no effect might have been anticipated, as otherwise the leaves would have excited into movement by every shower of rain.

Gum Arabic: Solution of four degrees of strength were made; one of six grains to the ounce of water (one part of 73); a second rather stronger, yet very thin, third moderately thick, and a fourth so thick that it would only just drop from a pointed instrument. These were tried on fourteen leaves; the drops being left on the discs from 24 hrs. to 44 hrs.; generally about 30 hrs. Inflection was never thus caused. It is necessary to try pure gum Arabic, for a friend tried a solution bought ready prepared, and this caused the tentacles to bend; but he afterwards ascertained that it contained much animal matter, probably glue.

Sugar: Drops of a solution of white sugar of three strengths (the weakest containing one part of sugar to 73 of water) were left on fourteen leaves from 32 hrs. to 48 hrs.; but no effect was produced.

Starch: A mixture about as thick as cream was dropped on six leaves and left on them for 30 hrs., no effect being produced. We are surprised at this fact, as we believe that the starch of commerce generally contains a trace of gluten, and this nitrogenous substance causes inflection, as we shall see in the next chapter.

Alcohol, Diluted: One part of alcohol was added to seven of water, and the usual drops were placed on the discs of three leaves. No inflection ensued in the course of 48 hrs. To ascertain whether these leaves had been at all injured, bits of meat were placed on them, and after 24 hrs. they were closely inflected. We also put drops of sherry wine on three other leaves; no inflection was caused, though two of them seemed somewhat injured. We shall hereafter see that cut-off leaves immersed diluted alcohol of the above strength do not become inflection.

Olive Oil: Drops were placed on the discs of eleven leaves, and no effect was produced in from 24 hrs. to 48 hrs. Four of these leaves were then tested by bits of meat on their discs, and three of them were found after 24 hrs. with all their tentacles and blades closely inflected, whilst the fourth had only a few tentacles inflected. It will, however, be shown in a future place, that cut-off leaves immersed in olive oil are powerfully affected.

Infusion and Decoction of Tea: Drops of a strong infusion and decoction, as well as of a rather weak decoction, of tea were placed on ten leaves, none of which became inflected. We afterwards tested three of them by adding bits of meat to the drops which still remained on their discs, and when we examined them after 24 hrs. they were closely inflected. The chemical principle of tea, namely, thence, was subsequently tried and produced no effect. The albumin's matter which the leaves must originally have contained, no doubt, had been rendered insoluble by their having been completely dried.

We thus see that, excluding the experiments with water, sixty-one leaves were tried with drops of the above-named non-nitrogenous fluids; and the tentacles were not in a single case inflected.

Effects of Nitrogenous Fluids

With respect to nitrogenous fluids, the first which came to hand were tried. The experiments were made at the same time and in exactly the same manner as the foregoing. As it was immediately evident that these fluids produced a great effect, we neglected in most cases to record how soon the tentacles became inflected, but this always occurred in less than 24 hrs.; whilst the drops of non-nitrogenous fluids which produced no effect were observed in every case during a considerably longer period.

Milk: Drops were placed on sixteen leaves, and the tentacles of all, as well as the blades of several, soon became greatly inflected, the periods were recorded in only three cases, namely, with leaves on which unusually small drops had been placed. Their tentacle were somewhat inflected in 45 m.; and after 7 hrs. 45 m. The blades of two were so much curved inwards that they formed little cups enclosing the drops. These leaves re-expanded on the third day. On another occasion the blade of a leaf was much inflected in 5 hrs. after a drop of milk had been placed on it.

Human Urine: Drops were placed on twelve leaves, and the tentacles of all, with a single exception, become greatly inflected. Owing to differences in the chemical nature of the urine on different occasion, the time required for the movements of the tentacles varied much, but was always effected in under 24 hrs. In two instances we recorded that all the exterior tentacles were completely inflected in 17 hrs., but not the blade of the leaf. In another case the edges of a leaf, after 25 hrs. 30 m., became so strongly inflected that it was converted into a cup. The power of urine does not lie in the urea, which, as we shall hereafter see, is inoperative.

Albumen (fresh from a hens egg): Drops were placed on seven leaves, caused the tentacles of six of them to be well inflected. In one case the edge of the leaf itself became much curled in after 20 hrs. The one leaf which was unaffected remained so for 26 hrs., and was then treated with a drop of milk, and this caused the tentacles to bend inwards in 12 hrs.

Cold Filtered Infusion of Raw Meat: This was tried only on a single leaf, which had most of its outer tentacles and the blade inflected in 19 hrs. During subsequent years we repeatedly used this infusion to test leaves which had been experimented on with other substances, and it was found to act most energetically, but as no exact account of these trials was kept, they are not here introduced.

Mucus: Thick and thin mucus from the bronchial tubes, placed on three leaves, caused inflection. A leaf with thin mucus had its marginal tentacles and blade somewhat curved inwards in 5 hrs. 30 m. and greatly so in 20 hrs. The action of this fluid no double is due either to the saliva or to some albumen's matter mingled with it, and not, as we shall see in the next chapter, to mucin or the chemical principle of mucus.

Saliva: Human saliva, when evaporated, yields from 1.14 to 1.19 per cent of residue; and this yields 0.25 per cent of ashes, so that the proportion of nitrogenous matter which saliva contains must be small. Nevertheless, drops placed on the discs of eight leaves acted on them all. In one case all the exterior tentacles, excepting nine, were inflected in 19 hrs. 30 m., in another case a few became so in 2 hrs., and after 7 hrs. 30 m. all those situated near where the drop lay, well as the blade, were acted on. Since making these trials, we have many scores of times just touched the gland with the handle of scalpel wetted with saliva, to ascertain whether a leaf was in an active condition; for this was shown in the course of a few minutes by the bending inwards of the tentacles. The edible nest of the Chinese swallow is formed of matter secreted by the salivary gland; two grains were added to one ounce of distilled water (one part to 218), which was boiled for several minutes, but did not dissolve the whole. The usual-sized drops were placed on three leaves, and these in 1 hr. 30 m. were well, and in 2 hrs. 15 m. closely inflected.

Isinglass: Drops of a solution about as thick as milk, and of a still thicker solution, were placed on eight leaves, and the tentacles of all became inflected. In one case the exterior tentacles were well curved in after 6 hrs. 30 m., and the blade of the leaf to partial extent after 24 hrs. As saliva acted so efficiently, and yet contains so small a proportion of nitrogenous matter, we tried how small a quantity of isinglass would act. One part was dissolved in 218 parts of distilled water, and drops were placed on four leaves. After 5 hrs. two of these were considerably and two moderately inflected; after 22 hrs. the former were greatly and the latter much more inflected. In the course of 48 hrs. from the time when the drops were placed on the leaves, all four had almost re-expanded. They were then given little bits of meat and these acted more powerful than the solution, one part of isinglass was next dissolved 437 of water; the fluid thus formed was so thin that it could not be distinguished from pure water. The usual-sized drops were placed on seven leaves, each of which thus received $\frac{1}{960}$ of grain (.0295 mg.). Three of them were observed for 41 hrs., but were in no way affected; the fourth and fifth had two or three of their exterior tentacles inflected after 18 hrs.; the sixth had a few more; and the seventh had in addition the edge of the leaf just perceptibly curved inwards. The tentacles of the four latter leaves began to re-expand after an additional interval of only 8 hrs. Hence the $\frac{1}{960}$ of a grain of isinglass is sufficient to affect very slightly the more sensitive or active leaves. On one of the leaves, which had not been acted on by the weak solution, and on another, which had only two of its tentacles inflected, drops of the solution as thick as milk were placed; and next morning, after an interval of 16 hr., both were found with all their tentacles strongly inflected.

Altogether we experimented on sixty-four leaves with the above nitrogenous fluids, the leaves tried only with the extremely weak solution of isinglass not being included, nor the numerous trials subsequently made, of which no exact account was kept. Of these sixty-four leaves, sixty-three had their tentacles and often their blands well inflected. The one which failed was probably too old and torpid. But to obtain so large a proportion of successful cases, care must be taken to select young and active leaves. Leaves in this condition were chosen with equal care for the sixty-one trials with non-nitrogenous fluids (water not included); and we have seen that not one of these was in the least affected. We may therefore safely conclude that in the sixty-four experiments with nitrogenous fluids the inflection of the experiments was due to the absorption of nitrogenous matter by the glands of the tentacles on the disc.

Difference in the Action of Two Sets of Fluids

Some of the leaves which were not affected by the non-nitrogenous fluids were, as above stated, immediately afterwards tested with bits of meat, and were thus proved to be in an active condition. But in addition these trials, twenty-three of the leaves, with drops of gum, syrup, or starch, still lying on their discs, which produced no effect in the course of between 24 hrs. and 48 hrs., were then tested with drops of milk, urine or albumen. Of the twenty-three leaves thus treated, seventeen had their tentacles, and in some cases their blades, well inflected; but their powers were somewhat impaired, for the rate of movement was decidedly slower than when fresh leaves were treated with these same nitrogenous fluids. The impairment, as well as the insensibility of six of the leaves, may be attributed to injury from exosmose, caused by the density of the fluids placed on their discs.

The results of few other experiments with nitrogenous fluids may be here conveniently given. Decoction of some vegetables known to be rich in nitrogen, were made, and these acted like animal fluids. Thus, a few green peas were boiled for some time in distilled water, and the moderately thick decoction thus made was allowed to settle. Drops of the super incumbent fluid were placed on four leaves, and when these were looked at after 16 hrs., the tentacles and blades of all were found strongly inflected. We infer from a remark by Gerhard that legumin is present in peas "in combination with an alkali, forming an incoagulable solution", and this would mingle with boiling water. We may mention, in relation to the above and following experiments, that according to Schiff certain forms of albumen exist, which are not coagulated by boiling water, but are converted into soluble peptones.

On three occasion chopped cabbage leaves were boiled in distilled water for 1 hr. or for 1¼ hr.; and by decanting the decoction after it had been allowed to rest, a pale dirty green fluid was obtained. The usual-sized drops were placed on thirteen leaves. Their tentacles and blades were inflected after 4 hrs. to quite extraordinary degree. Next day the protoplasm within the cells of the tentacles was fouled aggregated in the most strongly-marked manner. We also touched the viscid secretion round the glands of several tentacles with minute drops of the decoction on the head of a small pin, and they became well inflected in a few minutes. The fluid proving so powerful, one part was diluted with three of water, and drops were placed on the discs of five leaves; and these next morning were so much acted on that their blades were completely doubled over. We thus see that a decoction of cabbage leaves is nearly or quite as potent as an infusion of raw meat.

About the same quantity of chopped cabbage leaves and of distilled water as in the last experiment, were kept in a vessel for 20 hrs. in a hot closet, but not heated to near the boiling point. Drops of this infusion were placed on four leaves. One of these, after 23 hrs., was much inflected; a second slightly; a third had only the submarginal tentacles inflected; and the fourth was not at all affected. The power of this infusions therefore very much less than that of the decoction; and it is clear that the immersion cabbage leaves for an hour in water at the boiling temperature is much more efficient in extracting matter which excited Drosera than immersion during many hours in warm water. Perhaps the counts of the cells are protected (as Schiff remarks with respect to legumin) by the walls being formed of cellulose and that until these are ruptured by boiling-water, but little of the contained albumin's matter is dissolved. We know from the strong odour of cooked cabbage leaves that boiling-water produces some chemical change in them, and that they are thus rendered far more digestible and nutritious to man. It is therefore interesting that water at this temperature extracts matter from them which excites Drosera to an extraordinary degree.

Grasses contain far less nitrogenous matter than do peas or cabbages. The leaves and stalks of three common kinds were chopped and boiled for some time in distilled water. Drops of this decoction (after having stood for 24 hrs.) were placed on six leaves, and acted in a rather peculiar manner, of which other instances will be given in the seventh chapter on the salts of ammonia. After 2 hrs. 30 m. four of the leaves had their blades greatly inflected, but not their exterior tentacle; and so it was with all six leaves after 24 hrs. Two days afterwards the blades, as well as the few submarginal tentacles which had been inflected all re-expanded; and much of the fluid on their discs was by this time absorbed it appears that the decoction strongly excites the glands on the disc, causing the blade to be quickly and greatly inflected; but that the stimulus, differently from what occurs in ordinary cases, does not spread, or only in a feeble degree, to the exterior tentacles.

We may here add that one part of the extract of belladonna was dissolved in 437 of water, and drops were placed on six leaves. Next day all six were somewhat inflected, and after 48 hrs., were completely re-expanded. It was not the included atropine which produced this effect, for we subsequently ascertained that it is quite powerless. We also procured some extract of hyoscyamus from three shops, and made infusions of the same strength as before. Of these three infusion, only one acted on some of the leaves, which were tried. Though druggist's believe that all the albumen is precipitated in the preparation of these drugs, we cannot doubt that some is occasionally retained; and a trace would be sufficient to excite the more sensitive leaves of Drosera.

❑❑❑

6

The Digestive Power of the Secretion of Drosera

The Secretion rendered acid by the direct and indirect excitement of the glands – Nature of the acid – Digestible substances – Albumen, its digestion arrested by alkalies, recommences by the addition of an acid – Meat – Fibrin – Syntonin – Areolar tissue – Cartilage – Fibro-cartilage – Bone – Enamel and dentine – Phosphate of lime – Fibruos basis of bone – Gelatine – Chondrin – Milk, casein and cheese – Gluten – Legumin – Pollen – Globulin – Haematin – Indigestible substances – Epidermic productions – Fibro-elastic tissue – Mucin – Pepsin – Urea – Chitine – Cellulose – Guncotton – Chlorophyll – Fat and oil – Starch – Action of the secretion on living seeds – Summary and concluding remarks.

The Secretion and Excitement of the Glands

As we have seen that nitrogenous fluids act very differently on the leaves of Drosera from non-nitrogenous fluids, and as the leaves remain clasped for a much longer time over various organic bodies than over inorganic bodies, such as bits of glass, cinder, wood, & c., it becomes an interesting inquiry, whether they can only absorb matter already in solution, or render it soluble, that is, have the power of digestion. We shall immediately see that they certainly have this power, and that they act on albumen's compounds in exactly the same manner as does the gastric juice of mammals; the digested manner being afterwards absorbed. This fact, which will be clearly proved, is a wonderful one in the physiology of plants. We must here state that we have been aided through out all our later experiments by many valuable suggestions and assistance given with the greatest kindness by Dr. Burdon Sanderson.

It may be well to premise for the sake of any reader who knows nothing about the digestion of albumen's compounds by animals that this is effected by means of a ferment, pepsin, together with weak hydrochloric acid, though almost any acid will serve. Yet neither pepsin not an acid by itself has any such power. We have seen that when the glands of the disc are excited by the contact of any object especially of one containing nitrogenous matter, the outer tentacles and often the blade become inflected; the leaf being thus converted into a temporary cup or stomach. At the same time the discal glands secrete more copiously, and the secretion becomes acid. Moreover, they transmit some influence to the glands of the exterior tentacles, causing them to pour forth a more copious secretion, which also becomes acid or more acid than it was before.

Nature of the Acid

The secretion of many glands on thirty leaves, which had not been in any way excited, was tested with litmus paper; and the secretion of twenty-two of these leaves did not in the least affect the colour, whereas that of eight caused an exceedingly feeble and sometimes doubtful tinge of red. Two other old leaves, however, which appeared to have been inflected several times, acted much more decidedly on the paper. Particles of clean glass were then placed on five of the leaves, cubes of albumen on six, and bits of raw meat on three, on none of which was the secretion at this time in the least acid. After an interval of 24 hrs., when almost all the tentacles on these fourteen leaves had become more or less inflected, we again tested the secretion, selecting glands which had not as yet reached the centre or touched any object, and it was not plainly acid. The degree of acidity of the secretion varied somewhat on the glands of the same leaf. On some leaves, a few tentacles did not, from some unknown cause, become inflected as often happens; and in five instances their secretion was found not to be in the least acid; whilst the secretion of the adjoining and inflected tentacles on the same leaf was decidedly acid. With leaves excited by particles of glass placed on the central glands, the secretion which collects on the disc beneath them was much more strongly acid than that poured forth from the exterior tentacles, which were as yet only moderately inflected. When bits of albumen (and this is naturally alkaline), or bits of meat were placed on the disc, the secretion collected beneath them was likewise strongly acid. As raw moistened with water is slightly acid, we compared its action on litmus paper before it was placed on the leaves, and afterwards when bathed in the secretion; and there could not be the least doubt that the latter was very much more acid. We have indeed tried hundreds of times the state of the secretion on the discs of leaves which were inflected over various objects, and never failed to find it acid. We may, therefore, conclude that the secretion from unexcited leaves, though extremely viscid, is not acid or only slightly so, but that it becomes acid, or much more strongly so, after the tentacles have began to bend over any inorganic or organic object; and still more strongly acid after the tentacles have remained for some time closely clasped over any object.

We may here remind the reader that the secretion appears to be to a certain extent antiseptic, as it checks the apperance of mould and infusoria, thus preventing for a time the discolouration and decay of such substances as the white of an egg, cheese, &c. It therefore acts like the gastric juice of the higher animals, which is known to arrest putrefaction by destroying the microzymes.

As we were anxious to learn what acid the secretion contained 445 leaves were washed in distilled water given by Prof. Frankland; but the secretion is so viscid that it is scarcely possible to scrape or wash off the whole. The conditions were also unfavourable, as it was late in the year and the leaves were small. Prof. Frankland with great kindness undertook to test the fluid thus collected. The leaves were excited by clean particles of glass placed on them 24 hrs. previously. No doubt much more acid would have been secreted had the leaves been excited by animal matter, but this would have rendered the analysis more difficult. Prof. Frankland informed that the fluid contained no trace of hydrochloric, sulfuric, tartaric, oxalic, or formic acids. This having been ascertained, the remainder of the fluid was evaporated nearly to dryness, and acidified with sulfuric acid; it then evolved volatile acid vapour, which was condensed and digested with carbonate of silver. "The weight of the silver salt thus produced was only 37 gr., much too small a quantity for the accurate determination of the molecular weight of the acid. The number obtained, however, correspond nearly with that of propionic acid; and we believe that this, or a mixture of acetic and butyric acids, were present in the liquid. The acid doubtless belongs to the acetic or fatty series."

Prof. Frankland, as well as his assistant, observed (and this is an important fact) that the fluid, "when acidified with sulphuric acid, emitted a powerful odour like that of pepsin". The leaves from which the secretion had been washed were also sent to Prof. Frankland; they were macerated for some hours, then acidified with sulphuric acid and distilled, but no acid passed over. Therefore the acid which fresh leaves contain, as shown by their discolouring litmus paper when crushed, must be a different nature from that present in the secretion. Nor was any odour or pepsin emitted by them.

Although it has long been known that pepsin with acetic acid has the power of digesting albumen's compounds, it appeared advisable to ascertain whether acetic acid could be replaced without the loss of digestive power, by the allied acids which are believed to occur in the secretion of Drosera, namely, propionic, butyric, or valerian. Dr. Burden Sanderson did the following experiments, the results of which are valuable, independently of the present inquiry. Prof. Frankland supplied the acids.

1. The purpose of the following experiments was to determine the digestive activity of liquids containing pepsin, when acidulated with certain volatile acids belonging to the acetic series, in comparison with liquids acidulated with hydrochloric acid, in proportion similar to that in which it exists in gastric juice.

2. It has been determined empirically that the best results are obtained in artificial digestion when a liquid containing two per thousands of hydrochloric acid gas by weight is used. This corresponds to about 6.25 cubic centimetres per litre of ordinary strong hydrochloric acid. The quantities of propionic, butyric and valerian acids respectively which are required to neutralise as much base as 6.25 cubic centimetres of HCl, are in Grammies 4.04 of propoinic acid, 4.82 of butyric acid, and 5.68 of valerian acid. It was therefore judged expedient, in comparing the digestive powers of these acids with that of hydrochloric acid, to use them in these proportions.

3. Five hundred cub. cent. of a liquid containing about 8 cub. cent. of a glycerine extract of the mucous membrane of the stomach of a dog killed during digestion having been prepared, 10 cub. cent. of it were evaporated and dried at 110°C. This quantity yielded 0.0031 of residue.

4. Of this liquid four quantities were taken which were severally acidulated with hydrochloric, propionic, butyric, and valerian acids, in the

proportions above indicated. Each liquid was then placed in a tube, which allowed to float in a bath, containing a thermometer which indicated a temperature of 38° to 40° C. Into each, a quantity of unboiled fibrin was introduced, and the whole allowed to stand for four hours, the temperature being maintained during the whole time, and care being taken that each contained throughout an excess of fibrin. At the end of the period each liquid was filtered. Of the filtrate, which of course continued as much of the fibrin as had been digested during the four hours, 10 cub. cent. were measured out and evaporated, and dried at 110°C as before. The residues were respectively:

In the liquid containing hydrochloric acid	0.4079
In the liquid containing Propionic acid	0.0601
In the liquid containing butyric acid	0.1468
In the liquid containing valerianic acid	0.1254

Hence, deducting from each of these the above-mentioned residue, left when the digestive liquid itself was evaporated, *viz.*, 0.0031, we have,

For propionic acid	0.0570
For butyric acid	0.1437
For valerianic acid	0.1223

as compared with 0.4048 for hydrochloric acid; these several numbers expressing the quantities of fibrin by weight digested in presence of equivalent of the respective acids under identical conditions.

The results of the experiment may be stated thus—If 100 represent the digestive power of a liquid containing pepsin with the usual proportion of hydrochloric acid, 14.0, 35.4 and 30.2 will represent respectively the digestive powers of the three acids under investigation.

5. In a second experiment in which the procedure was in every respect the same, excepting that all the tubes were pluged into the same water-bath, and the residues dried at 115°C, the results were as follows:

Quantity of fibrin dissolved in four hours by 10 cub. cent. of the liquid:

Propionic acid	0.0563
Butyric acid	0.0835
Valerianic acid	0.0615

The quantity digested by a similar liquid containing hydrochloric acid was 0.3376. Hence, taking this as 100, the following numbers represent the relative quantities digested by the other acids:

Propionic acid	16.5
Butyric Acid	24.7
Valerianic acid	16.1

6. A third experiment of the same kind gave:

Quantity of fibrin digested in four hours by 10 cub. cent. of the liquid:

Hydrochloric acid	0.2915
Propionic acid	0.1490
Butyric Acid	0.1044
Valerianic acid	0.0520

Comparing, as before, the three last numbers with the first taken as 100, the digestive power of propionic acid is represented by 16.8; that of butyric by 35.8; and that of valerianic by 17.8.

The mean of these three sets of observations (hydochloric acid being taken as 100) given for

Propionic acid	15.8
Butyric acid	32.0
Valerianic acid	21.4

7. A further experiment was made to ascertain whether the digestive activity of butyric acid (which was selected as being apparently the most efficacious) was relatively greater at ordinary temperatures than at the temperature of the body. It was found that whereas 10 cub. cent. of liquid containing the ordinary proportion of hydrochloric acid digested 0.1311 gramme, a similar liquid prepared with butyric acid digested 0.0455 gramme of fibrin.

"Hence, taking the quantities with hydrochloric acid at the temperature of the body as 100° F, we have the digestive power of hydrochloric acid at the temperature of 16° to 18° C represented by 44.9; that of butyric acid at the same temperature being 15.6."

We here see that at the lower of these two temperatures, hydrochloric acid with pepsin digests, within the same time, rather less than half the quantity of fibrin compared with, what it digests at the higher temperature: and the power of butyric acid is reduced in the same proportion under similar condition and temperatures. We have also seen that butyric acid, which is much more efficacious than propionic or valerian acids, digests with pepsin at the higher temperature less than a third of the fibrin which is digested at the same temperature by hydrochloric acid.

We will now give in detail my experiments on the digestive power of the secretion of Drosera, dividing the substances tried into two series, namely, those which are digested more or less completely, and those which are not digested. We shall presently see that all these substances are acted on by the gastric juice of the higher animals in the same manner. We beg leave to call attention to the experiments under the head albumen, showing that the secretion loses its power when neutralised by an alkali, and recovers it when an acid is added.

Substances which are Completely or Partially Digested by the Secretion of Drosera

Albumen: After having tried substances, Dr. Burdon Sanderson suggested the use of cubes of coagulated albumen of hard-boiled egg. We may premise that five cubes of the same size as those used in the following experiments were placed for the sake of comparison at the same time on wet moss close to the plants of Drosera. The weather was hot, and after four days some of the cubes were discoulored and mouldy, with their angles a little rounded; but they were not surrounded by a zone of transparent fluid as in the case of those undergoing digestion. Other cubes retained their angles and whitle colour. After eight days all were somewhat reduced in size, discoloured, with their angles much rounded. Nevertheless in four out of the five specimens, the central parts were still white and opaque, so that their state differed widely, as we shall see, from that of the cubes subjected to the action.

Experiment 1: Rather large cubes of albumen were first tried; the tentacles were well inflected in 24 hrs.; after an additional day the angles of the cubes were dissolved and rounded; but the cubes were too large, so that the leaves were injured, and after seven days one died and the others were dying. Albumen which has been kept for four or five days, and which, it may be presumed, has begun to decay slightly, seems to act more quickly than freshly boiled eggs. As the latter were generally used, we often moistened them with a little saliva, to make the tentacles close more quickly.

Experiment 2: A cube of $\frac{1}{10}$ of an inch (*i.e.*, with each side $\frac{1}{10}$ of an inch, or 2.54 mm., in length) was placed on a leaf, and after 50 hrs. it as converted into a sphere about $\frac{3}{40}$ of an inch (1.905 mm.) in diameter, surrounded by perfectly transparent fluid. After ten days the leaf re-expanded, but there was still left on the disc a minute bit of albumen now rendered transparent. More albumen had been given to the leaf than could be dissolved or digested.

Experiment 3: Two cubes of albumen of $\frac{1}{20}$ of an inch (1.27 mm.) were placed on two leaves. After 46 hrs. every atom of one was dissolved, and most of the liquefied matter was absorbed, the fluid which remained being in this, as in all other cases, very acid and viscid. The other cube was acted on at a rather slower rate.

Experiment 4: Two cubes of albumen of the same size as the last were placed on two leaves, and were converted in 50 hrs. into two large drops of transparent fluid; but when these were removed from beneath the inflected tentacles, and viewed by reflected light under the microscope, fine streaks of white opaque matter could be seen in the one, and traces of similar streaks in the other. The drops wee replaced on the leaves, which re-expanded after 10 days; and now nothing was left except a very little transparent acid fluid.

Experiment 5: This experiment was slightly varied, so that the albumen might be more quickly exposed to the action of the secretion. Two cubes, each of about 1/10 of an inch (.635 mm.) were placed on the same leaf, and two similar cubes on another leaf. These were examined after 21 hrs. 30 m., and all four were found rounded. After 46 hrs. the two cubes on the one leaf were completely liquefied, the fluid being perfectly transparent; on the other leaf some opaque white streaks could still be seen in the midst of the fluid. After 72, hrs. these streaks disappeared, but there was still a little viscid fluid left on the disc; whereas it was almost all absorbed on the first leaf. Both leaves were now beginning to re-expand.

The best and almost sole test of the presence of some ferment analogous to pepsin in the secretion appeared to be to neutralise the acid of the secretion with an alkali, and to observe weather the process of digestion ceased; and then to add a little acid and observe whether the process recommenced. This was done, and as, we shall see, with success, but it was necessary first to try two control experiments; namely, whether the addition of minute drops of water of the same size as those of the dissolved alkalis to be used would stop the process of digestion; and, secondly, whether minute drops of weak hydrochloric acid, of the same strength and size as those to be used, would injure the leaves. The two following experiments were therefore tried:

Experiment 6: Small cubes of albumen were put on there leaves, and minute drops of distilled water on the head of a pin were added two or three times daily. These did not in the least delay the process; for, after 48 hrs., the cubes were completely dissolved on all there leaves. On the third day the leaves to re-expand, and on the fourth day all the fluid was absorbed.

Experiment 7: Small cubes of albumen were put on two leaves, and minute drops of hydrochloric acid, of the strength of one part of 437 of water, were added two or three times. This did not in the least delay, but seemed rather to hasten, the process of digestion; for every trace of the albumen disappeared in 24 hrs. 30 m. After three days the leaves partially re-expanded, and by this time almost all the viscid fluid on their discs was absorbed. It is almost superfluous to state that cubes of albumen of the same size as those above used, left for seven days in a little hydrochloric acid of the above strength, retained all their angles as perfect as ever.

Experiment 8: Cubes of albumen (of $\frac{1}{20}$ of an inch, or 1.27 mm.) were placed on five leaves, and minute drops of a solution of one part of carbonate of soda to 437 of water were added at intervals to three of them, and drops of carbonate of potash, of the same strength to the other two. The drops were given on the head of a rather large pin, and we ascertained that each was equal of about $\frac{1}{10}$ of a minim (.0059 cc), so that each contained only $\frac{1}{4800}$ of a grain (.0135 mg.) of the alkali. This was not sufficient, for after 46 hrs. all five cubes were dissolved.

Experiment 9: The last experiment was repeated on four leaves, with this difference, that drops of the same solution of carbonate of soda were added rather oftener, as often as the secretion became acid, so that it was much more effectually neutralised. And now after 24 hrs. the angles of three of the cubes were not in the least rounded, those of the fourth being so in a very slight degree. Drops of extremely weak hydrochloric acid (*viz.*, one part to 847 of water) were then added, just enough to neutralise the alkali which was still present; and now digestion immediately recommenced, so that after 23 hrs. 30 m., three of the cubes were completely dissolved, whilst the fourth was converted into a minute sphere, surrounded by transparent fluid; and this sphere next day disappeared.

Experiment 10: Stronger solutions of carbonate of soda and of potash were next used, *viz.*, one part to 109 of water; and as the same sized drops were given as before, each drop contained $\frac{1}{1200}$ of a grain (.0539 mg.) of either salt. Two cubes of albumen (each about $\frac{1}{40}$ of an inch, or .635 mm.) were placed on the same leaf, and two on another. Each leaf received, as soon as the secretion became slightly acid (and this occurred four time within 24 hrs.) drops either of the soda or potash, and the acid was thus effectually neutralised. The experiment now succeeded perfectly, for after 22 hrs. the angles of the cubes were as sharp as they were at first, and we know from experiment 5 that such small cubes would have been completely rounded within this time by the secretion in its natural state. Some of the fluid was now removed with blotting-paper from the discs of the leaves, and minute drops of hydrochloric acid of the strength of one part to 200 water was added. Acid of this greater strength was used as the solutions of the alkalis were stronger. The process of digestion now commenced, so that within 48 hrs. from the time when the acid was given the four cubes were not only completely dissolved, but much of the liquefied albumen was absorbed.

Experiment 11. Two cubes of albumen ($\frac{1}{40}$ of an inch, or .635 mm.) were placed on two leaves, and were treated with alkalis of the last experiment, and with the same result; for after 22 hrs. they had their angles perfectly sharp, showing that the digestive process had been completely arrested. We then wished to ascertain what would be the effect of using stronger hydrochloric acid; so we added minute drops of the strength of 1 per cent. This proved rather too strong, for after 48 hrs. from the time when the acid was added one cube was still almost perfect, and the other only very slightly rounded, and leaves were not thus coloured, and we can thus understand why the cubes were not dissolved.

From these experiments we clearly see that the secretion has the power of dissolving albumen, and we further see that if an alkali is added, the process of digestion is stopped, but immediately recommences as soon as the alkali is neutralised by weak hydrochloric acid. Even if we had tried no other experiments than these, they would have almost sufficed to prove that the glands of Drosera secrete some ferment analogous to pepsin, which in presence of an acid gives to the secretion its power of dissolving albumin's compounds.

Splinters of clean glass were scattered on a large number of leaves, and these became moderately inflected. They were cut off and divided into three lots; two of them, after being left for some time in a little distilled water, were strained, and some discoloured, viscid, slightly acid fluid was thus obtained. The third lot was well soaked in a few drops of glycerine, which is well known to dissolve pepsin. Cubes of albumen ($\frac{1}{20}$ of an inch) were not placed in the three fluids in watch-glasses, some of which were kept for several days at about 90° F (32°.2 C) and others at the temperature of my room; but none of the cubes were dissolved, the angles remaining as sharp as ever. This fact probably indicates that the ferment is not secreted until the glands are excited by the absorption of a minute quantity of already soluble animal matter—a conclusion which is supported by what we shall hereafter see with respect to Dionaea. Dr. Hooker likewise found that, although the fluid within the pitchers of Nepenthes possess extraordinary power of digestion, yet when removed from the pitchers before they have been excited and placed in a vessel, it has no such power, although it is already acid; and we can account for this fact only on the supposition that the proper ferment is not secreted until some exciting matter is absorbed.

On three other occasions eight leaves were strongly excited with albumen moistened with saliva; they were then cut off, and allowed to soak for several hours or for a whole day in few drops of glycerine. Some of this extract was added to a little hydrochloric acid of various strengths (generally one to 400 of water), and minute cubes of albumen were placed in the mixture. In two of these trails the cubes were not in the least acted on; but in the third the experiment was successful. For in a vessel containing two cubes, both were reduced in size in 3 hrs.; and after 24 hrs. mere streaks of undissolved albumen were left. In a second vessel, containing two minute ragged bits of albumen, both were likewise reduced in size in 3 hrs., and after 24 hrs., completely disappeared. We then added a little weak hydrochloric acid to both vessels, and placed fresh cubes of albumen in them; but these were not acted on. This latter fact is intelligible according to the high authority of Schiff, who has demonstrated, as he believes, in opposition to the view held by some physiologists, that a certain small amount of pepsin is destroyed during the act of digestion. So that if my solution contained, as is probable, an extremely small amount of the ferment, this would have been consumed by the dissolution of the cubes of albumen first given: none being left when the hydrochloric acid was added. The destruction of the ferment during the process of digestion, or its absorption after the albumen had been converted into a peptone, will also account for only one out of the three latter sets of experiments having been successful.

Digestion of Roast Meat: Cubes of about $\frac{1}{20}$ of an inch (1.27 mm.) of moderately roasted meat were placed on five leaves which became in 12 hrs. closely inflected. After 48 hrs. We gently opened one leaf, and the meat now consisted of a minute central sphere, partially digested and surrounded by a thick envelope of transparent viscid fluid. The whole, without being much disturbed, was removed and placed under the microscope. In the central part the transverse striae on the muscular fibres were quite distinct; and it was interesting to observe how gradually they disappeared, when the same fibre was traced into the surrounding fluid. They disappeared by the striae being replaced by transverse lines formed of excessively minute dark points, which towards the exterior could be seen only under a very high power; and ultimately these points were lost.When we made these observation, we had not read Schiff a account of the digestion of meat by gastric juice, and we did not understand the meaning of the dark points.

It the viscid fluid surrounding the central sphere of undigested meat there were globules of fat and little bits of fibro-elastic tissue; neither of which were in the least digested. There were also little free parallelograms of yellowish, highly translucent matter. Schiff, in speaking of the digestion of meat by gastric juice, alludes to such parallelograms.

After an interval of 72 hrs. from the time when the five cubes were placed on the leaves, we opened the four remaining ones. On two nothing could be seen by little masses of transparent viscid fluid; but when these were examined under a high power, fat-globules, bits of fibro-elastic tissue, and some few parallelograms of sarcous matter, could be distinguished, but not a vestige of transverse striae. On the two leaves there were minute spheres of only partially digested meat in the centre of much transparent fluid.

Fibrin: Bits of fibrin were left in water during four days, whilst the following experiments were tried, but they were not in the least acted on. The fibrin which we first used was not pure, and included dark particles: it had either not been well prepared or had subsequently undergone some change. Thin portions, about $\frac{1}{10}$ of an inch square, were placed on several leaves, and though the fibrin was soon liquefied, the whole was never dissolved. Smaller particles were then placed on four leaves, and minute drops of hydrochloric acid (one part to 437 of water) were added; this seemed to hasten the process of digestion, for on one leaf all was liquefied and absorbed after 20 hrs.; but on the three other leaves some undissolved residue was left after 48 hrs. It is remarkable that in all the above and following experiments, as well as when much larger bits of fibrin were used, the leaves were very little excited; and it was sometimes necessary to add a little saliva to induce complete inflection. The leaves, moreover, began to re-expand after only 48 hrs. Whereas they would have remained inflected for much longer time had insects, meat, cartilage, albumen, & c., been placed on them.

We then tried some pure white fibrin, sent by Dr. Burdon Sanderson.

Experiment 1: Two particles, barely 1/20 of an inch (1.27 mm.) square, were placed on opposite sides of the same leaf. One of these did not excite the surrounding tentacles, and the gland on which it rested soon dried. The other particles caused a few of the short adjoining tentacles to be inflected, the more distant ones not being affected. After 24 hrs. both were almost, and after 72 hrs. completely, dissolved.

Experiment 2: The same experiment with the same result, only one of the two bits of fibrin exciting the short surrounding tentacles. This bit was so slowing acted on that after a day we pushed it on to some fresh glands. In three days from the time when it was first placed on the leaf it was completely dissolved.

Experiment 3: Bits of fibrin of about the same size as before were placed on the discs of two leaves; these caused very little inflection in 23 hrs., but after 48 hrs. both were well clasped by the surrounding short tentacles, and after an additional 24 hrs. were completely dissolved. On the disc of one of these leaves much clear acid fluid was left.

Experiment 4: Similar bits of fibrin were placed on the discs of two leaves; as after 2 hrs. the glands seemed rather dry, they were freely moistened with saliva; this soon caused strong inflection both of the tentacles and blades, with copious secretion from the glands. In 18 hrs., the fibrin was completely liquefied, but undigested atoms still floated in the liquid; these, however, disappeared in under two additional days.

From these experiments it is clear that the secretion completely dissolves pure fibrin. The rate of dissolution is rather slow; but this depends merely on this substance not exciting the leaves sufficiently, so that only the immediately adjoining tentacles are inflected, and the supply of secretion is small.

Syntonin: This substance, extracted from muscle, was kindly prepared by Dr. Moore. Very differently from fibrin, it acts quickly and energetically. Small portions placed on the discs of three leaves caused their tentacles an blades to be strongly inflected within 8 hrs. But no further observation were made. It is probably due to the presence of this substance that raw meat is to powerful a stimulant, often injuring or even killing the leaves.

Areolar Tissue: Small portions of this tissue from a sheep were placed on the discs of three leaves; these became moderately well inflected in 24 hrs., but began to re-expand after 48 hrs., and were fully re-expanded in 72 hrs., always reckoning from the time when the bits were first given. This substance, therefore, like fibrin, excites the leaves for only a short time. The residue left on the leaves, after they were fully re-expanded, was examined under a high power and found much altered, but, owing to the presence of a quantity of elastic tissue, which is never acted on, could hardly be said to be in a liquefied condition.

Some areolar tissue free from elastic tissue was next procured from the visceral cavity of a toad, and moderately sized, as well as very small, bits were placed on five leaves. After 24 hrs. two of the bits were completely liquefied; two others were rendered transparent, but not quite liquefied; whilst the fifth was but little affected. Several glands on the three latter leaves were not moistened with a little saliva, which soon caused much inflection and secretion, with the results that in the course of 12 additional hrs., one leaf alone showed a remnant of undigested tissue. On the discs of the four other leaves (to one of which a rather large bit had been given) nothing was left except some transparent viscid fluid. We may add that some of this tissue indulged points of black pigment, and these were not at all affected. As a control experiment, small portions of this tissue were left in water and on wet moss for the same length of time, and remained white and opaque. From these facts it is clear that areolar tissue is easily and quickly digested by the secretion; but that it does not greatly excited the leaves.

Cartilage: Three cubes ($\frac{1}{20}$ of an inch or 1.27 mm.) of white, translucent, extremely tough cartilage were cut from the end of a slightly roasted leg-bone of a sheep. These were placed on three leaves, borne by poor, small plants in my greenhouse during November; and it seemed in the highest degree improbable that so hard a substance would be digested under such unfavourable circumstances. Nevertheless, after 48 hrs., the cubes were largely dissolved and converted into minute spheres, surrounded by transparent, very acid fluid. Two of these spheres were completely softened to their centres; whilst the third still contained a very small irregularly shaped core of solid cartilage. Their surfaces were seen under the microscope to be curiously marked by prominent ridges, showing that the cartilage had been unequally corroded by the secretion. We need hardly say that cubes of the same cartilage, kept in water for the same length of time, were not in the least affected.

During a more favourable season, moderately sized bits of the skinned ear of a cat, which includes cartilage, areolar and elastic tissue, were placed on three leaves. Some of the glands were touched with saliva, which caused prompt inflection. Two of the leaves began to re-expaned after three days, and the third on the fifth day. The fluid residue left on their discs was now examined, and consisted in one case of perfectly transparent, viscid matter; in the other two cases, it contained some elastic tissue and apparently remnants of half digested areolar tissue.

Fibro-cartilage *(from between the vertebrae of the tail of a sheep)*: Moderately sized and small bits (the latter about $\frac{1}{20}$ of an inch) were placed on nine leaves. Some of the these were well and some very little inflected. In the latter case the bits were dragged over the discs, so that they were well bedaubed with the secretion, and many glands thus irritated. All the leaves re-expanded after only two days; so that they were but little excited by this substance. The bits were not liquefied, but were certainly in an altered condition, being swollen, much more transparent, and so tender as to disintegrate very easily. Francis prepared some artificial gastric juice, which was proved efficient by quickly dissolving fibrin, and suspended portions of the fibro-cartilage in it. These swelled and became hyaline, exactly like those exposed to the secretion of Drosera, but were not dissolved. This result surprised much, as two physiologists were of opinion that fibro-cartilage would be easily digested by gastric juice. Dr. Klein examined and reported that the two which had been subjected to artificial gastric juice were "in that state of digestion in which we find connective tissue when treated with an acid, *viz.*, swollen, more or less hyaline, the fibular bundles having become homogeneous and lost their fibriller structure". In the specimens which had been left on the leaves of Drosera, until they re-expanded, "parts were altered, though only slightly so, in the same manner as those subjected to the gastric juice, as they had become more transparent, almost hyaline, with the fibrillation of the bundles indistinct". Fibro-cartilage is therefore acted on in nearly the same manner by gastric juice and by the secretion of Drosera.

Bone: Small smooth bits of the dried hyoidal bone of a fowl moistened with saliva were placed on two leaves, and a similarly moistened splinter of an extremely hard, broiled mutton-chop on a third leaf. These leaves soon became strongly inflected, and remained so for an unusual length of time; namely, one leaf for ten and the other two for nine days. The bits of bone were surrounded all the time by acid secretion. When examined under a weak power, they were found quite softened, so that they were readily penetrated by a blunt needle, torn into fibre, or compressed. Dr. Klein was so kind as to make section of both bones and examine them. He informed that both presented the normal appearance of decalcified bone, with traces of the earthy salts occasionally left. The corpuscles with their processes were very distinct in most parts; but in some parts, especially near the periphery of the hyoidal bone, none could be seen. Other parts again appeared amorphous, with even the longitudinal striation of bone not distinguishable. This amorphous structure, as Dr. Klein thinks, may be the result either of the incipient digestion of the fibrous basis or of all the earthy matter having been removed, the corpuscles being thus rendered invisible. A hard, brittle, yellowish substance occupied the position of the medulla in the fragments of the hyoidal bone.

As the angles and little projections of the fibrous basis were not in the least rounded or corroded, two of the bits were placed on fresh leaves. These by the next morning were closely inflected, and remained so – the one for six and the other for seven days—therefore for not so long a time as on the first occasion, but for a much longer time than ever occurs with leaves inflected over inorganic or even over many organic bodies. The secretion during the whole time coloured litmus paper of a bright red; but this may have been due to the presence of the acid super phosphate of lime. When the leaves re-expanded, the angles and projection of the fibrous basis were as sharp as ever. We therefore concluded, falsely, as we shall presently see, that the secretion cannot touch the fibrous basis of bone. The more probable explanations that the acid was all consumed in decomposing the phosphate of lime which still remained; so that none was left in a free state to act in conjunction with the ferment on the fibrous basis.

Enamel and Dentine: As the secretion decalcified ordinary bone, we determined to try whether it would act on enamel and dentine, but did not expect that it would succeed with so hard a substance as enamel. Dr. Klein gave some thin transverse slices of the canine tooth of a dog; small angular fragments of which were placed on four leaves; and these were examined each succeeding day at the same hour. The results are worth giving in detail.

Experiment 1: May 1st, fragment placed on leaf 3rd, tentacles but little inflected, so a little saliva was added; 6th, as the tentacles were not strongly inflected, the fragment was transferred to another leaf, which acted at first slowly, but by the 9th closely embraced it. On the 11th this second leaf began to re-expand; the fragment was manifestly softened, and Dr. Klein reported, "great deal of enamel and the greater part of the part of the dentine decalcified".

Experiment 2: May 1st, fragment placed on leaf; 2nd, tentacles fairly well inflected, with much secretion on the disc, and remained so until the 7th, when the leaf re-expanded. The fragment was now transferred to a fresh leaf, which next day (8th) was inflected in the strongest manner, and thus remained until the 11th,when it expanded. Dr. Klein reported, "a great deal of enamel and the greater part of the dentine decalcified".

Experiment 3: May 1st, fragment moistened with saliva and placed on a leaf, which remained well inflected until 5th, when it re-expanded. The enamel was not at all, and the dentine only slightly, softened. The fragment was now transferred to a fresh leaf, which next, morning (6th) was strongly inflected, and remained so until the 11th. The enamel and dentine both now somewhat softened; and Dr. Klein reported, "less than half the enamel, but the greater part of the dentine decalcified".

Experiment 4: May 1st, a minute and thin bit of dentine, moistened with saliva, was placed on a leaf, which was soon inflected, and re-expanded on the 5th. The dentine had become as flexible as thin paper. It was then transferred to a fresh leaf, which next morning (6th) was strongly inflected, and reopened on the 10th. The decalcified was now so tender than it was torn into shreds merely by the force of the re-expanding tentacles.

From these experiments it appears that enamel is attached by the secretion with more difficulty than dentine, as might have been expected from its extreme harness; and both with more difficulty than ordinary bone. After the process of dissolution has once commenced, it is carried on with greater ease; this may be inferred from the leaves, to which the fragment were transferred, becoming in all four cases strongly inflected in the course of a single day; whereas the first set of leaves acted much less quickly and energetically. The angles of projection of the fibrous basis of the enamel and dentine (except, perhaps, in no. 4, which could not be well observed) were not in the least rebounded; and Dr. Klein remarks that their microscopical structure was not altered. But this could not have been expected, as the decalcification was not complete in the three specimens, which were carefully examined.

Fibrous Bass of Bone: We at first concluded, as already stated, that the secretion could not digest this substance. We therefore asked Dr. Burdon Sanderson to try bone, enamel, and dentine, in artificial gastric juice, and he found that they were after a considerable time completely dissolved. Dr. Klein examined some of the small lamellae, into which part of the skull of cat became broken up after abut a week's immersion in the fluid, and he found that towards the edges the "matrix appeared rarified, thus producing the appearance as if the canaliculi of the bone-corpuscles had become larger. Otherwise the corpuscles and their canaliculi were very distinct." So that with bone subject to artificial gastric juice complete decalcification precedes the dissolution of the fibrous basis. Dr. Burdon Sanderson suggested to that the failure of Drosera to digestive fibrous basis of bone, enamel, and dentine, might be due to the acid being consumed in the decomposition of the earthy salts, so that there was none left for the work of digestion. Accordingly, my son thoroughly decalcified the bone of a sheep with weak hydrochloric acid; and seven minute fragments of the fibrous basis were placed on so many leaves, four of the garments being first damped with saliva to aid prompt inflection. All seven leaves became inflected, but only very moderately, in the course of a day. They quickly began to re expand; five of them of the second day, and the other two the third day. On all seven leaves the fibrous tissue was converted into perfectly transparent, viscid, more or less liquefied little masses. In the middle, however, of one, saw under a high power a few corpuscles, with traces of fibrillation in the surrouding transparent matter. From these facts it is clear that the leaves are very little excited by the fibrous basis of bone, but that the secretion easily and quickly liquefies it, if thoroughly decalcified. The glands which had remained in contact for two or three days with the viscid masses were not discoloured, and apparently had absorbed little of the liquefied tissue, or had been little affected by it.

Phosphate of Lime: As we have seen that the tentacles of the first set of leaves remained clasped for nine or ten days over minute fragments of bone, and the tentacles of the second set for six or seven days over the same fragments, we were led to suppose that it was the phosphate of lime, and not any included animal matter, which caused such long continued inflection. It is at least certain from what has just been shown that this cannot have been due to the presence of the fibrous basis. With enamel and dentine(the former of which contains only 4 per cent of organic mater) the tentacles of two successive sets of leaves remained inflected altogether for eleven days. In order to test my belief in the potency of phosphate of lime, we procured some from Prof. Frankland absolutely free of animal matter and of any acid. A small quantity moistened with water was placed on the disc of two leaves. One of these was only slightly affected; the other remained closely inflected for ten days, when a few of the tentacles began to re-expand, the rest being much injured or killed. We repeated the experiment, but in moistened the phosphate with saliva to insure prompt inflection; one leaf remained inflected for six days (the little saliva used would not have acted for nearly so long a time) and then dies, the other leaf tried to re-expand on the sixth day, but after nine days failed to do so, and likewise died. Although the quantity of phosphate given to the above four leaves was extremely small, much was left in every case undissolved. A larger quantity wetted with water was next placed on the disc of the three leaves; and these became most strongly inflected in the course of 24 hrs. They never re-expanded; on the fourth day they looked sickly, and on the sixth were almost dead. Large drops of not very viscid fluid hung from their edges during the six days. This fluid was tested each day with litmus paper, but never coloured it; and this circumstance we do not understand, as the super phosphate of lime is acid. We suppose that some super phosphate must have been formed by the acid of the secretion acting on he phosphate, but that it was all absorbed and injured the leaves; the large drops which hung from their edges being an abnormal and dropsical secretion, any how, it is manifest that the phosphate of lime is a most powerful stimulant. Even small doses are more or less poisonous, probably on the same principle that raw meat and other nutrition's substances, given in excess, kill the leaves hence the conclusion, that the long-continued inflection of the tentacles over fragments of bone, enamel dentine, is caused by the presence of phosphate of lime, and not of any included animal matter, is no doubt correct.

Gelatine: For comparsion, squares of the same size as those placed on the leaves were left close by on three moss. These soon swelled, but retained their angles for three days, after five days they formed rounded, softened masses, but even on the eighth day a trace of gelatine could still be detected. Other squares were immersed in water, and these, though much swollen, retained their angles for six days. Squats of $\frac{1}{10}$ of an inch (2.54 mm), just moistened with water, were placed on two leaves; and after two or three days nothing was left on them but some acid viscid fluid which in this and other cases never showed any tendency to regelatinise; so that the section must act on the gelatine manner as gastric juice. Four squares of the same size as before were then soaked for three days in water, and placed in two days, but did not excite much inflection. The leaves began to re-expand after four or five days, much viscid fluid being left on their discs, as if but little had been absorbed. One of these leaves as soon as it re-expanded, caught a small fly, and after 24 hrs. was closely inflected, showing how much more potent than gelatine is the animal matter absorbed from an insect. Some larger pieces of gelatine, soaked for five days in water, were next placed on three leaves, but these did not become much inflected until the third day, nor was the gelatine completely liquefied until the fourth day. On this day one leaf began to re-expand; the second on the fifth; and third on the sixth. These several facts prove that gelatine is far from acting energetically on Drosera.

In the last chapter it was shown that a solution of isinglass of commerce, as thick as milk or cream, induces strong inflection, we therefore wished to compare its action with that of pure gelatine. Solution of one part of both substances to 218 of water were made; and half-minim drops (.0296 o.e) were placed on the discs of eight leaves, so that each received 1/480 of a grain, or .135mg. The four with the isinglass were much more strongly inflected than the other four. We conclude, therefore, that isinglass contains some, though perhaps very little, soluble albumin's matter. As soon as these eight leaves re-expanded, they were given bits of roast meat, and in some hours all become greatly inflected; again showing how much more meat excites Drosera than does gelatine or isinglass. This is an interesting fact, as it is well known that gelatine by itself has little power of nourishing animals.

Chondrin: Some chondrin was slowly dried, and a small chip was placed on a leaf, and a much larger chip on a second leaf. The first was liquefied in a day; the larger piece was much swollen and softened, but was not completely liquefied until the third day. The undried jelly was next tried, and as a control experiment small cubes were left water for four days and retained their angles. Cubes of the same size were placed on two leaves, and larger cubes on two other leaves. The tentacles and laminae of the latter were closely inflected after 22 hrs. But those of the two leaves with the smaller cubes only to a moderate degree. The jelly on all four was by this time liquefied, and rendered very acid. The glands were blackened from the aggregation of their protoplasmic contents. In 46 hrs. from the time when the jelly was given, the leaves had almost re-expanded, and completely so after 70 hrs.; and now only a little slightly adhesive fluid was left unabsorbed on their discs.

One part of chondrin jelly was dissolved in 218 parts of boiling water, and half-minim drops were given to four leaves; so that each received about 1/460 of a grain (.135 mg) of the jelly; and, of course, much less of dry chondrin. This acted most powerful, for after only 3 hrs. 30 m. all four leaves were strongly inflected. Three of them began to re-expand after 24 hrs., and in 48 hrs. were completely open; but the fourth had only partially re-expanded. All the liquefied chondrin was by this time absorbed. Hence a solution of chondrin seems to act far more quickly and energetically than gelatine or isinglass; but we were assured by good authorities that it is most difficult, or impossible, to know whether chondrin is pure, and if it contained any albumin's compound, this would have produced the above effects. Nevertheless, we have thought these facts worth giving, as there is so much doubt on the nutritious value of gelatine; and Dr. Lauder Brunton does not know of any experiments with respect to animals on the relative value of gelatine and chrondrin.

Milk: We have seen in the last chapter that milk acts most powerfully on the leaves; but whether this is due to the contained casein or albumen, we know not. Rather large drops of milk excite so much secretion (which is very acid) that it sometimes trickles down from the leaves, and this is likewise characteristic of chemically prepared casein. Minute drops of milk, placed on leaves, were coagulated in about ten minutes. Schiff denies that the coagulation of milk by gastric juice is exclusively due to the acid which is present, but attributes it in part to the pepsin; and it seems doubtful whether with Drosera the coagulation can be wholly due to the acid, as the secretion does not commonly colour litmus paper until the tentacles have become well inflected; whereas the coagulation commences, as we have seen, in about ten minutes. Minute drops of skimmed milk were placed on the disc of five leaves; and a large proportion of the coagulated matter or curd was dissolved in 6 hrs., and still more completely in 8 hrs. These leaves re-expanded after two days, and the viscid fluid left on their discs was then carefully scrapped off and examined. It seemed at first sight as if all the casein had not been dissolved, for a little matter was left which appeared of a whitish colour by reflected light. But this matter, when examined under a high power, and when compared with a minute drop of skimmed milk coagulated by acetic acid, was seen to consist exclusively of oil-globules, more or less aggregated together, with no trace of casein. As we was not familiar with the microscopical appearance of milk, we asked Dr. Lauder Brunton to examine the slides, and he tested the globules with ether, and found that they were dissolved. We may therefore conclude that the secretion quickly dissolves casein, in the state in which it exists in milk.

Chemically Prepared Casein: This substance, which is insoluble in water, is supposed by many chemists to differ from the casein of fresh milk. We procured some, consisting of hard globules, and tried many experiments with it. Small particles and the powder, both in a dry state and moistened with water, caused the leaves on which they were placed to be inflected very slowly, generally not until two days had elapsed. Other particles, wetted with weak hydrochloric acid (one part to 437 of water) acted in a single day, as did some casein freshly prepared for by Dr. Moore. The tentacles commonly remained inflected from seven to nine days; and during the whole of this time the secretion was strongly acid. Even on the eleventh day some secretion left on the discs of a fully re-expanded leaf was strongly acid. The acid seems to be secreted quickly, for in one case the secretion from the discal glands, on which a little powdered casein had been strewed, coloured litmus paper, before any of the exterior tentacles were inflected.

Some cubes of hard casein, moistened with water, were placed on two leaves; after three days one cube had its angles a little rounded, and after seven days both consisted of rounded softened masses, in the midst of much viscid and acid secretion; but it must not be inferred from this fact that the angles were dissolved, for cubes immersed in water were similarly acted on. After nine days these leaves began to re-expand, but in this and other cases the casein did not appear, as far as could be judged by the eye, much if at all reduced in bulk. According to Hoppe-Seyler and Lubavin, casein consists of an albuminous, with a non-albuminous, substance; and the absorption of a very small quantity of the former would excite the leaves, and yet not decrease the casein to a perceptible degree. So that here we have another point of accordance between the secretion of Drosera and gastric juice, as both act so differently on the fresh casein of milk, and on that prepared by chemists.

A few trials were made with cheese; cubes of 1/20 of an inch (1.27mm.) were placed on four leaves, and these after one or two days became well inflected, their glands pouring fourth much acid secretion. After five days they began to re-expand, but one dies, and some of the glands of the other leaves were injured. Judging by the eye, the softened and subsided masses of cheese, left on the discs, were very little or not at all reduced in bulk. We may however, infer from the time during which the tentacles remind inflected, from the changed colour of some of the glands, and from the injury done to others, that matter had been absorbed from the cheese.

Legumin: There can hardly be a doubt that it would be easily digested, judging from the powerful effect produced by drops of a decoction of green peas, as described in the last chapter. This slices of dried pea, after being soaked in water, were placed on two leaves; these became somewhat inflected in the course of single hour, and most strongly so in 21 hrs. They re-expanded after three or four days. The slices were not liquefied, for the walls of the cells, composed of cellulose, are not in the least acted on by the secretion.

Pollen: A little fresh pollen from the common pea was placed on the discs of five leaves, which soon became closely inflected, and remained so for two or three days.

The grains being then removed, and examined under the microscope, were found discoloured, with the oil-globules remarkably aggregated. Many had their contents much shrunk, and some were almost empty. There could be no doubt that the secretion has penetrated the outer coats of the grains, and had partially digested their contents. So it must be with the gastric juice of the insects which feed on pollen, without masticating it. Drosera in state of nature cannot fail to profit to a certain extent by this power of digesting pollen, as innumerable grains from the carices, grasses, rumices, fir-trees, and other wind-fertilised plants, which commonly grow in the same neighbourhood, will be inevitably caught by the viscid secretion surrounding the many glands.

Gluten: This substance is composed of two albuminoids, one soluble, the other insoluble in alcohol. Some was prepared by merely washing wheaten flour in water. A provisional trial was made with rather large pieces placed on two leaves; these, after 21 hrs., were closely inflected, and remained so for four days, when one was killed and the other had its glands extremely blackened, but was not afterward observed. Smaller bits were placed on two leaves; these were only slightly inflected in two days, but afterwards became much more so. Their secretion was not so strongly acid as that of leaves excited by casein. The bits of gluten, after lying for three days on the leaves, were more transparent than other bits left for the same time in water. After seven days both leaves re-expanded, but the gluten seemed hardly at all reduced in bulk. The glands which had been in contact with it were extremely black. Still smaller bits of half putrid fluted were now tried on two leaves; these were well inflected in 24 hrs., and thoroughly in four days, the glands in contact being much blackened. After five days one leaf began to re-expanded, and after eight days both were fully re-expanded, some gluten being still left on their disc. Four little chips of dried gluten, just dipped in water, were next tried, and these acted rather differently from fresh gluten. One leaf was almost fully re-expanded in three days, and the other three leaves in four days. The chips were greatly softened, almost liquefied, but not nearly all dissolved. The glands which had been in contact with them, instead of being much blackened, were of a very pale colour, and many of them were evidently killed.

In not one of these ten cases was the whole of the gluten dissolved, even when very small bits were given. We therefore asked Dr. Burdon Sanderson to try gluten in artificial digestive fluid of pepsin with hydrochloric acid; and this dissolved the whole. The gluten, however was acted on much more slowly than fibrin; the proportion dissolved within four hours being as 40.8 of gluten to 100 of fibrin. Gluten was also tried in two other digestive fluids, in which hydrochloric acid was replaced by propionic and butyric acids, and it was completely dissolved by these fluids at the ordinary temperature of room. Here, then, at last, we have a case in which it appears that there exists an essential difference in digestive power between the secretion of Drosera and gastric juice; the difference being confined to the ferment, for, as we have just seen, pepsin in combination with acids of the acetic series acts perfectly on gluten. We believe that the explanation lies simply in the fact that gluten is too powerful stimulant (like raw meat, or phosphate of lime, or even too large a piece of albumen), and that it injures or kills the glands before they have had time to pour forth a sufficient supply of the proper secretion. That some matter is absorbed from the gluten, we have clear evidence in the length of time during which the tentacles remain inflected, and in the greatly changed colour of the glands.

At the suggestion of Dr. Sanderson, some gluten was left for 15 hrs. in weak hydrochloric acid (.20 per cent) in order to remove the starch. It became colourless, more transparent, and swollen. Small portions were washed and placed on five leaves, which were soon closely inflected, but to my surprise re-expanded completely in 43 hrs. A mere vestige of gluten was left on two of the leaves, and not a vestige on the other three. The viscid and acid secretion, which remained on the discs of the three latter leaves, was scraped off and examined by my son under a high power; but nothing could be seen except a little dirt, and a good many starch grains which had not been dissolved by the hydrochloric acid. Some of the glands were rather pale. We thus learn that gluten, treated with weak hydrochloric acid, is not so powerful or so enduring a stimulant as fresh gluten, and does not much injure the glands; and we further learn that it can be digested quickly and completely by the secretion.

Globulin or Crystallin: This substance was kindly prepared by Dr. Moore, and consisted of hard, colourless, transparent fragment. It is said that globulin ought to "swell up in water and dissolve, for the most part forming a gummy liquid;" but this did not occur with the above fragments, though kept in water for four days. Particles, some moistened with water, others with weak hydrochloric acid, others soaked in water for one or two days, were placed on nineteen leaves. Most of these leaves, especially those with the long soaked particles, became strongly inflected in a few hours. The greater number re-expanded after three or four days; but three of the leaves remained inflected during one, two, or three additional days. Hence, some exciting matter, must have been absorbed; but the fragments, though perhaps softened in a greater degree than those kept for the same time in water, retained all their angles as sharp as ever. As globulin is an albuminious substance, we were astonished at this result; and my object being to compare the action of the secretion with that of gastric juice, we asked Dr. Burdon Sanderson to try some of the globulin used by us. He reported that "it was subjected to a liquid containing 0.2 per cent of hydrochloric acid, and about 1 per cent of glycerine extract of the stomach of dog. It was then ascertained that this liquid was capable of digesting 1.31 of its weight of unboiled fibrin in 1 hr.; whereas, during the hour, only 0.141 of the above globulin was dissolved. In both cases an excess of the substance to be digested was subjected to the liquids." We thus see that within the same time less than one ninth by weight of globulin than of fibrin was dissolved: and bearing in mind that pepsin with acids of the acetic series had only about one-third of the digestive power of pepsin with hydrochloric acid, it is not surprising that the fragments of globulin were not corroded or rounded by the secretion of Drosera, though some soluble matter was certainly extracted from them and absorbed by the glands.

Haematin: Some dark red granules from bullock's blood, were taken; these were found by Dr. Sanderson to be insoluble in water, acids, and alcohol, so that they were probably hematite, together with other bodies derived from the blood. Particles with little drops of water were placed on four leaves, three of which were pretty closely inflected in two days; the fourth only moderately so. On the third day the glands in contact with the haematin were blackened, and some of the tentacles seemed injured. After five days two leaves dies, and the third was drying; the fourth was beginning to re-expand, but many of its glands were blackened and injured. It is therefore clear that matter had been absorbed which was either actually poisonous or of too simulating a nature. The particles were much more softened than those kept for the same time in water, but judging by the eye, very little reduced in bulk. Dr. Sanderson tried this substance with artificial digestive fluid, in the manner described under globulin, and found that whilst 1.31 of fibrin, only 0.546 of the hematite was dissolved in an hour; but the dissolution by the secretion of even a less amount would account for its action on Drosera. The residue left by the artificial digestive fluid at first yielded nothing more to it during several succeeding days.

Substances which are not Digested by the Secretion

All the substances hitherto mentioned cause prolonged inflection of the tentacles, and are either completely or at least partially dissolved by the secretion. But there are many other substances, some of them containing nitrogen, which are not in the least acted on by the secretion, and do not induce inflection for a longer time than do inorganic and insoluble objects. These unexciting and indigestible substances are, as far as we have observed, epidermic production (such as bits of human nails, balls of hair, the quills of feathers), fibro-elastic tissue, mucin, pepsin, urea, chitine, chlorophyll, cellulose, gun-cotton, fat, oil, and starch.

To these may be added dissolved sugar and gum, diluted alcohol and vegetable infusions not containing albumen, for none of these, as shown in the last chapter, excite inflection. Now, it is a remarkable fact, which affords additional and important evidence, that the ferment of Drosera is closely similar to or identical with pepsin, that none of these same substances are, as far as it is known, digested by the gastric juice of animals, though some of them are acted on by the other secretions of the alimentary canal. Nothing more need be said about some of the above enumerated substances, excepting that they were repeatedly tried on the leaves of Drosera, and were not in the least affected by the secretion.

Fibro-elastic Tissue: We have already seen that when little cubes of meat, &c., placed on leaves, the muscles, areolar tissue, and cartilage was completely dissolved, but the fibro-elastic tissue, even the most delicate threads, were left without the least signs of having been attacked. And it is well known that this tissue cannot be digested by the gastric juice of animals.

Mucin: As this substance contains about 7 per cent of nitrogen, we expected that it would have excited the leaves greatly and been digested by the secretion, but in this we were mistaken. From what is stated in chemical works, it appears extremely doubtful whether mucin can be prepared as a pure principle. That which we used (prepared by Dr. Moore) was dry and hard. Particles moistened with water were placed on four leaves, but after two days there was only a trace of inflection in the immediately adjoining tentacles. These leaves were then tried with bits of meat, and all four soon became strongly inflected. Some of the dried mucin was then soaked in water for two days, and little cubes of the proper size were placed on three leaves. After four days the tentacles round the margins of the disc were a little inflected, and the secretion collected on the disc was acid, but the exterior tentacles were not affected. One leaf began to re-expand on the fourth day, and all were fully re-expanded on the sixth. The glands which had been in contact with the mucin were a little darkened. We may therefore conclude that a small amount of some impurity of a moderately exciting nature had been absorbed. That the mucin employed did contain some soluble matter was proved by Dr. Sanderson, who on subjecting it to artificial gastric juice found that in 1 hr. some was dissolved, but only in the proportion of 23 to 100 of fibrin during the same time. The cubes, though perhaps rather softer than those left in water for the same time, retained their angles as sharp as ever. We may therefore infer that the mucin itself was not dissolved or digested. Nor is it digested by the gastric juice of living animals, and according to Schiff, it is a layer of this substance which protects the coats of the stomach from being corroded during digestion.

Pepsin: We were curious to ascertain, as far as that was possible, whether the ferment of the secretion of Drosera would act on the ferment of the gastric juice of animals. We first used the common pepsin sold for medicinal purposes, and afterwards some which was much purer, prepared by Dr. Moore. Five leaves to which a considerable quantity of the former was given remained inflected for five days; four of them then dies, apparently from too great stimulation. We then tried Dr. Moore's pepsin, making it into a paste with water, and placing much small particles on the discs of five leaves that all would have been quickly dissolved had it been meat or albumen. The leaves were soon inflected; two of them began to re-expanded after 44 hrs. Some of the glands which had been in contact with the particles of pepsin, or with the acid secretion surrouding them, were singularly pale, whereas others were singularly dark-coloured. Some of the secretion was scraped off the examined under a high power; and it abounded with granules undistinguishable from those of pepsin left in water for the same length of time. We may therefore infer, as highly probable (remembering what small quantities were given),that the ferment of Drosera does not act on or digest pepsin, but absorbs from it some albumen's impurity which induces inflection, and which in large quantity is highly injurious. Dr. Lauder Brunton request endeavoured to ascertain whether pepsin with hydrochloric acid would digest pepsin, and as far as he could judge, it had no such power. Gastric juice, therefore, apparently agrees in this respect with the secretion of Drosera.

Urea: It seemed to an interesting inquiry whether this refuse of the living body, which contains much nitrogen, would, like so many other animal fluids and substances, be absorbed by the glands of Drosera and cause inflection. Half-minim drops of a solution of one part of 437 of water were placed on the discs of four leaves, each drop containing the quantity usually employed, namely, 1/960 of a grain, or .0674 mg.; but the leaves were hardly at all affected. They were then tested with bits of meat, and soon became closely inflected. We repeated the same experiment on four leaves with some fresh urea prepared by Dr. Moore; after two days there was no inflection; we then gave them another dose, but still there was no inflection. These leaves were afterwards tested with similarly sized drops of an infusion of raw meat, and in 6 hrs. there was considerable inflection, which became excessive in 24 hrs. But the urea apparently was not quite pure, for when four leaves immersed in 2 hr. (7.1 c.c.) of the solution, so that all the glands, instead of merely those on the disc, were enabled to absorb any small amount of impurity in solution, there was considerable inflection, after 24 hrs., certainly more than would have followed from a similar immersion in pure water. That the urea, which was not perfectly white, should have contained a sufficient quantity of albuminous matter, or of some salt of ammonia, to have caused the above effect, is far from surprising, for, as we shall see in the next chapter, astonishingly small doses of ammonia are highly efficient. We may therefore conclude that the urea itself is not exciting or nutritious to Drosera; nor is it modified by the secretion, so as to be rendered nutritious, for, had this been the case, all the leaves with drops on their disc assuredly would have been well inflected. Dr. Lauder Brunton informed that from experiments made it appears that urea is not acted on by artificial gastric juice, that is by pepsin with hydrochloric acid.

Chitine: The chitionous coats of insects naturally captured by the leaves do not appear in the least corroded. Small square pieces of the delicate wing and of the elytron of a staphylinus were placed on some leaves, and after these had re-expanded, the pieces were carefully examined. Their angles were as sharp as ever, and they did not differ in appearance form the other wing and elytron of the same insects which had been left in water. The elytron, however, had evidently yielded some nutrition's matter, for the leaf remained clasped over it for four days; whereas the leaves with bits of the true wing re-expanded on the second day. Any one who will examine the excrement of insect-eating animals will see how powerless their gastric-juice is on chitine.

Cellulose: We did not obtain this substance in a separate state, but tried angular bits of dry wood, cork, sphagnum moss, linen, and cotton thread. None of these bodies were in the least attacked by the secretion, and they caused only that moderate amount of inflection which is common to all inorganic objects. Gun-cotton, which consists of cellulose, with the hydrogen replaced by nitrogen, was tried with the same result. We have seen that a decoction of cabbage leaves excites the most powerful inflection. We therefore placed two little square bits of the blade of cabbage leaf, and four little cubes cut from the midrib, on six leaves of Drosera. These became well inflected in 12 hrs., and remained so for between two and four days; the bits of cabbage being bathed all the time by acid secretion. This shows that some exciting matter, to which we shall presently refer, had been absorbed; but the angles of the squares and cubes remained as sharp as ever, proving that the framework of cellulose had not been attacked. Small square bits of spinach leaves were tried with the same result; the glands pouring forth a moderate supply of acid secretion, and the tentacles pouring forth a moderate supply of acid secretion, and the tentacles remaining inflected for three days. We have also seen that the delicate coats of pollen grains are not dissolved by the secretion. It is well known that the gastric juice of animals does not attack cellulose.

Chlorophyll: This substance was tried, as it contains nitrogen. It was dried, but soon deliquesced. Particles were placed on four leaves; after 3 hrs. the secretion was acid; after 8 hrs., there was a good deal of inflection which in 24 hrs. became fairly well marked. After four days two of the leaves began to open, and the other two were then almost fully expanded. It is therefore clear that this chlorophyll contained matter which excited the leaves to a moderate degree; so that in a pure state it would not probably have been attacked by the secretion. Dr. Sanderson tried that which we used, as well as some freshly prepared, with artificial digested liquid, and found that it was not digested. Dr. Lauder Brunton likewise tried some prepared by the process given in the British pharmacopoeia, and exposed it for five days at the temperature of 37° C. to digestive liquid, but it was not diminished in bulk, though the fluid acquired a slightly brown colour. It is also tried with the glycerine extract of pancreas with a negative result. Nor does chlorophyll seem affected by the intestinal secretions of various animals, judging by the colour of their excrement.

It must not be supposed from these facts that the grains of chlorophyll, as they exist in living plants, cannot be attacked by the secretion; for these grains consist of protoplasm merely coloured by chlorophyll. We placed a thin slice of spinach leaf moistened with saliva, on a leaf of Drosera, and other slices on damp cotton-wool, all exposed to the same temperature. After 19 hrs. the slice on the leaf of the Drosera was bathed in much secretion from the inflected tentacles, and was now examined under the microscope. No perfect grains of chlorophyll could be distinguished; some were shrunken, of a yellowish-green colour, and collected in the middle of the cells; others were disintegrated and formed a yellowish mass, likewise in the middle of the cells. On the other hand, in the slices surrounded by damp cotton-wool, the grains of chlorophyll were green and as perfect as ever. We also placed some slices in artificial gastric juice, and these were acted on in nearly the same manner as by the secretion. We have seen that bits of fresh cabbage and spinach leaves cause the tentacles to be inflected and the glands to pour forth much acid secretion; and there can be little doubt that it is the protoplasm forming the grains of chlorophyll, as well as that lining the walls of the cells, which excited the leaves.

Fat and Oil: Cubes of almost pure uncooked fat, placed on several leaves, did not have their angles in the least rounded. We have also seen that the oil-globules in milk are not digested. Nor does olive oil dropped on the discs of leaves cause any inflection; but when they are immersed in olive oil they become strongly inflected. Oily substances are not digested by the gastric juice of animals.

Starch: Rather large bits of dry starch caused well-marked inflection, and the leaves did not re-expand until the fourth day; but we have no doubt that this was due to the prolonged irritation of the glands, as the starch continued to absorb the secretion. The particles were not in the least reduced in size; and we know that leaves immersed in an emulsion of starch not at all affected. We need hardly say that starch is not digested by the gastric juice of animals.

Action of the Secretion on Living Seeds

The results of some experiments on living seeds, selected by hazard, may here be given, though they bear only indirectly on our present subject of digestion.

Seven cabbage seeds of the previous year were placed on the same number of leaves. Some of these leaves were moderately, but the greater number only slightly inflected, and most of them re-expanded on the third day. One, however, remained clasped till the fourth, and another till the fifth day. These leaves therefore were excited somewhat more by the seeds than by inorganic objects of the same size. After they re-expanded, the seeds were placed under favourable conditions on damp sand; other deeds of the same lot being tried at the same time in the same manner, and found to germinate well. Of the seven seeds which had been exposed to the secretion, only three germinated;

and one of the three seedlings soon perished, the tip of its radical being from the first decayed, and the edges of its cotyledons of a dark brown colour; so that altogether five out of the seven seeds ultimately perished.

Radish seeds *(Raphanus sativus)* of the previous year were placed on three leaves, which became moderately inflected, and re-expanded on the third or fourth day. Two of these seeds were transferred to damp sand; only one germinated, and that very slowly. This seedling had on extremely short, crooked, diseased, radical, with no absorbent hairs; and the cotyledons were oddly mottled with purple, with the edges blackened and partly withered.

Cress seeds *(Lepidium sativum)* of the previous year were placed on four leaves; two of these next morning were moderately and two strongly inflected, and remained so for four, five, and even six days. Soon after these seeds were placed on the leaves and had become damp, they secreted in the usual manner a layer of tenacious mucus; and to ascertain whether it was the absorption of this substance by the glands which caused so much inflection, two seeds were put into water, and as much of the mucus as possible scraped off. They were then placed on leaves, which became very strongly inflected in the course of 3 hrs., and were still closely inflected on the third day; so that it evidently was not the mucus which excited so much inflection; on the contrary, this served to a certain extent as a protection to the seeds. Two of the six seeds germinated whilst still lying on the leaves, but the seedlings, when transferred to damp sand, soon died; of the other four seeds, only one germinated.

Two seeds of mustard (*Sinapis nigra*), two of celery (*Apium graveolens*)—both of the previous year, two seeds well soaked of caraway (*Carum carui*), and two of wheat, did not excite the leaves more than inorganic objects often do. Five seeds, hardly ripe, of a buttercup (*Ranunclus*), and two fresh seeds of *Anemone nemorosa*, induced only a little more effect. On the other hand, four seeds, perhaps not quite ripe, of *Carex sylvatica* caused the leaves in which they were placed to be very strongly inflected; and these only began to re-expand on the day, one remaining inflected for seven days.

It follows from these few facts that different kinds of seeds excite the leaves in very different degrees; whether this is solely due to the nature of their coats is not clear. In the case of the cress seeds, the particle removal of the layer of mucus hastened the inflection of the tentacles. Whenever the leaves remain inflected during several days over seeds, it is clear that they absorb some matter from them. That the secretion penetrates their coats is also evident from the large proportion of cabbage, radish, and cress seeds which were killed, and from several of the seedlings being greatly injured. This injury to the seeds and seedlings may, however, be due solely to the acid of the secretion, and not to any process of digestion; for Mr. Traherne Moggridge has shown that very weak acids of the acetic series are highly injurious to seeds. It never occurred to me to observe whether seeds are often blown on to the viscid leaves of plants growing in a state of nature; but this can hardly fail sometimes to occur, as we shall hereafter see in the case of Pinguicula. If so, Drosera will profit to a slight degree by absorbing matter from such seeds.

Summary and Concluding Remarks

When the glands on the disc are excited either by the absorption of nitrogenous matter or by mechanical irritation, their secretion increase in quantity and becomes acid. They likewise transmit some inflect to the glands of the exterior tentacles, causing them to secrete more copiously; and their secretion likewise becomes acid. With animals, according to Schiff, mechanical irritation excites the glands of the stomach to secrete an acid, but not pepsin. Now, we have every reason to believe (though the fact is not fully established), that although the glands of Drosera are continually secreting viscid fluid to replace the lost by evaporation, yet they do not secrete the ferment proper for digestion when mechanically irritated, but only after absorbing certain matter, probably of nitrogenous nature. We infer that this is the case, as the secretion from a large number of leaves which had been irritated by particles of glass placed on their discs did not digest albumen; and more especially from the analogy of Dionaea and Nepenthes. In like manner, the glands of the stomach of animals secrete pepsin, as Schiff asserts, only after they have absorbed certain soluble substances, which he designates as peptones. There is, therefore, a remarkable parallels between the glands of Drosera and those of the stomach in the secretion of their proper acid and ferment.

The secretion, as we have seen, completely dissolves albumen, muscle, fibrin, areola tissue, cartilage, the fibrous basis of bone, gelatine, chrondrin, casein in the state in which it exists in milk, and gluten which has been subject to weak hydrochloric acid. Syntonin and legumin excite the leaves so powerfully and quickly that there can hardly be a doubt that both would be dissolved by the secretion. The secretion failed to digest fresh gluten, apparently from its injuring the glands, though some was absorbed. Raw meat, unless in very small bits, and large pieces albumen, &c., likewise injure the leaves, which seem to suffer, like animals, from a surfeit. We know not whether the analogy is real one, but it is worth notice that a decoction of cabbage leaves is far more exciting and probably nutritious to Drosera than an infusion made with tepid water; and boiled cabbages are far more nutritious, at least to man, than the uncooked leaves. The most striking of all the cases, thought not really more remarkable than many others, is the digestion of so hard and tough a substance as cartilage. The dissolution of pure phosphate of lime, of bone, dentine, and especially enamel, seems wonderful; but it depends merely on the long-continued secretion of an acid; and this is secreted for a longer time under these circumstances than under any other. It was interesting to observe that as long as the acid was consumed in dissolving the phosphate of lime, no true digestion occurred; but that as soon as the bone was completely decalcified, the fibrous basis was attacked and liquefied with the greatest ease. The twelve substances above enumerated, which are completely dissolved by the secretion, are likewise dissolved by the gastric juice of the higher animals; and they are acted on in the same manner, as shown by the rounding of the angles of the albumen, and more especially by the manner in which the transverse striae of the fibres of muscle disappear.

The secretion of Drosera and gastric juice were both able to dissolve some element or impurity out of the globulin and haematin. The secretion also dissolved something out of chemically prepared casein which is said to consist of two substances; and although Schiff asserts that casein in this state is not attacked by gastric juice, he might easily have overlooked a minute quantity of some albumin's matter, which Drosera would detect and absorb. Again, fibro-cartilage, though not properly dissolved, is acted on in the same manner, both by the secretion of Drosera and gastric juice. But this substance, as well as the so-called haematin, ought perhaps to have been classed with indigestible substances.

The gastric juice acts by means of its ferment, pepsin, solely in the presence of an acid, is well established; and we have excellent evidence that a ferment is present in the secretion of Drosera, which likewise acts only in the presence of an acid; for we have seen that when the secretion is neutralised by minute drops of the solution of an alkail, the digestion of albumen is completely stopped, and that on the addition of a minute does of hydrochloric acid it immediately recommences.

The nine followings substances, or classes of substances, namely, epidemic production, fibro-elastic tissue, mucin, pepsin, urea, chitine, cellulose, gun-cotton, chlorophyll, starch, fat and oil, are not acted on by the secretion of Drosera; not are they, as far as is known, by the gastric juice of animals. Some soluble matter, however, was extracted from the mucin, pepsin, and chlorophyll, both by the secretion and by artificial gastric juice.

The several substances, which are completely dissolved by the secretion, and which are afterwards absorbed by the glands, affect the leaves rather differently. They induce inflection at very different rates, and in very different degrees; and the tentacles remain inflected for very different periods of time. Quick inflection depends partly on the quantity of the substance given, so that many glands are simultaneously affected, partly on the facility with which it is penetrated, as liquefied by the secretion, and partly on its nature, but chiefly on the presence of exciting matter already in solution. Thus saliva, or a weak solution of raw meat, acts much more quickly than even a strongly solution of gelatine. So again leaves which have re-expanded, after absorbing drops of solution of pure gelatine or isinglass (the latter being the more powerful of the two), if given bits of meat, are inflected much more energetically and quickly then they were before, not withstanding that some rest is generally requisite between two acts of inflection. We probably see the influence of texture in gelatine and globulin when softened by having been soaked in water acting more quickly than when merely wetted. It may be partly due to changed texture, and partly to changes chemical nature, that albumen, which has been kept for some time, and gluten which has been subjected to weak hydrochloric acid, act more quickly than these substances in their fresh state.

The length of time during which the tentacles remain inflected largely depends on the quantity of the substance given, partly on the facility with which it is penetrated or acted on by the secretion, and partly on its essential nature. The tentacles always remain inflected much longer over large bits or large drops than over small bits or drops. Texture probably plays a part in determining the extra ordinary length of time during which the tentacles remain inflected over the hard grains of chemically prepared casein. But the tentacles remain inflected for an equally long time over finely powdered, precipitated phosphate of lime; phosphorus in this latter case evidently being the attraction, and animal matter in case of casein. The leaves remain long inflected over insects, but it is doubtful how far this is due to the protection afforded by their chitinous integuments; for animal matter is soon extracted from insects (probably by exosmose from their bodies into the dense surrounding secretion), as shown by the prompt inflection of the leaves. We see the influence of the nature of different substances in bits of meat, albumen, and flesh gluten acting very differently from equal-sized bits of gelatine, areolar tissue, and the fibrous basis of bone. The former cause not only far more prompt and energetic, but more prolonged, inflection than do the latter. Hence we are justified in believing that gelatine, areolar tissue, and the fibrous basis of bone, would be far less nutritious to Drosera than such substances as insects, meat, albumen, &c. This is an interesting conclusion, as it is known the gelatine affords but little nutriment to animals; and so, probably would areolar tissue and the fibrous basis of bone. The chondrin which we used acted more powerfully than gelatine, but then we do not know that it was pure. It is a more remarkable fact that fibrin, which belongs to the great class of proteins, including albumen in one of its sub-groups, does not excite the tentacles in a greater degree, or keep them inflected for a longer time, than does gelatine, or areolar tissue, or the fibrous basis of bone. It is not known how long an animal would survive if fed on fibrin alone, but Dr. Sanderson has no doubt longer than on gelatine, and it would be hardly rash to predict, judging from the effects on Drosera, that albumen would be found more nutritious than fibrin. Globulin likewise belongs to the proteins, forming another sub-group, and this substance, though containing some matter which excited Drosera rather strongly, was hardly attacked by the secretion, and was very little or very slowly attacked by gastric juice. How far globulin would be nutritious to animals is not known. We thus see how differently the above specified several digestible substances act on Drosera; and we may infer, as highly probable, that they would in like manner be nutritious in very different degrees both to Drosera and to animals.

The glands of Drosera absorb matter from living seeds, which are injured or killed by the secretion. They likewise absorb matter from pollen, and from fresh leaves; and this is notoriously the case with the stomachs of vegetable-feeding animals. Drosera is properly an insectivorous plant; but as pollen cannot fail to be often blown on to the glands, as will occasionally the seeds and leaves of surrounding plants, Drosera is, to a certain extent, a vegetable feeder.

Finally, the experiments recorded in this chapter show us that there is a remarkable accordance in the power of digestion between the gastric juice of animals with its pepsin and hydrochloric acid and the secretion of Drosera with its ferment and acid belonging to the acetic series. We can therefore hardly doubt that the ferment in both cases is closely similar, if not identically the same. That a plant and an animal should pour forth the same, or nearby the same, complex secretion, adapted for the same purpose of digestion, is a new and wonderful fact in physiology.

7

The Effects of Salts of Ammonia

Manner of performing the experiments—Action of distilled water in comparison with the solution—Carbonate of ammonia, absorbed by the roots—The vapour absorbed by the gland—Drops on the disc—Minute drops applied to separate gland—Leaves immersed in weak solution—Minuteness of the doses which induce aggregation of the protoplasm—Nitrate of ammonia, analogous experiments with—Phosphate of ammonia, analogous experiments with—Other salts of ammonia—Summary and concluding remarks on the action of the salts of ammonia.

Manner of Performing the Experiments

The chief object in this chapter is to show how powerfully the salts of ammonia act on the leaves of Drosera, and more especially, to show what an extraordinarily small quantity suffices to excite infection. We shall therefore be compelled to enter into full details. Doubly distilled water was always used; and for the more delicate experiments, water had been prepared with the utmost possible care. The graduated measures were tested, and found as accurate as such measure can be. The salts were carefully weighed, and in all the more delicate experiments, by Borda's Double method. But extreme accuracy would have been superfluous as the leaves differ greatly in irritability, according to age, condition, constitution. Even the tentacles on the same leaf differ in irritability to a marked degree. Experiments were tried in the following several ways.

Firstly, drops which were ascertained by repeated trials to be on an average about half a minim, or the $\frac{1}{960}$ of a fluid ounce (0296 c.c.), were placed by the same pointed instrument on the discs of the leaves, and the inflection of the exterior rows of tentacles observed at successive intervals of time. It was first ascertained, from between thirty and forty trials, that distilled water dropped in this manner produces no effect, that sometimes, though rarely, two or three tentacles become inflected. In fact all the many trials with solutions which were so weak as to produce no effect lead to the same result that water is inefficient.

Secondly, the head of a small pin, fixed into a handle, was dipped into the solution under trial. The small drop which adhered to it, and which was much to small to fall off, was cautiously placed, by the aid of a lens, in contact with the secretion surrounding the glands of one, two, three, or four of the exterior tentacles of the same leaf. Great care was taken that the glands themselves should not be touched. We had supposed that the drops were of nearly the same size; but on trial this proved a great mistake. We first measured some water, and removed 300 drops, touching the pin's head each time on blotting paper; and on again measuring the water, a drop was found to equal on an average about the $\frac{1}{60}$ of a minim. Some water in small vessel was weighed (and this is a more accurate method), and 300 drops removed as before; and on again weighing the water, a drop was found to equal on an average only the 189 of a minim. We repeated the operation, but endeavoured this time, by taking the pin's head out of the water obliquely and rather quickly, to remove as large drops as possible; and the result showed that we had succeeded, for each drop on an average equalled $\frac{1}{19.4}$ of a minim. We repeated the operation in exactly the same manner, and now the drops averaged $\frac{1}{23.5}$ of a minim. Bearing in mind that on these two latter occasions special pains were taken to remove as large drops as possible, we may safely conclude that the drops used in the experiments were at least equal to the $\frac{1}{20}$ of a minim, or .0029c.c. One of these drops could be applied to three or even four glands, and if the tentacles became inflected, some of the solution must have been absorbed by all; for drops of pure water, spilled in the same manner, never produced any effect. We were able to hold the drop in steady contact with the secretion only for ten to fifteen seconds; and this was not time enough for the diffusion of all the salt in solution, as was evident, from three or four tentacles treated successively with the same drop, often becoming inflected. All the matter in solution was even then probably not exhausted.

Thirdly, leaves were cut off and immersed in a measured quantity of the solution under trial; the same number of leaves being immersed at the same time, in the same quantity of the distilled water which had been used in making the solution. The leaves in the two lots were compared at short intervals of time, up to 24 hrs., and sometimes to 48 hrs. They were immersed by being laid as gently as possible in numbered watch glasses, and thirty minims (1.775 c.c.) of the solution or of water was poured over each.

Some solutions, for instance that of carbonate of ammonia, quickly discolour the glands; and as all on the same leaf were discoloured simultaneously, they must all have absorbed some of the salt within the same short period of time. This was likewise shown by the simultaneous inflection of the several exterior rows of tentacles. If we had no such evidence as this, it might have been supposed that only the glands of the exterior and inflected tentacles and absorbed the salt; or that only those on the disc had absored it, and had then transmitted a motor impulse to the exterior tentacles; but in this latter case the exterior tentacles would not have become inflected until some time had elapsed, instead of within half an hour, or even within a few minutes, as usually occurred. All the glands on the same leaf are of nearly the same size, as may best be seen by cutting off a narrow transverse strip, and laying it on its side; hence their absorbing surfaces are nearly equally. The long-headed glands on the extreme margin must be excepted, as they are much longer than the others; but only the upper surface is capable of absorption. Besides the glands, both surfaces of the leaves and the pedicels of the tentacles bear numerous minute papilla, which absorb carbonate of ammonia, an infusion of raw meat, metallic salts, and probably many other substances, but the absorption of matter by these papillae never induces inflection. We must remember that the movement of each separate tentacle depends on its gland being excited, except when a motor impulse is transmitted from the glands of the disc, and then the movement, as just stated, does not place until some little time has elapsed. We have made these remarks because they show us that when a leaf is immersed in a solution, and the tentacles are inflected, we can judge with some accuracy how much of the salt each gland has absorbed. For instance, if a leaf bearing 212 glands, be immersed in a measured quantity of a solution, containing $\frac{1}{10}$ of a grain of salt; and all the exterior tentacles, except twelve, are inflected, we may feel sure that each of the 200 glands can on an average have absorbed at most $\frac{1}{2000}$ of a grain of the salt. We say at most, for the papillae will have absorbed some small amount, and so will perhaps the glands of the twelve excluded tentacles which did not become inflected. The application of this principle leads to remarkable conclusions with respect to the minuteness of the doses inflection.

Action of Distilled Water in Causing Inflection

Although in all the more important experiments the difference between the leaves simultaneously immersed in water and in the several solutions will be described, nevertheless it may be will have to give a summary of the effects of water. The fact, moreover, of pure water acting on the glands deserves in itself some notice. Leaves to the numner of 141 were immersed in water at the same time with those in the solutions, and their state recorded at short intervals of time. Thirty-two other leaves were separately observed in water, making altogether 173 experiments. Many scores of leaves were also immersed in water at other times, but no exact record of the effects produced was kept; yet these cursory observations support the conclusions arrived at in this chapter. A few of the long-headed tentacles, namely from one to about six, were commonly inflected within half an hour after immersion; as were occasionally a few, and rarely a considerable number of the exterior round-headed tentacles. After an immersion on from 5 to 8 hrs. the short tentacles surrounding the outer parts of the disc generally become inflected, so that their glands form a small dark ring on the disc; the exterior tentacles not partaking of this movement. Hence, excepting in a few cases hereafter to be specified, we can judge whether a solution produces any effect only by observing the exterior tentacles within the first 3 or 4 hrs. after immersion.

Now for a summary of the state of the 173 leaves after an immersion of 3 or 4 hrs. in pure water. One leaf had almost all its tentacles inflected; three leaves had most of them sub-inflected; and thirteen had on average 36.5 tentacles inflected. Thus seventeen leaves out of the 173 were acted on in a marked manner. Eighteen leaves had from seven to nineteen tentacles inflected, the average being 9.3 tentacles for each leaf. Forty-four leaves had from one to six tentacles inflected generally the long-headed ones. So that altogether of the 173 leaves carefully observed, seventy-nine were affected by the water in some degree, though commonly to a very slight degree; and ninety-four were not affected in the least degree. This amount of inflection is utterly insignificant, as we shall hereafter see, compared with solutions of several salts of ammonia.

Plants which have lived for some time in a rather high temperature are far more sensitive to the action of water then those grown out of doors, or recently brought into a warm greenhouse. Thus in the above seventeen cases, in which the immersed leaves had a considerable number of tentacles inflected, the plants had been kept during the winter in very warm greenhouse; and they bore in the early spring remarkably fine leaves of a light red colour. Had then known that the sensitiveness of plants was thus increased, perhaps we should not have used the leaves for my experiments with the very weak solutions of phosphate of leaves ammonia; but my experiments are not thus vitiated, as we invariably used leaves from the same plants for simultaneous immersion in water. It often happened that some leaves on the same plant, and some tentacles on the same leaf, were more sensitive than others.

Fig. 9

(Drosera rotundifolia)

Leaf (enlarged) with all the tentacles closely inflected, from immersion in a solution of phosphate of ammonia (one part to 87,500 of water).

Besides the differences just indicated between the leaves immersed in water and in weak solutions of ammonia, the tentacles of the latter are in most cases much more closely inflected. The appearance of a leaf after immersion in a few drops of a solution of one grain of phosphate of ammonia to 200 oz. of water (*i.e.*, one part to 87,500) is here reproduced: such energetic inflection is never caused by water alone. With leaves in the weak solutions, the blade or lamina often becomes inflected; and this is so rare a circumstance with leaves in water that we have seen only two instances; and in both of these the inflection was very feeble. Again, with leaves in the weak solutions, the inflection of the tentacles and blade often goes on steadily, thought slowly, increasing during many hours; and this again is so rare a circumstance with leaves in water that we have seen only three instances of any such increase after the first 8 to 12; and in these three instances the two outer rows of tentacles were not at all affected. Hence there is sometimes a much greater difference between the leaves in water and in the weak solution, after from 8 hrs. to 24 hrs., than there was within the first 3 hrs.; though as general rule it is best to trust to the difference observed within the shorter time.

With respect to the period of the re-expansion of the leaves, when left immersed either in water or in the weak solution, nothing could be more variable. In both cases the exterior tentacles not rarely begin to re-expand, after an interval of only from 6 to 8 hrs.; that is just about the time when the short tentacles round the borders of the disc become inflected. On the other hand, the tentacles sometimes remain inflected for a longer period in very weak solution than in water. In solutions which are not extremely weak, they never re-expand within nearly so short a period as six or eight hours. From these statements it might be though difficult to distinguish between the effects of water and the weak solutions are tried; and then the distinction, as might be expected, becomes very doubtful, and at last disappears. But as in all, except the simplest, cases the state of leaves simultaneously immersed for an equal length of time in water and in the solutions will be described, the reader can judge for himself.

Carbonate of Ammonia

This salt, when absorbed by the roots, does not cause the tentacles to be inflected. A plant was so placed in a solution of one part of the carbonate to 146 of water that the young uninjured roots could be observed. The terminal cells, which were of pink colour, instantly become colourless, and their limpid contents cloudy, like a mezzo-tinto engraving. So that some degree of aggregation was almost instantly caused; but no further change ensued, and the absorbent hairs were not visibly affected. The tentacles did not bend. Two other plants were placed with their roots surrounded by damp moss, in half an ounce (14.198 c.c.) of a solution of one part of the carbonate to 218 of water, and were observed for 24 hrs.; but not a single tentacle was inflected . In order to produce this effect, the carbonate must be absorbed by the glands.

The vapour produce effect on the glands, and induces inflection. There plants with their roots in bottles, so that the surrounding air could not have become very humid, were placed under a bell-glass (holding 122 fluid ounces), together with 4 grains of carbonate of ammonia in a watch-glass. After an interval of 6 hrs. 15 m. the leaves appeared unaffected; but next morning, after 20 hrs., the blackened glands were secreting copiously, and most of the tentacles were strongly inflected. These plants soon dies. Two other plants were placed under the same bell-glass together with half a grain of the carbonate, the air being rendered as damp as possible; and in 2 hrs. most of the leaves were affected, many of the glands being blackened and the tentacles inflected. But it is a curious fact that some of the closely adjoining tentacles in the same leaf, both on the disc and round the margins, were much, and some, apparently, not in the least affected. The plants were kept under the bell-glass for 24 hrs., but no further change ensued. One healthy leaf was hardly at all affected, though other leaves on the same plant were much affected. On some leaves all the tentacles on one side, but not those on the opposite side, were inflected. We doubt whether this extremely unequal action can be explained by supposing that the more active glands absorb all the vapour as quickly as it is generated, so that none is left for the others; for we shall meet with analogous cases with air thoroughly permeated with the vapours of chloroform and ether.

Minute particles of the carbonate were added to the secretion surrounding several glands. These instantly became black and secreted copiously; but, except in two instances, when extremely minute particles were given, there was no inflection. This result is analogous to that which follows from the immersion of leaves in a strong solutions of one part of the carbonate to 109, or 146, or even 218 of water, for the leaves are then paralysed and no inflection ensues, though the glands are blackend, and the protoplasm in the cells of the tentacles undergoes strong aggregation.

We will now turn to the effects of solutions of the carbonate. Half minims of a solution of one part to 437 of water were placed on the discs to twelve leaves; so the each received $\frac{1}{960}$ of a grain of. 0675 mg. Ten of these had their exterior tentacles well inflected; the blades of some being also much curved inwards. In two cases several of the exterior tentacles were inflected in 35 m.; but the movement was generally slower. These ten leaves re-expanded in periods varying between 21 hrs., and 45 hrs., but in one case not until 67 hrs. had elapsed; so that they re-expanded much more quickly than leaves which have caught insects.

The same-sized drops of a solution of one part of 875 of water were placed on the discs of eleven leaves; six or eight of their exterior tentacles inflected; but this degree of movement can hardly be considered as trustworthy. Each of these leaves received $\frac{1}{1920}$ of a grain (.0337 mg.) distributed between the glands of the disc, but this was too small an amount to produce any decided effect on the exterior tentacles, the glands of which had not themselves received any of the salt.

Minute drops on the head of small pin, of a solution of one part of the carbonate to 218 of water, were next tried in the manner above described. A drop of this kind equals on an average $\frac{1}{20}$ of a minim, and therefore contain $\frac{1}{4800}$ of a grain (.0135 mg.) of the carbonate. We touched with it the viscid secretion round three glands, so that each gland received only $\frac{1}{14400}$ of a grain (0.00445 mg). Nevertheless, in two trails all the glands were plainly blackened; in one case all three tentacles were well inflected after an interval of 2 hrs. 40 m.; and in another case two of the three tentacles were inflected. We then tried drops of a weaker solution of one part of 292 of water of twenty-four glands, always touching the viscid secretion round three glands with the same little drop. Each gland thus received only the $\frac{1}{19200}$ of a grain (.00337 mg.). Yet some of them were a little darkened; but in no one instances were any of the tentacles inflected, though they were watched for 12 hrs. when a still weaker solution (*viz.*, one part to 437 of water) was tried on six glands, no effect whatever was perceptible. We thus learn that the $\frac{1}{14400}$ of a grain (.00445 mg.) of carbonate of ammonia, if absorbed by a gland, suffices to induce inflection in the basal part of the same tentacle; but as already stated, we were able to hold with a steady hand the minute drops in contact with the secretion only for a few seconds; and if more time had been allowed for diffusion and absorption, much weaker solution would certainly have acted.

Some experiments were made by immersing cut-off leaves in solution of different strengths. Thus four leaves were left for about 3 hrs. each in a drachm (3.549 c.c.) of a solution of one part of the carbonate of 5250 of water; two of these had almost every tentacle inflected, the third had about half the tentacles and the fourth about one-third inflected; and all the glands were blackened. Another leaf was placed in the same quantity of a solutions of one part to 7000 of water, and in 1 hr. 16 m. every single tentacle was well inflected, and all the glands blackened. Six leaves were immersed, each in thirty minims (1.774 c.c.) of a solution of one part to 4375 of water, and the glands were all blackened in 31 m. All six leaves exhibits some slight inflection, and one was strongly inflected. Four leaves were then immersed in thirty minims of a solution of one part to 8750 of water, so that each leaf received the $\frac{1}{320}$ of grain (.2025 mg.). Only one became strongly inflected; but all the glands on all the leaves were of so dark a red after one hour as almost to deserve to be called black, whereas this did not occur with the leaves which were at the same time immersed in water; nor did water produce this effect on any other occasion in nearly so short a time as an hour. These cases of the simultaneous darkening or blackening of the glands from the action of weak solutions are important, as they show that all the glands absorbed the carbonate with in the same time, which fact indeed there was not the least reason to doubt. So again, whenever all the tentacles become inflected with in the same time, we have evidence, as before remarked, of simultaneous absorption. We did not count the number of glands of these four leaves; but as they were fine ones and as we know that the average number of glands on thirty-one leaves was 192, we may safely assume that each bore on an average at least 170; and if so, each blackened gland could have absorbed only $\frac{1}{54400}$ of a grain (.00119 mg.) of the carbonate.

Nitrate and Phosphate of Ammonia—A large number of trials had been previously made with solutions of one part of the nitrate and phosphate of ammonia to 43750 of water (*i.e.*, one grain to 100 ounces), and these were found highly efficient. Fourteen leaves were therefore placed, each in thirty minims of a solution of one part of the carbonate to the above quantity of water; so that each leaf received $\frac{1}{1600}$ of a grain (.0405). The glands were not much darkened. Ten of the leaves were not affected, or only very slightly so. Four, however, were strongly affected; the first having all the tentacles, except forty, inflected in 47 m.; in 6 hrs. 30 m. all except eight; and after 4 hrs. the blade itself. The second leaf after 9 m. had all its tentacles except nine inflected; after 6 hrs. 30 m. these nine were sub-inflected; the blade having become much inflected in 4 hrs. The third leaf after 1 hr. 6,. had all but forty tentacles inflected. The fourth, after 2 hrs. 5 m., had about half its tentacles and after 4 hrs. all but forty-five inflected. Leaves which were immersed in water at the same time were not at all affected, with the exception of one; and this not until 8 hrs. had elapsed. Hence, there can be no doubt that a lightly sensitive leaf, if immersed in a solution, so that glands are able to absorb, is acted on by $\frac{1}{1600}$ of a grain of the carbonate. Assuming that the leaf, which was a large one, and which had all its tentacles excepting eight inflected, bore 170 glands, each gland could have absorbed only $\frac{1}{268800}$ of a grain (.00024 mg.); yet this sufficed to act on each of the 162 tentacles which were inflected. But as only four out of the above fourteen leaves were plainly affected, this is nearly the minimum dose which is efficient.

Aggregation of the Protoplasm from the Action of Carbonate of Ammonia— We have fully described in the third chapter the remarkable effects of moderately strong doses of this salt in causing in aggregation of the protoplasm within the cells of the glands and tentacles and here my object is merely to show what small doses suffice. A leaf was immersed in twenty minims (1.183 c.c.) of a solution of one part of 1750 of water, and another leaf in the same quantity of a solution of one part of 3062; in the former case aggregation occurred in 4m. in the latter in 11 m. A leaf was then immersed in twenty minims of solution of one part to 4375 of water, so that it received $\frac{1}{240}$ of grain (.27 mg); in 5 m. there was a slight change of colour in the glands, and in 15 m. small spheres of protoplasm were formed in the cells beneath glands of all the tentacles. In these cases there could not be a shadow of doubt about the action of the solution.

A solution was then made of one part to 5250 of water, and we experimented on fourteen leaves, but will give only a few of the cases. Eight young leaves were selected and examined with care, and they showed no trace of aggregation. Four of these were placed in a drachm (3.549 c.c.) of distilled water; and four in similar vessel, with a drachm of the solution. After a time the leaves were examined under a high power, being taken alternately from the solution and the water. The first leaf was taken out of the solution after an immersion of 2 hrs. 40 m., and the last leaf out of the water after 3 hrs. 50 m.; the examination lasting for 1 hr. 40. In the four leaves out of the water there was no trace of aggregation except in one specimen, in which a very few extremely minute spheres of protoplasm were present beneath some of the round glands. All the gland were translucent and red. The four leaves which had been immersed in the solution, besides being inflected, presented a widely different appearance; for the contents of the cell so ever single tentacle on all for 4 leaves were conspicucusly aggregated; the spheres and elongated masses of protoplasm in many cases extending halfway down the tentacles. All the glands, both those of the central and exterior tentacles, were opaque and blackened; and this shows that all had absorbed some of the carbonate. These four leaves were of very nearly the same size, and the glands were counted on one and found to be 167. This being the case, and the four leaves having been immersed in a drachm of the solution, each gland could have received on an average only $\frac{1}{64128}$ of a grain (.001009 mg.) of the salt: and this quantity sufficed to induce within a short time conspicuous aggregation in the cells beneath all the glands.

A vigorous but rather small red leaf was placed in six minims of the same solution (*viz.*, one part to 5250 of water), so that it received $\frac{1}{960}$ of a grain (.0675 mg.). In 40 m. the glands appeared rather darker; and in 1 hr. from four to six spheres of protoplasm were formed in the cells beneath the glands of all the tentacles. We did not count the tentacles; but we may safely assume that there were at east 140; and if so, each gland could have received only the $\frac{1}{134400}$ of grain, or .00048 mg.

A weaker solution was then made of one part to 7000 of water, and four leaves were immersed in it; but we will give only one case. A leaf was placed in ten minims of this solutions; after 1 hr. 37 m. the glands became somewhat darker, and the cells beneath all of them now contained many spheres of aggregated protoplasm. This leaf received $\frac{1}{768}$ of a grain, and bore 166 glands. Each could, therefore, have received only $\frac{1}{127488}$ of grain (.000507 mg.) of the carbonate.

Two other experiments are worth giving. A leaf was immersed for 4 hrs. 15 m. in distilled water, and there was no aggregation; it was then placed for 1 hr. 15 m. in a little solution of one part to 5250 of water; and this excited well-marked aggregation and inflection. Another leaf, after having been immersed for 21 hrs. 15 m. in distilled water, had its glands blackened. But there was no aggregation in the cells beneath them; it was then left in six minims of the same solution,. and in 1 hr. there was much aggregation in may of the tentacle; in 2 hrs. all the tentacles (146 in number) were affected—the aggregation extending down for a length equal to half or the whole of the glands. It is extremely improbable that these two leaves would have undergone aggregation if they had been left for a little longer in the water, namely, for 1 hr. and 1 hr. 15 m., during which time they were immersed in the solution; for the process of aggregation seems invariably to supervene slowly and very gradually in water.

Summary of the Results with Carbonate of Ammonia—The roots absorb the solution, as shown by their changed colour, and by the aggregation of the contents of their cells. The vapour is absorbed by the glands; these are blackened, and the tentacles are inflected. The glands of the disc, when excited by a half minim drop (.0296 c.c.), containing $\frac{1}{960}$ of a grain (.0675 mg.), transmit a motor impulse to the exterior tentacles, causing them to bend inwards. A minute drop, containing $\frac{1}{14400}$ of a grain (.00445 mg.), if held for a few seconds in contact with a gland, soon causes the tentacle bearing it to be inflected. If a leaf is left immersed for a few hours in a solution, and a gland absorbs the $\frac{1}{13400}$ of a grain (.00048 mg.) its colour becomes darker, though not actually black; and the contents of the cells beneath the gland are plainly aggregated. Lastly, under the same circumstances, the absorption by a gland of the $\frac{1}{26800}$ of a grain (.00024 mg.) suffices to excite the tentacle bearing this gland into movement.

Nitrate of Ammonia

With this salt we attended only to the inflection of the leaves, for it is far less efficient than the carbonate in causing aggregation, although considerably more potent in causing inflection. We experimented with half-minims (.296 c.c.) on the discs of fifty-two leaves, but will give only a few cases. A solution of one part to 109 of water was too strong, causing little inflection, and after 24 hrs. killing, or nearly killing, four out of six leaves which were thus tried; each of which received the $\frac{1}{240}$ of a grain (or .27 mg.). A solution of one part to 218 of water acted most energetically, causing not only the tentacles of all the leaves, but the blades of some to be strongly inflected. Fourteen leaves were tried with drops of solution of one part to 875 of water, so that the disc of each received the $\frac{1}{1920}$ of a grain (.0337 mg.). Of these inflected; two were moderately acted on; and five not at all. We subsequently tried three of these latter five leaves with urine, saliva, and mucus, but they were only slightly affected; and this proves that they were not in an active condition. We mention this fact to show how necessary it is to experiment on several leaves. Two of the leaves, which were well inflected, re-expanded after 51 hrs.

In the following experiment we happened to select very sensitive leaves. Half-minims of a solution of one part to 1094 of water (*i.e.*, 1 ge. to 2½ oz.)were placed on the discs of nine leaves, so that each received the $\frac{1}{2400}$ of a grain (.027 mg). Three of them had their tentacles strongly inflected and their blades curled inwards; five were slightly and somewhat doubtfully affected, having from three to eight of their exterior tentacles inflected; one leaf was not at all affected, yet was afterwards acted on by saliva. In six of these cases, trace of action was perceptible in 7 hrs., but the full effect was not produced until from 24 hrs. to 30 hrs. had elapsed. Two of the leaves, which were only slightly inflected, re-expanded after an additional interval of 19 hrs.

Half-minims of a rather weaker solution, *viz.*, of one part to 1312 of water (1 gr. to 3 oz.) were tried on fourteen leaves; so that each received $\frac{1}{2880}$ of grain (.0225 mg.) instead of, as in the last experiment, $\frac{1}{2400}$ of a grain. The blade of one was plainly inflected, as were six of the exterior tentacles; the blade of a second was lightly, and two of the exterior tentacles well inflected. All the other tentacles being curled in at right angles to the disc; three other leaves had from five to eight tentacles inflected; five others only two or three, and occasionally, thought very rarely, drops or pure water cause this such action; the four remaining leaves were in no way affected, yet three of them, when subsequently tried with urine, became greatly inflected. In most of these cases a slight effect was perceptible in from 6 hrs. to 7 hrs., but the full effect was not produced until from 24 hrs. to 30 hrs. had elapsed. It is obvious that we have reached very nearly the minimum amount, which, distributed between the glands of the disc, acts on the exterior tentacles; these having themselves not received any of the solution.

In the next place, the viscid secretion round three of the exterior glands was touched with the same little drop ($\frac{1}{20}$ of minim) of a solution of one part to 437 of water; and after an interval of 2 hrs. 50m. all three tentacles were well infected. Each of these glands could have received only the $\frac{1}{28800}$ of a grain, or .00225 mg. A little drop of the same size and strength was also applied to four other glands, and in 1 hr. two became inflected, whilst the other two never moved. We here see, as in the case of the half-minims placed on the disc, that the nitrate of ammonia is more potent in causing inflection than the carbonate; for minute drops of the latter salt of this strength produced no effect. We tried minute drops of a still weaker solution of the nitrate, *viz.*, one part to 875 of water on twenty-one glands, but no effect whatever was produced, except perhaps in one instance.

Sixty-three leaves were immersed in solutions of various strengths; other leaves being immersed at the same time in the same pure water used in marking the solution. The results are so remarkable, though less so than with phosphate of ammonia, that we must describe the experiments in detail, but we will give only a few. In speaking of the successive periods when inflection occurred, we always reckon from the time of first immersion.

Having made some preliminary trials as guide, five leaves were placed in the same little vessel in thirty minims of a solution of one part of the nitrate to 7875 of water (1 gr. to 18 oz.); and this amount of fluid just sufficed to cover them. After 2 hrs. 10 m. three of the leaves were considerably inflected, and the other two moderately. The glands of all became of so dark a red as almost to deserve to be called blade. After 8 hrs. four of the leaves had all their tentacles more or less inflected; whilst the fifth, which was then perceived to be an old leaf, had only thirty tentacles inflected. Next morning, after 23 hrs. 40 m., all the leaves were in the same state, excepting that the old leaf had a few more tentacles inflected. Five leaves which had been placed at the same time in water were observed at the same intervals of time; after 2 hrs. 10 m., two of them had four, one had seven, ne had ten, of the long-marginal tentacles, and the fifth had four round-headed tentacles, inflected. After 8 hrs. there was no change in these leaves, and after 24 hrs. all the marginal tentacles had re-expanded; but in one leaf, a dozen, and in a second leaf, half a dozen, submarginal tentacles had become inflected. As the glands of the five leaves in the solution wee simultaneously darkened, no doubt they had all absorbed a nearly equal amount of the salt: and as $\frac{1}{288}$ of a grain was given to the five leaves together, each got $\frac{1}{1440}$ of a grain (.045 mg.). I did not count the tentacles on these leaves, which were in moderately fine ones, but as the average number on thirty-one leaves was 192, it would be safe to assume that each bore on an average at least 160. If so each of the darkened glands could have received only $\frac{1}{230400}$ of a grain of nitrate; and this caused the inflection of a great majority of the tentacles.

This plan of immersing several leaves in the same vessel is a bad one, as it is impossible to fell sure that the more vigorous leaves do not rob the weaker ones of their share of the salt. The glands, moreover, must often touch one another or the sides of the vessel, and movement may have been thus excited; but the corresponding leaves in water, which were little inflected, though rather more so than commonly occurs, were exposed in an almost equal degree to these same sources of error. We will, therefore, give only one other experiment made in this manner, though many were tried and all confirmed the foregoing and following results. Four leaves were placed in forty minims of a solution of one part to 10500 of water; and assuming that they absorbed equally, each leaf received $\frac{1}{1152}$ of grain (.0562mg.). After 1 hr. 20 many of the tentacles on all four leaves were somewhat inflected. After 5 hr. 30m. two leaves had all their tentacles inflected; a third leaf all except the extreme marginal, which seemed old and torpid; and the fourth a large number. After 21 hrs. every single tentacles, on all four leaves, was closely inflected. Of the four leaves placed at the same time in water, one had, after 5 hrs. 45 m., five marginal tentacles inflected; a second, ten; a third, nine marginal and submarginals; and the fourth, twelve, chiefly submarginals inflected. After 21 hrs. all these marginal tentacles re-expanded, but a few of the submarginals on two of the leaves remained slightly curved inwards. The contrast was wonderfully great between these four leaves in water and those in the solution, the latter having every one of their tentacles closely inflected. Making the moderate assumption that each of these leaves bore 160 tentacles, each gland could have absorbed only $\frac{1}{184320}$ of a grain (.000351mg.). This experiment was repeated on three leaves with the same relative amount of the solution; and after 6 hrs. 15 m. all the tentacles except nine, on all three leaves taken together, were closely inflected. In this case the tentacles on each leaf were counted, and gave zn average of 162 per leaf.

The following experiments were tried during the summer, by placing the leaves, such in a separate watch-glass and pouring over it thirty minims (1.775 c.c.) of the solution; other leaves being treated in exactly the same manner with the doubly distilled water used in making the solutions. Six leaves were first immersed, each in thirty minims of a solution of one part of the nitrate to 8750 of water (1 gr. to 20 oz.), so that each received $\frac{1}{320}$ of a grain (.2025 mg.). Before 30 m. had elapsed, four of these leaves were immensely, and two of them moderately, inflected. The glands were rendered of a dark red. The four corresponding leaves in water were not at all affected until 6 hrs. had elapsed, and then only the short tentacles on the borders of the disc; and their inflection, as previously explained, is never of any significance.

Four leaves were immersed, each in thirty minims of a solutions of one part to 17500 of water (1 gr. to 40 oz), so that each received $\frac{1}{640}$ of a grain (.101 mg.); and in less then 45 m. three of them had all their tentacles, except from four to ten, inflected; the blade of one being inflected after 6 hrs. and the blade of second after 21 hrs. The fourth leaf was not at all affected. The glands of none were darkened. Of the corresponding leaves in water, only one had any of its exterior tentacles, namely five, inflected; after 6 hrs. In one case, and after 21 hrs. in two other cases, the short tentacles on the borders of disc formed a ring, in the usual manner.

Four leaves were immersed, each in thirty minims of solution of one part to 43750 of water (1 ge. to 100 oz), so that each leaf got $\frac{1}{1600}$ of grain (.405 mg,). Of these, one was much inflected in 8, and after 2 hrs. 7m. had all the tentacles, except thirteen, inflected. The second leaf, after 10 m., had all except three inflected. The third and fourth were hardly at all affected, scarcely more than the corresponding leaves in water. Of the latter, only one was affected, this having two tentacles inflected, with those on the outer parts of the disc forming a ring in the usual manner. In the leaf which had all its tentacles except three inflected in 10 m., each gland (assuming that the leaf bore 160 tentacles) could have absorbed only $\frac{1}{251200}$ of a grain, or .000258 mg.

Four leaves were separately immersed as before in a solution on one part to 131250 of water (1 gr. to 300 oz), so that each received $\frac{1}{4800}$ of a grain, or .0135 mg. After 50m. one leaf had all its tentacles except sixteen, and after 8 hrs. 20 m., all but fourteen, infected. The second leaf, after 40 m., had all but twenty inflected; and after 8 hrs. 10 m. began to re-expand. The third, in 3 hrs. had about half its tentacles inflected, which began to re-expand after 8 hrs. 15 m. The fourth leaf, after 3 hrs. 7 m., had only twenty-nine tentacles more or less inflected. Thus three out of the four leaves were strongly acted on. It is clear that very sensitive leaves had been accidentally selected. The day moreover was hot. The four corresponding leaves in water were likewise acted on rather more than is usual; for after 3 hrs. one had nine tentacles, another four, and another two, and the fourth none, inflected, with respect to the leaf of which all the tentacles, except sixteen, were inflected after 50 m., each gland (assuming that the leaf bore 160 tentacles) could have absorbed only $\frac{1}{691200}$ of a grain (.0000937 mg.) and this appears to be about the least quantity of the nitrate which suffices to induce the inflection of a single tentacle.

As negative results are important in confirming the foregoing positive ones, eight leaves were immersed as before, each in thirty minims of a solution of one part in 175000 of water (1 gr. to 400 oz.) so that each received only $\frac{1}{6400}$ of a grain (.0101 mg). This minute quantity produced a slight effect on only four of the eight leaves. One had fifty-six tentacles inflected after 2 hrs. 13 m.; a second, twenty-six inflected, or sub-inflected, after 33 m.; a third, eighteen inflected, after 1 hr.; and a fourth, ten inflected, after 35 m. The four other leaves were not in the least affected. Of the eight corresponding leaves in water, one had, after 2 hrs. 10 m., nine tentacles, and four others from one to four long-headed tentacles, inflected; the remaining three being unaffected. Hence, the 1/6400 of a grain given to a sensitive leaf during warm weather perhaps produces a slight effect; but we must bear in mind that occasionally water causes as great an amount of inflection as occurred in this last experiment.

Summary of the Results with Nitrate of Ammonia—The glands of the disc, when excited by a half-minim drop (.0296 c.c.), containing $\frac{1}{2400}$ of a grain of the nitrate (.027 mg.), transmit a motor impulse to the exterior tentacles, causing them to bend inwards. A minute drop, containing $\frac{1}{28800}$ of a grain (.000225 mg.), if held for a few seconds in contact with a gland, causes the tentacle bearing this gland to be inflected. If a leaf is left immersed for a few hours, and sometimes for only a few minutes, in a solutions of such strengths that each gland can absorb only the $\frac{1}{691200}$ of a grain (.0000937 mg.) this small amount is enough to excite each tentacle into movement, and it becomes closely insected.

Phosphate of Ammonia

This salt is more powerful than the nitrate, even in a greater degree than the nitrate is more powerful than the carbonate.This is shown by weaker solutions of the phosphate acting when dropped on the discs, or applied to the glands of the exterior tentacles or when leaves are immersed. The difference in the power of these three salts, as tried in three different ways, supports the results presently to be given, which are so surprising that their credibility required every kind of support. Earlier, we experimented on twelve immersed leaves, giving each only ten minims of a solution; but this was a bad method, for so small a quantity hardly covered them. None of these experiments will, therefore, be given, though they indicate that excessively minute doses are efficient. Later, we determined to make another set of experiments with scrupulous care, on the same plan as those made with the nitrate, namely, by placing leaves in watch glasses, and pouring over each thirty minims of the solutions under trial, treating at the same time and in the same manner other leaves with the distilled water used in making the solutions. Seventy-one leaves were thus tried in solution of various strengths, as the same number in water. Notwithstanding the care taken and the number of the trials made, when in the following year we looked merely at the results, without reading over observations, we again thought that there must have been some error, and thirty-five fresh trials were made with the weakest solution; but the results were as plainly marked as before. Altogether, 106 carefully selected leaves were tried, both in water and in solutions on the phosphate. Hence, after the most anxious consideration, we can entertain no doubt of the substantial accuracy of my results.

Before giving the experiments, it may be well to premise that crystallised of ammonia contains 35.33 per cent of water of crystallisation; so that in all the following trails the efficient elements formed only 64.67 per cent of the salt used.

Extremely minute particles of the dry phosphate were placed with the point of a needle on the secretion surrounding several glands. These poured fourth much secretion, were blackened, and ultimately died; but the tentacles moved only slightly. The dose, small as it was, evidently was too great, and the result was the same as with particles of the carbonate of ammonia.

Half-minims of a solution of one part to 437 of water were placed on the disc of three leaves and acted most energetically, causing the tentacles of one to be inflected in 15 m., and the blades of all three to be much curved inwards in 2 hrs. 15 m., similar drops of a solutions of one part to 1312 of water (1 gr .

to 3 oz.) were then placed on the discs of five leaves, so that each received the $\frac{1}{2880}$ of a grain (.0225 mg.) after 8 hrs. the tentacles of four of them were considerably inflected, and after 24 hrs. the blades of three. After 48 hrs. all five were almost fully re-expanded. We may mention with respect to one of these leaves, that a drop of water had been left during the previous 24 hrs. on its disc, but produced no effect; and that this was hardly dry when the solutions was added.

Similar drops of a solution of one part to 1750 of water (1 gr. to 4 oz.) were next placed on the discs of six leaves; so that each received $\frac{1}{3840}$ of grain (.0169mg); after 8 hrs. three of them had many tentacles and their blades inflected; two others had only a few tentacles slightly inflected, and the sixth was not at all affected. After 24 hrs. most of the leaves had a few more tentacles inflected, but one had began to re-expand. We thus see that with the more sensitive leaves the $\frac{1}{3840}$ of grain, absorbed by the central glands, is enough to make many of the exterior tentacle and the blades bend, whereas the $\frac{1}{1920}$ of a grain of the carbonate similarly given produced no effect; and $\frac{1}{2880}$ of a grain of the nitrate was only just sufficient to produce a well marked effect.

A minute drop, about equal to $\frac{1}{20}$ of a minim, of a solution of one part of the phosphate to 875 of water, was applied to the secretion on three glands, each of which thus received only $\frac{1}{57600}$ of a grain (.00112 mg.) and all three tentacles become inflected. Similar drops of a solution of one part to 1312 of water (1 gr. to 3 oz.) were now tried on three leaves; a drop being applied to four glands on the same leaf. On the first leaf three of the tentacles became slightly inflected to 6 m., and re-expanded after 8 hrs. 45 m. On the second, two tentacles became sub-inflected in 12 m. and on the third all four tentacles were decidedly inflected in 12 m.; they remained so for 8 hrs. 30 m., but by the next morning were fully re-expanded. In this latter case each gland could have received only the $\frac{1}{115200}$ (or .000563 mg.) of a grain. Lastly, similar drops of a solution of one part to 1750 of water (1 gr. to 4 oz.) were tried on five leaves; a drop being applied to four glands on the same leaf. The tentacles on three of these leaves were not in the least affected; on the fourth leaf two became inflected; whilst on the fifth, which happened to be a very sensitive one, all four tentacles were plainly inflected in 6 hrs. 15 m; but only one remained inflected after 24 hrs. we should however, state that in this case an unusually large drop adhered to the head of the pin. Each of these glands could have received very little more than $\frac{1}{153600}$ of a grain (or .000423 mg.); but this small quantity sufficed to cause inflection. We must bear in mind that these drops were applied to the viscid secretion for only from 10 to 15 seconds, and we have good reason to believe that all the phosphate in the solution would not be diffused and absorbed in this time. We have seen under the same circumstances that the adsorption by a gland of the $\frac{1}{19200}$ of a grain of the carbonate, and of $\frac{1}{57600}$ of grain of the nitrate, did not cause the tentacle bearing the gland in question to be inflected; so that here again the phosphate is much more powerful than the other two salts.

We will now turn to the 106 experiments with immersed leaves. Having ascertained by repeated trails that moderately strong solution were highly efficient, we commenced with sixteen leaves, each placed in thirty minims of a solutions of one part of 43750 of after (1 gr. to 100 oz.); so that each received $\frac{1}{1600}$ of a grain, of .04058 mg. Of these leaves, eleven had nearly all or a great number of their tentacles inflected in 1 hr., and the twelfth leaf in 23 hrs. One of the eleven had every single tentacle closely inflected in 50m. Two leaves out of the sixteen were only moderately affected, yet more so than any of those simultaneously immersed in water; and there remaining two, which were pale leaves, were hardly at all affected. Of the sixteen corresponding leaves in water, one had nine tentacles, another six, and two others two tentacles inflected, in the course of 5 hrs. so that the contrast in appearance between the two lots was extremely great.

Eighteen leaves were immersed, each in thirty minims of a solution of one part to 87500 of water (1 gr. to 200 oz.)so that each received $\frac{1}{3200}$ of a grain (.0200 mg.) Fourteen of these were strongly inflected within 2 hrs., and some of them within 15 m., three our of the eighteen were only slightly affected, having twenty-one, nineteen, and twelve tentacles inflected; and one was not at all acted on. By an accident only fifteen, instead of eighteen leaves were immersed at the same time in water; these were observed for 24 hrs.; one had six, another four, and a third two of their outer tentacles inflected; the remainder being quite unaffected.

The next experiment was tried under very favourable circumstances, for the day was very warm, and we happened to have unusually fine leaves.

Five were immersed as before in solution of one part to 131250 of a grain, or .0135 mg. After an immersion of 25 m. all five leaves were much inflected. After 1 hr. 25 m. one leaf had all but eight tentacles inflected; the second, all but three; the third, all but five; the fourth, all but twenty-three; the fifth, on the other hand, never had more than twenty-four inflected. Of the corresponding five leaves in water, one had seven, a second two, a third ten, a fourth-one, and a fifteen one inflected. Let it be observed what a contrast is presented between these latter leaves and those in the solution. We counted the glands on the second leaf in the solution, and the number was 217; assuming that the three tentacles which did not become inflected absorbed nothing, we find that each of the 214 remaining glands could have absorbed only $\frac{1}{1027200}$ of a grain, or .0000631 mg. The third leaf bore 236 glands, and subtracting the five which did not become inflected, each of the remaining 231 glands could have absorbed only $\frac{1}{1108800}$ of a grain (or .0000584mg.) and this amount sufficed to cause the tentacles to bend.

Twelve leaves were tried as before in a solution of one part to 175000 of water (1 gr. to 400 oz.) so that each leaf received $\frac{1}{6400}$ of a grain (.0101 mg.). The plants were not at the time in a good state, and many of the leaves were young and pale. Nevertheless, two of them had all their tentacles, except three or four, closely inflected in under 1 hr. Seven were considerably affected, some within 1 hr., and others not until 3 hrs., 4 hrs., 30, and 8 hrs. had elapsed; and this slow action may be attributed to the leaves being young and pale. Of these nine leaves, four had their blades well inflected, and a fifth slightly so. The three remaining leaves were not affected. With respects to the twelve corresponding leaves in water, not one had its blade inflected; after from 1 to 2 hrs. one had thirteen of its outer tentacles inflected; a second six and four others either one or two inflected. After 8 hrs., the outer tentacles did not become more inflected; whereas this occurred with the leaves in the solution. After the 8 hrs. it was impossible to compare the two lots, and doubt for an instant the power of the solutions.

Two of the above leaves in the solution had all their tentacles, except three and four, inflected within an hour. We counted their glands, and, on the same principle as before, each gland on one leaf could have absorbed only $\frac{1}{1164800}$ and on the other leaf only $\frac{1}{1472000}$ of a grain of the phosphate.

Twenty leaves were immersed in the usual manner, each in thirty minims of a solution of one part to 218750 of water (1 gr. to 500 oz.). So many leaves were tried because I was then under the false impression that it was incredible that any weaker solution could produce an effect. Each leaf received $\frac{1}{8000}$ of a grain,, or .0081 mg. The first eight leaves which I tried both in the solution and water were either young and pale or too old; and the weather was not hot. They were hardly at all affected; nevertheless, it would be unfair to exclude them. We then waited until we had got eight pairs of fine leaves, and the weather was favourable, the temperature of the room where the leaves were immersed varying from 75° to 81° F (23.8° to 27.2° C). In another trail with four pairs (included in the above twenty pairs), the temperature in my room was rather low, about 60° F (15.5° C); but the plants had been kept for several days in a very warm greenhouse and thus rendered extremely sensitive. Special precaution were taken for this set of experiments; the leaves were selected form a large number of plants in the following manner: the four finest were immersed in water, and the next four finest in the solution, and so on till the twenty pairs were complete. The water specimens were thus a little favoured, but they did not undergo more inflection than in the previous cases, comparatively with those in the solutions.

Of the twenty leaves in the solution, eleven became inflected within 40 m.; eight of them plainly and three rather doubtfully; but the latter had at east twenty of their outer tentacles inflected. Owing to the weakness of the solutions, inflection occurred, except in No. 1, much more slowly than in the previous trials. The condition of the eleven leaves which were considerably inflected will now be given at stated intervals, always reckoning from the time of immersion:

1. After only 8 m. a large number of tentacles inflected, and after 17 m. all but fifteen; after 2 hrs. all but eight inflected, or plainly sub-inflected. After 4 hrs. the tentacles began to re-expand, and such prompt re-expansions unusual; after 7 hrs. 30 m. they were almost fully re-expanded.

2. After 39m. a large number of tentacles inflected; after 2 hrs. 18 m. all but twenty-five inflected; after 4 hrs. 17 m. all but sixteen inflected. The leaf remained in this state for many hours.

3. After 12 m. a considerable amount of inflection; after 4 hrs. all the tentacles

inflected except those of the two outer rows, and the leaf remained in this state for some time; after 23 hrs. began to re-expand.

4. After 40 m. much inflection; after 4 hrs. 13 m. fully half the tentacles inflected; after 23 hrs. still slightly inflected.

5. After 40 m. much inflection; after 4 hrs. 22 m. fully half the tentacles inflected; after 23 hrs. still slightly inflected.

6. After 40 m. some inflection; after 2 hrs. 18 m. about twenty-eight outer tentacles inflected; after 5 hrs. 20 m. about third of the tentacles inflected; after 8 hrs. much re-expanded.

7. After 20 m. some inflection; after 2 hrs. a considerable number of tentacles inflected; after 7 hrs. 45 began to re-expand.

8. After 38 m. twenty-eight tentacles inflected; after 3 hrs. 45 m. thirty-three inflected, with most of the submarginal tentacles sub-inflected; continued so for two days and then partially re-expanded.

9. After 38 m. forty-two tentacles inflected; after 3 hrs. 12 m. sixty–six inflected or sub-inflected; after 6 hrs. 40 m. all but twenty four inflected or sub-inflected; after 9 hrs. 40 m. all but seventeen inflected; after 24 hrs. all but four inflected or sub-inflected, only a few being closely inflected; after 27 hrs. all but four inflected or sub-inflected, only a few being closely inflected; after 27 hrs. 40 m. the blade inflected. The leaf remained in this state for two days, and then began to re-expand.

10. After 38 m. twenty-one tentacles; after 3 hrs. 12 m. forty-six tentacles inflected or sub-inflected; after 6 hrs. 40 m. all but seventeen inflected, though none closely; after 24 hrs. every tentacle slightly curved inwards; after 27 hrs. 40 m. blade strongly inflected, and so continued for two days, and then the tentacles and blade very slowly re-expanded.

11. This fine dark red and rather old leaf, though not very large, bore an extraordinary number of tentacles (*viz.*, 252), and behaved in an anomalous manner. After 6 hrs. 40 m. only the short tentacles round the outer part of the disc were inflected, forming a ring as so often occurs in from 8 to 24 hrs. with leaves both in water and the weaker solutions. But after 9 hrs. 40 m. all the outer tentacles except twenty-five were inflected, as was the blade in a strongly marked manner. After 24 hrs. every tentacle

except one was closely inflected, and the blade was completely doubled over. Thus the leaf remained for two days, when it began to re-expand. We may add that the three latter leaves (nos. 9, 10 and 11) were still somewhat inflected after three days. The tentacles in but few of these eleven leaves became closely inflected within so short a time as in the previous experiments with stronger solutions.

We will now turn to the corresponding leaves in water: nine had none of their outer tentacles inflected; nine others had from one to three inflected; and these re-expanded after 8 hrs. The remaining two leaves were moderately affected; one having six tentacles inflected in 34 m.; the other, twenty-three inflected in 2 hrs. 12 m. and both thus remained for 24 hrs. None of these leaves had their blades inflected. So that the contrast between the twenty leaves in water and the twenty in the solution was very great, both within the first hour and after from 8 to 12 hrs. had elapsed.

Of the leaves in the solutions, the glands on leaf No. 1 which in 2 hrs. had all its tentacles except eight inflected, were counted and found to be 202. Subtracting the eight, each gland could have received only the $\frac{1}{1552000}$ of a grain (.0000411 mg.) of the phosphate. Leaf No. 9 had 213 tentacles, all of which, with the exception of four, were inflected after 24 hrs., but none of them closely; the blade was also inflected; each gland could have received only the $\frac{1}{1672000}$ of a grain, or .000387 mg. Lastly, leaf No. 11, which had after 24 hrs. all its tentacles, except one, closely inflected, as well as the blade, bore the unusually large number of 252 tentacles; and, on the same principle as before, each gland could have absorbed only the $\frac{1}{2008000}$ of a grain, or .0000322 mg.

With respect to the following experiments, we must premise that the leaves, both those placed in the solutions and in water, were taken from plants, which had been kept in a very warm greenhouse during the winter. They were thus rendered extremely sensitive, as was shown by water exciting them much more than in the previous experiments. Before giving my observations, it may be well to remind the reader that, judging from thirty-one fine leaves, the average number of tentacles is 192, and that the outer or exterior ones, the movements of which are alone significant, are to the short ones on the disc in the proportion of about sixteen to nine.

Four leaves were immersed as before, each in thirty minims of a solutions of one part of 328125 of water (1 gr. to 750 oz.). Each leaf thus received $\frac{1}{12000}$ of a grain (.0054 mg.)of the salt; and all four were greatly inflected.

1. After 1 hr. all the outer tentacles but one inflected, and the blade greatly so; after 7 hrs. began to re-expand.

2. After 1 hr. all the outer tentacles but eight inflected; after 12 hrs. all re-expanded.

3. After 1 hr. much inflection; after 2 hrs. 30 m. all the tentacles but thirty-six inflected; after 6 hrs. all but twenty-two inflected; after 12 hrs. partly re-expanded.

4. After 1 hr. all the tentacles but thirty-two inflected; after 2 hrs. 30 m. all but twenty-one inflected; after 6 hrs., almost re-expanded.

Of the four corresponding leaves in water:

1. After 1 hr. forty-five tentacles inflected; but after 7 hrs. so many had re-expanded that only ten remained much inflected.
2. After 1 hr. seven tentacles inflected; these were almost re-expanded in 6 hrs.
3. and (4) not affected, except that, as usual, after 11 hrs. the short tentacles on the borders of the disc formed a ring.

There can, therefore, be no doubt about the efficiency of the above solution; and it follows as before that each gland of No. 1 could have absorbed only $\frac{1}{2412000}$ of a grain (.0000268 mg.) and of No. 2 only $\frac{1}{2460000}$ of a grain (.0000263 mg.) of the phosphate.

Seven leaves were immersed, each in thirty minims of a solution of one part to 437500 of water (1 gr. to 1000 oz.). Each leaf thus received $\frac{1}{16000}$ of a grain (.00405 mg.). The day was warm, and the leaves were very fine, so that all circumstances were favourable.

1. After 30 m. all the outer tentacles except five inflected, and most of them closely; after 1 hr. blade slightly inflected; after 9 hrs. 30 m. began re-expand.

2. After 33 m. all the outer tentacles but twenty-five inflected, and blade slightly so; after 1 hrs. 30 m. blade strongly inflected and remained so for 24 hrs.; but some of the tentacles had then re-expanded.

3. After 1 hr. all but twelve tentacles inflected; after 2 hrs. 30 m. all but nine

inflected; and of the inflected tentacles all excepting four closely; blade slightly inflected. After 8 hrs. blade quite double up, and now all the tentacles excepting eight closely inflected. The leaf remained in this state for two days.

4. After 2 hrs. 20 m. only fifty-nine tentacles inflected; but after 5 hrs. all the tentacles closely inflected excepting two which were not affected, and eleven which were only sub-inflected; after 7 hrs. blade considerably inflected; after 12 hrs. much re-expansion.
5. After 4 hrs. all the tentacles but fourteen inflected; after 9 hrs. 30 m. beginning to re-expand.
6. After 1 hr. thirty-six tentacles inflected; after 5 hrs. all but fifty-four inflected; after 12 hrs. considerable re-expansion.
7. After 4 hrs. 30 m. only thirty-five tentacles inflected or sub inflected, and this small amount of inflection never increased.

Now for the seven corresponding leaves in water:

1. After 4 hrs. thirty-eight tentacles inflected; but after 7 hrs. these, with the exception on six, re-expanded.
2. After 4 hrs. 20 m. twenty inflected; these after 9 hrs. partially re-expanded.
3. After 4 hrs. five inflected, which began to re-expand after 7 hrs.
4. After 24 hrs. one inflected.
5., (6) and (7) not at all affected, though observed for 24 hrs., excepting the short tentacles on the borders of the disc, which as usual formed a ring.

A comparison of the leaves in the solutions, especially of the first five or even six on the list, with those in the water, after 1 hr. or after 4 hrs., and in a still more marked degree after 7 hrs. or 8 hrs. could not leaves the least doubt that the solutions had produced a great effect. This was shown, not only by the vastly greater number of inflected tentacles, but by the degree or closeness of their inflection, and by that of their blades. Yet each gland on leaf No. 1 (which bore 255 glands, all of which, excepting five, were inflected in 30 m.) could not have received more than one-four-millionth of a grain (.0000162 mg.) of the salt. Again, each gland on leaf No. 3 (which bore 233 glands, all of which, except nine were inflected in 2 hrs. 30 m.) could have received at most only the $\frac{1}{3584000}$ of grain, or .0000181 mg.

Four leaves were immersed as before in a solution of one part to 656,250 of water (1 gr. to 1500 oz.); but on this occasion we happened to select leaves which were very little sensitive, as on other occasions we chanced to select unusually sensitive leaves. The leaves were not more affected after 12 hrs. then the four corresponding ones in water; but after 24 hrs. they were slightly more inflected. Such evidence, however, is not at all trustworthy.

Twelve leaves were immersed, each in thirty minims of a solution of one part to 1312500 of water (1 gr. to 3000 oz.); so that each leaf received $\frac{1}{48000}$ of a grain (.00135 mg.). The leaves were not in very good condition; four of them were too old and of a dark red colour; four were too pale, yet one of these latter acted well; the four others, as far as could be told by the eye, seemed in excellent condition. The result was as follows:

1. This was a pale leaf; after 40 m. about thirty-eight tentacles inflected; after 3 hrs. 30 m. the blade and many of the outer tentacles inflected; after 10 hrs. 15 m. all the tentacles but seventeen inflected, and the blade quite doubled up after 24 hrs. all the tentacles but ten more are less inflected. Most of them were closely inflected, but twenty-five were only sub-inflected.
2. After 1 hr. 40 m. twenty-five inflected; after 6 hrs. all but twenty-one inflected; after 10 hrs. all but sixteen more or less inflected; after 24 hrs. re-expanded.
3. After 1 hr. 40 m. thirty-five inflected; after 6 hrs. "a large number" inflected, but from want of time they were not counted; after 24 hrs. re-expanded.
4. After 1 hr. 40 m. about thirty inflected; after 6 hrs. "a large number all round the leaf" inflected, but they were not counted; after 10 hrs., began to re-expand.
5. to (12) these were not more inflected than leaves often are in water, having respectively 16, 8, 10, 8, 4, 9, 14 and 0 tentacles inflected. Two of these leaves however, were remarkable from having their blades slightly inflected after 6 hrs.

With respect to the twelve corresponding leaves in water, (1) had, after 1 hr. 35 m., fifty tentacles inflected, but after 11 hrs. only twenty-two remained so, and these formed a group, with the blade at this point slightly inflected. It appeared as if this leaf had been in some manner accidentally excited, for instance, by a particle of animal matter which was dissolved by the water; (2)

After 1 hr. 45 m. thirty-two tentacles inflected, but after 5 hrs. 30 m. only twenty-five inflected, and these after 10 hrs. all re-expanded; (3) after 1 hr. twenty-five inflected, which after 10 hrs. all re-expanded; (4) and (5) after 1 hr. 35 m. six and seven tentacles inflected, which re-expanded after 11 hrs.; (6), (7) and (8) from one to three inflected, which soon re-expanded; (9), (10), (11) and (12) none inflected, though observed for 24 hrs.

Comparing the states of the twelve leaves in water with those in the solutions, there could be no doubt that in the latter a larger number of tentacles were inflected, and these to a greater degree; but the evidence was by no means so clear as in the former experiments with stronger solutions. It deserves attention that the inflection of four of the leaves in the solutions went on increasing during the first 6 hrs., and with some of them for a longer time; whereas in the water the inflection of the three leaves which were the most affected, as well as of all the others, began to decrease during this same interval. It is also remarkable that the blades of three of the leaves in the solution were slightly inflected, and this is a most rare event with leaves water, though it occurred to a slight extent in one (No. 1), which seemed to have been in some manner accidentally excited. All this shows that the solutions produced some effect, though less and at a much slower rate than in the previous cases. The small effect produced may, however, be accounted for in large part by the majority of the leaves having been in a poor condition.

Of the leaves in the solution, No. 1 bore 200 glands and received $\frac{1}{48000}$ of a grain of the salt. Subtracting the seventeen tentacles, which were not inflected, each gland could have absorbed only the $\frac{1}{8784000}$ of a grain (.000000738 mg.). This amount caused the tentacles bearing each gland to be greatly inflected. The blade was also inflected.

Lastly, eight leaves were immersed, each in thirty minims of a solution of one part of the phosphate 21875000 of water (1 gr. to 5000 oz.). Each leaf thus received 1/8000 of a grain of the salt, or .00081 mg. we took especial pains in selecting the finest leaves from the hot-house for immersion, both in the solution and the water, and almost all proved extremely sensitive. Beginning as before with those in the solutions:

1. After 2 hrs. 30 m. all the tentacles but twenty-two inflected, but some only sub-inflected; the blade much inflected; after 6 hrs. 30 m. all but thirteen inflected, with the blade immense inflected; and remained so for 48 hrs.

2. No change for the first 12 hrs., but after 24 hrs. all the tentacles inflected, excepting those of the outermost row, of which only eleven were inflected. The inflection continued to increase, and after 48 hrs. all the tentacles except three were inflected, and most of them rather closely, four or five being only sub-inflected.

3. No change for the first 12 hrs.; but after 24 hrs. all the tentacles excepting those of the outermost row were sub-inflected, with the blade inflected. After 36 hrs. blade strongly inflected, with all the tentacles, except three, inflected or sub-inflected. After 48 hrs. in the same state.

4. to (8) These leaves, after 2 hrs. 30m., had respectively 32, 17, 7, 4, and 0, tentacles inflected, most of which, after a few hours, re-expanded, with the exception of No. 4, which retained its thirty-two tentacles inflected for 48 hrs.

Now for the eight corresponding leaves in water:

1. After 2 hrs. 40 m. this had twenty of its outer tentacles inflected, five of which re-expanded after 6 hrs. 30 m. After 10 hrs. 15 m. a most unusual circumstance occurred, namely, the whole blade became slightly bowed towards the footstalk, and so remained for 48 hrs. The exterior tentacles, excepting those of the three or four outermost rows, were now also inflected to an unusual degree.

2. to (8) These leaves, after 2 hrs. 40 m., had respectively 42, 12, 9, 8, 2, 1 and 0 tentacles inflected, which all re-expanded within 24 hrs., and most of them within a much shorter time.

When the two lots of eight leaves in the solutions and in the water were compared after the lapse of 24 hrs., they undoubtedly differed much in appearance. The few tentacles on the leaves in water which were inflected had after this interval re-expanded, with the exception of one leaf; and this presented the very unusual case of the blade being somewhat inflected, though in a degree hardly approaching that of the two leaves in the solution. Of these latter leaves, No. 1 had almost all its tentacles, together with its blade, inflected after an immersion of 2 hrs. 30 m. Leaves No. 2 and 3 were affected at a much slower rate; but after from 24 hrs. to 48 hrs. almost all their tentacles were closely inflected, and the blade of one quite doubled up. We must therefore admit, incredible as the fact may at first appear, that this extremely weak solutions acted on the more sensitive leaves; each of which received only the $\frac{1}{80000}$ of a grain (.00081 mg.) of the phosphate. Now, leaf No. 3 bore 178 tentacles, and, subtracting the three which were not inflected, each gland could have absorbed only the $\frac{1}{14000000}$ of grain, or .00000463 mg. Leaf No. 1, which was strongly acted on within 2 hrs. 30 m., and had all its outer tentacles, except thirteen, inflected within 6 hrs. 30 m., bore 260 tentacles; and, on the same principle as before, each gland could have observed only $\frac{1}{19760000}$ of a grain,or .00000328 mg.; and this excessively minute amount sufficed to cause all the tentacles bearing these glands to be greatly inflected. The blade was also inflected.

Summary of the Results with Phosphate of Ammonia- The glands of the disc, when excited by a half-minim drop (.0296 c.c.), containing 1/3840 of a grain (.0169 mg.) of this salt, transmit a motor impulse to the exterior tentacles, causing them to bend inwards. A minute drop, containing $\frac{1}{153600}$ of a grain (.000423 mg.), if held for a few seconds in contact with a gland, causes the tentacles bearing this gland to be inflected. If a leaf is left immersed for a few hours, and sometimes for a shorter time, in a solution so weak that each gland can absorb only the $\frac{1}{19760000}$ of a grain (0.00000328 mg.) this is enough to excite the tentacle into movement, so that it becomes closely inflected, as does sometimes the blade. In the general summary to this chapter a few remarks will be added, showing that the efficiency of such extremely minute doses is not so incredible as it much at first appear.

Other Salts of Ammonia

Sulphate of Ammonia—The few trials made with this and the following five salt of ammonia were undertaken merely to ascertain whether they induced inflection. Half-minims of a solution of one part of the sulphate of ammonia to 437 of water were placed on the discs of seven leaves, so that each received $\frac{1}{960}$ of a grain, or .0675 mg. After 1 hr. the tentacles of five of them, as well as the blade of one, were strongly inflected. The leaves were not afterwards observed.

Citrate of Ammonia—Half-minims of a solution of one part to 437 of water were placed on the discs of six leaves. In 1 hr. the short outer tentacles round the discs were a little inflected, with the glands on the disc blackened. After 3 hrs. 25 m. one leaf had its blade inflected, but none of the exterior tentacles. All six leaves remained in nearly the same state during the day, the submarginal tentacles, however, becoming more and more inflected. After 23 hrs. three of the leaves had their blades somewhat inflected, and the submarginal tentacles of all considerably inflected, but in none were the two, three, or four outer row, affected. We have rarely seen cases like this, except from the action of a decoction of grass. The glands on the discs of the above leaves, instead of being almost black, as after the first hour, were now, after 23 hrs., very pale. We next tried on four leaves half-minims of a weaker solutions of one part to 1312 of water (1 grain in 3 oz.); so that each received $\frac{1}{2880}$ of a grain (.0225 mg.). After 2 hrs. 18 m. the glands on the disc were very dark-coloured; after 24 hrs. two of the leaves were slightly affected; the other two not at all.

Acetate of Ammonia—Half-minims of a solution of about one part to 109 of water were placed on the discs of two leaves, both of which were acted on in 5 hrs. 30 m., and after 23 hrs. had every single tentacle closely inflected.

Oxalate of Ammonia—Half-minims of a solution of one part to 218 of water were placed on two leaves, which, after 7 hrs., became moderately, and after 23 hrs. strongly, inflected. Two other leaves were tried with a weaker solution of one part to 437 of water; one was strongly inflected in 7 hrs.; the other not until 30 hrs. had elapsed.

Tartrate of Ammonia—Half-minims of a solution of one part to 437 of water were placed on the discs of five leaves. In 31 m. there was a trace of inflection in the exterior tentacles of some of the leaves, and this became more decided after 1 hr. with all the leaves; but the tentacles were never closely inflected. After 8 hrs. 30 m. they began to re-expand. Next morning, after 23 hrs., all were fully re-expanded, excepting one which was still slightly inflected. The shortness of the period of inflection in this and the following case is remarkable.

Chloride of Ammonia—Half-minims of a solution of one part to 437 of water were placed on the discs of six leaves. A decided degree of inflection in the outer and submarginal tentacles was perceptible in 25 m.; and this increased during the next three or four hours, but never became strongly marked. After only 8 hrs. 30 m. the tentacles began to re-expand, and by the next morning, after 24 hrs., were fully re-expanded on four of the leaves, but still slightly inflected on two.

General Summary and Concluding Remarks

We have now seen that the nine salts of ammonia which were tried all cause the inflection of the tentacles, and often of the blade of the leaf. As far as can be ascertained from the superficial trials with the last six salts, the citrate the least powerful, and the phosphate certainly by far the most. The tartrate and chloride are remarkable from the short duration of their action. The relative efficiency of the carbonate, nitrate, and phosphate, is shown in the following table by the smallest amount, which suffice to cause the infection of the tentacles.

Solutions, How Applied	Carbonate of Ammonia	Nitrate of Ammonia	Phosphate of Ammonia
Placed on the glands of the disc, so as to act indirectly on the outer tentacles	$\frac{1}{960}$ *of* a grain, or .0675 mg.	$\frac{1}{2400}$ *of* a grain, or .027 mg.	$\frac{1}{3810}$ *of* a grain, or .0169 mg.
Applied for a few seconds directly to the gland of an outer tentacle	$\frac{1}{14400}$ *of* a grain, or .00445 mg.	$\frac{1}{28800}$ *of* a grain, or .0025 mg.	$\frac{1}{153600}$ *of* a grain, or .000423 mg.
Leaf immersed, with time allowed for each gland to absorb all that it can . . .	$\frac{1}{268800}$ *of* a grain, or .00024 mg.	$\frac{1}{691200}$ *of* a grain, or .0000967 mg.	$\frac{1}{19760000}$ of a grain, or .00000328 mg.
Amount absorbed by a gland which suffices to cause the aggregation of the protoplasm in the adjoinng cells of the tentacles.	$\frac{1}{134400}$ *of* a grain, or .00048 mg.		

From the experiments tried in these three different ways, we see that the carbonate, which contains 23.7 per cent of nitrogen, is less efficient than the nitrate, which contains 35 per cent. The phosphate contains less nitrogen than either of these salts, namely, only 21.2 per cent, and yet is far more efficient; its power, no doubt, depending quite as much on the phosphorus as on the nitrogen which it contains. We may infer that this is the case, from the energetic manner in which bits of bone and phosphate of lime affect the leaves. The inflection excited by the other salts of ammonia is probably due solely to their nitrogen, on the same principle that nitrogenous organic fluids act powerfully, whilst non-nitrogenous organic fluids are powerless. As such minute doses of the salts of ammonia affect the leaves, we may fell almost sure than Drosera absorbs and profits by the amount, though small, which is present in rain-water, in same manner as other plants absorb these same salts by their roots. The smallness of the doses of the nitrate, and more especially of the phosphate of ammonia, which cause the tentacles of immersed leaves to be inflected, is perhaps the most remarkable fact recorded in this volume. When we see that much less than the millionth of a grain of the phosphate, absorbed by a gland of one of the exterior tentacles, causes it to bend, it may be thought that the effects of the solutions on the gland of the disc have been overlooked, namely, the transmission of a motor impulse from them to the exterior tentacles. No doubt the movements of the latter are thus aided; but the aid thus rendered must be insignificant; for we know that a drop containing as much as the $\frac{1}{3840}$ of a grain placed on the disc is only just able to cause the outer tentacles of a highly sensitive leaf to bend. It is certainly a most surprising fact that the $\frac{1}{19760000}$ of a grain or in round numbers the one-twenty millionth of a grain (.0000033 mg.) of the phosphate should affect any plant or indeed any animal; and as this salt contains 35.33 per cent of water of crystallisation, the efficient elements are reduced to $\frac{1}{30555126}$ of a grain, or in round number to one-thirty-millionth of a grain (.00000216 mg). The solution, moreover, in these experiments was diluted in the proportion of one part of the salt to 2187500 of water, or one grain to 5000 oz. The reader will perhaps best realise this degree of dilution by remembering that 5000 oz. would more than fill a 31-galon cask; and that to this large body of water one grain of the salt was added; only half a drachm, or thirty minims, of the solutions being poured over the leaf. Yet this amount sufficed to cause the inflection of almost every tentacle, and often of the blade of the leaf.

Drosera is far from rivalling the power of the spectroscope, but it can detect, as shown by the movement of its leaves, a very much smaller quantity of the phosphate of ammonia than the most skilful chemist can of any substance. The salt was in some cases weighed in an excellent balance; and fresh water was measured many times with care. The observations were repeated during several years. We compared several lots of leaves simultaneously immersed in the weaker solutions and in water, and declared that there could be no doubt about the difference in their appearance. We hope that someone may hereafter be induced to repeat my experiments; in this case he should select young and vigorous leaves, with the glands surrounded by abundant secretion. The leaves should be carefully cut off and laid gently in watch-glasses, and a measured quantity of the solutions and of water poured over each. The water used must be as absolutely pure as it can be made. It is to be especially observed that the experiments with the weaker solutions ought to be tried after several days of very warm weather. Those with the weakest solutions should be made on plants, which have been kept for a considerable time in a warm greenhouse, or cool hothouse; but this is by no means necessary for trials with solutions of moderate strength.

We beg the reader to observe that the sensitiveness or irritability of the tentacles was ascertained by three different methods—indirectly by drops placed on the disc, directly by drops applied to the glands of the outer tentacles, and by the immersion of whole leaves—and it was found by these three methods that the nitrate was more powerful than the carbonate, and the phosphate much more powerful than the nitrate; this result being intelligible from the difference in the amount of nitrogen in the first two salts, and from the presence of phosphorus in the third. It may aid the reader's faith to turn to the experiments with a solution of one grain of the phosphate to 1000 oz. of water, and he will there find decisive evidence that the one-four-millionth of a grain is sufficient to cause the inflection of a single tentacle. There is therefore, nothing very improbable in the fifth of this weight, or the one-twenty-millionth of a grain, acting on the tentacle of a highly sensitive leaf. Again, two of the leaves in the solution of one grain to 3000 oz., and three of the leaves in the solution of one grain to 5000 oz., were affected, not only far more than leaves tried at the same time in water, but incomparably more than any five leaves which can be picked out of the 173 observed at different times in water.

There is nothing remarkable in the mere fact of the one-twenty-millionth of a grain of the phosphate, dissolved in about two million times its weight of water, being absorbed by a gland. All physiologists admit that the roots of plants absorb the salts of ammonia brought to them by the rain; and fourteen gallons of rain-water contain a grain of ammonia, therefore, only a little more than twice as much as in the weakest solution employed by me. The fact which appears truly wonderful is, that the one-twenty-millionth of a grain of the phosphate of ammonia including less than the one-thirty-millionth of efficient matter, when absorbed by a gland, should induce some change in it, which leads to a motor impulse being transmitted down the whole length of the tentacle, causing the basal part to bend, often through an angle of above 180 degrees.

Astonishing as is this result, there is no sound reason why we should reject it as incredible. Prof. Donders, of Utrecht, informed that, from experiments formerly made by him and Dr. De Ruyter, he inferred that less than the one-millionth of a grain of sulphate of atropine, in an extremely diluted state, if applied directly to the iris of a dog, paralyses the muscles of this organ. but, in fact, every time that we perceive an odour, we have evidence that infinitely smaller particles act on our nerves. When a dog stands a quarter of a mile to leeward of a deer or other animal, and perceived its presence, the odorous particles produce some change in the olfactory nerves; yet these particles must be infinitely smaller than those of the phosphate of ammonia weighing the one-twenty-millionth of a grain. These nerves then transmit some influence to the brain of the dog, which leads to action on its part . With Drosera, the really marvellous fact is, that a plant without any specialised nervous system should be affected by such minute particles; but we have no grounds for assuming that other tissues could be rendered as exquisitely susceptible to impressions from without, if this were beneficial to the organism, as is the nervous system of the higher animals.

❑❑❑

8

The Effects of Various Salts and Acids on the Leaves

Salts of sodium, potassium, and other alkaline, earthy, and metallic salts—Summary on the action of these salts—Various acids—Summary on their action.

Action of Other Salts: An Introduction

Having found that the salts of ammonia were so powerful, we were led to investigate the action of some other salts. It will be convenient, first, to give a list of the substances tried (including forty-nine salts and two metallic acids), divided into two columns, showing those which cause inflection, and those which do not do so, or only doubtfully. The experiments were made by placing half-minim drops on the discs of leaves, or, more commonly, by immersing them in the solutions; and sometimes by both methods. A summary of the results, with some concluding remarks, will then be given. The action of various acids will afterwards be described.

Salts Causing Inflection	Salts not Causing Inflection
(Arranged in Groups according to the Chemical Classification in Watts' 'Dictionary of Chemistry')	
Sodium carbonate, rapid inflection.	Potassium carbonate; slowly poisonous.
Sodium nitrate, rapid inflection.	Potassium nitrate: somewhat poisonous.
Sodium sulphate, moderately rapid inflection.	Potassium sulphate.
Sodium phosphate, very rapid inflection.	Potassium phosphate.
Sodium citrate, rapid inflection.	Potassium citrate.
Sodium oxalate, rapid inflection.	
Sodium chloride, moderately rapid inflection.	Potassium chloride.

(Arranged in Groups according to the Chemical Classification in Watts' 'Dictionary of Chemistry')

Salts Cuasing Inflection	Salts not Causing Inflection
Sodium iodide, rather slow inflection.	Potassium iodide, a slight and doubtful amount of inflection.
Sodium bromide, moderately rapid inflection.	Potassium bromide.
Potassium oxalate, slow and doubtful inflection.	
Lithium nitrate, moderately rapid inflection.	Lithium acetate.
Caesium chloride, rather slow inflection.	Rubidium chloride.
Silver nitrate, rapid inflection: quick poison.	
Cadmium chloride, slow inflection.	Calcium acetate.
Mercury perchloride, rapid inflection: quick poison.	Calcium nitrate.
	Magnesium acetate.
	Magnesium nitrate.
	Magnesium chloride.
	Magnesium sulphate.
	Barium acetate.
	Barium nitrate.
	Strontium acetate.
	Strontium nitrate.
	Zinc chloride.
Aluminium chloride, slow and doubtful inflection.	Aluminium nitrate, a trace of inflection.
Gold chloride, rapid inflection: quick poison.	Aluminium and potassium sulphate.
Tin chloride, slow inflection: poisonous.	Lead chloride.
Antimony tartrate, slow inflection: probably poisonous.	
Arsenious acid, quick inflection: poisonous.	
Iron chloride, slow inflection: probably poisonous.	Manganese chloride.
Chromic acid, quick inflection: highly poisonous.	
Copper chloride, rather slow inflection: inflection: poisonous.	Cobalt chloride
Nickel chloride, rapid inflection: probably poisonous.	
Platinum chloride, rapid inflection: poisonous.	

Salts of Sodium

Sodium Carbonate *(pure)*—Half-minims (.296 c.c.) of a solution of one part to 218 of water (2 grs. to 1 oz.) were placed on the discs of twelve leaves. Seven of these became well inflected; three had only two or three of their outer tentacles inflected, and the remaining two were quite unaffected. But the dose, though only the $\frac{1}{480}$ of a grain (.135 mg.), was evidently too strong, for three of the seven well-inflected leaves were killed. On the other hand, one of the seven, which had only a few tentacles inflected, re-expanded and seemed quite healthy after 48 hrs. By employing a weaker solution (*viz.*, one part to 437 of water, of 1 gr. to 1 oz.), doses of $\frac{1}{960}$ of a grain (.0675 mg.) were given to six leaves. Some of these were affected in 37 m.; and in 8 hrs. the outer tentacles of all, as well as the blades of two, were considerably inflected. After 23 hrs. 15 m. the tentacles had almost re-expanded, but the blades of the two were still just perceptibly curved inwards. After 48 hrs. all six leaves were fully re-expanded, and appeared perfectly healthy.

Three leaves were immersed, each in thirty minims of a solution of one part to 875 of water (1 gr. to 2 oz.), so that each received $\frac{1}{32}$ of a grain (2.02 mg.); after 40 m. the three were much affected and after 6 hrs .45 m the tentacles of all and the blade of one closely inflected.

Sodium Nitrate *(pure)*—Half minims of a solution of one part to 437 of water, containing $\frac{1}{960}$ of a grain (.0675 mg.) were placed on the discs of five leaves. After 1 hr. 25 m. the tentacles of nearly all, and the blade of one, were somewhat inflected. The inflection continued to increase, and in 21 hrs. 15m. the tentacles and the blades of four of them were greatly affected, and the blade of the fifth to a slight extent. After an additional 24 hrs. the four leaves still remained closely inflected, whilst the fifth was beginning to expand. Four days after the solution had been applied, two of the leaves had quite, and one had partially, re-expanded; whilst the remaining two remained closely inflected and appeared injured.

Three leaves were immersed, each in thirty minims of a solution of one part 875 of water; in 1 hr. there was great inflection, and after 8 hrs. 15 m. every tentacle and the blades of all three were most strongly inflected.

Sodium Sulphate—Half-minims of a solution of one part to 437 of water were placed on the discs of six leaves. After 5 hrs. 30 m. the tentacles of three of them (with the blade of one) were considerably, and those of the other three slightly, inflected. After 21 hrs. the inflection had a little decreased, and in 45 hrs. the leaves were fully expanded, appearing quite healthy.

Sodium Phosphate – Half-minims of a solution of one part to 437 of water were placed on the discs of six leaves. The solutions acted with extraordinary rapidity, for in 8 m. the outer tentacles on several of the leaves were much incurved. After 6 hrs. the tentacles of all six leaves, and the blades of two, were closely inflected. This state of things continued for 24 hr., excepting that the blade of a third leaves became incurved. After 48 hrs. all the leaves re-expanded. It is clear that $\frac{1}{960}$ of a grain of phosphate of soda has great power in causing inflection.

Sodium Citrate – Half-minims of a solution of one part of 437 of water were placed on the discs of six leaves, but these were not observed until 22 hrs. had elapsed. The submarginal tentacles of five of them, and the blades of four, were then found inflected; but the outer rows of tentacles were not affected. One leaf, which appeared older than the others, was very little affected in any way. After 46 hrs. four of the leaves were almost re-expanded, including their blades. Three leaves were also immersed, each in thirty minims of a solution of one part of the citrate to 875 of water; they were much acted on in 25 m.; and after 6 hrs. 35 m. almost all the tentacles, including those of the outer rows, were inflected, but not the blades.

Sodium Oxalate – Half-minims of a solution of one part to 437 of water were placed on the disc of seven leaves; after 5 hrs. 30 m. the tentacles of all, and the blades of most of them, were much affected. In 22 hrs., besides the inflection of the tentacles, the blades of all seven leaves were so much doubled over that their tips and bases almost touched. On no other occasion we have seen the blades so strongly affected. Three leaves were also immersed, each in thirty minims of a solution of one part to 875 of water; after 30 m. there was much inflection, and after 6 hrs. 35 m. the blades of two and the tentacles or all were closely inflected.

Sodium Chloride *(best culinary salt)* – Half-minims of a solution of one part to 218 of water were placed on the discs of four leaves. Two, apparently, were not all affected in 48 hrs; the third had its tentacles slightly inflected;

whilst the fourth had almost all its tentacles inflected in 24 hrs., and these did not began to re-expand until the fourth day, and were not perfectly expanded on the seventh day. We presume that this leaf was injured by the salt. Half-minims of a weaker solution, of one part to 437 of water, were then dropped on the discs of six leaves, so that each received $\frac{1}{960}$ of a grain. In 1 hr. 33 m. there was slight inflection; and after 5 hrs. 30 m. the tentacles of all six leaves were considerably, but not closely, inflected. After 23 hrs. 15 m. all had completely re-expanded, and did not appear in the least injured.

Three leaves were immersed, each in thirty minims of a solution of one part to 875 of water, so that each received $\frac{1}{132}$ of a grain, or 2.02 mg. After 1 hr. there was much inflection; after 8 hrs. 30 m., all the tentacles and the blades of all three were closely inflected. Four other leaves were also immersed in the solutions, each receiving the same amount of salt as before, *viz.*, $\frac{1}{132}$ of a grain. They all soon became inflected; after 48 hrs. they began to re-expand, and appeared quite uninjured, though the solution was sufficiently strong to taste saline.

Sodium Iodide—Half-minims of a solution of one part to 437 of water were placed on the discs of six leaves. After 24 hrs. four of them had their blades and many tentacles inflected. The other two had only their submarginal tentacles inflected; the outer ones in most of the leaves being but little affected. After 46 hrs., the leaves had nearly re-expanded. There leaves were also immersed, each in thirty minims of a solution of one part to 875 of water. After 6 hrs. 30 m. almost all the tentacles, and the blade of one leaf, were closely inflected.

Sodium Bromide—Half-minims of a solution of one part to 437 of water placed on six leaves. After 7 hrs. there was some inflection; after 22 hrs. three of the leaves had their blades and most of their tentacles inflected; the fourth leaf was very slightly, and the fifth and sixth hardly at all, affected. There leaves were also immersed, each in thirty minims of a solution of one part of 875 of water; after 40 m. there was some inflection; after 4 hrs. the tentacles of all three leaves and the blades of two were inflected. These leaves were then placed in water, and after 17 hrs. 30 m. two of them were almost completely, and the third partially, re-expanded; so that apparently they were not injured.

Salts of Potassium

Potassium Carbonate *(pure)*—Half-minims of a solution of one part to 437 of water were placed on six leaves. No effect was produced in 24 hrs. but after 48 hrs. Some of the leaves had their tentacles and one blade, considerably inflected. This, however, seemed the result of their being injured; for, on the third day after the solution was given, there of the leaves were dead, and one was very unhealthy; the other two were recovering, but with several of their tentacles apparently injured, and these remained permanently inflected. It is evident that the $\frac{1}{960}$ of a grain of this salt acts as a poison. Three leaves were also immersed, each in thirty minims of a solution of one part to 875 of water, though only for 9 hrs.; and very differently from what occurs with the salts of soda, no inflection ensued.

Potassium Nitrate—Half-minims of a strong solution, if one part to 109 of water (4 grg. to oz.) were placed on the discs of four leaves; two were much injured, but no inflection ensued. Eight leaves were treated in the same manner, with drops of weaker solutions, of one part to 218 of water. After 50 hrs. there was no inflection, but two of the leaves seemed injured. Five of these leaves were subsequently tested with drops of milk and a solution of gelatine on their discs, and only one became inflected; so that the solution of the nitrate of the above strength, acting for 50 hrs., apparently had injured or paralysed the leaves. Six leaves were then treated in the same manner with a still weaker solution, of one part to 437 of water, and these, after 48 hrs. were in no way affected, with the exception of perhaps a single leaf. Three leaves were next immersed for 25 hrs., each in thirty minims of solution of one part to 875 of water, and this produced no apparent affect. They were then put into a solution of one part of carbonate of ammonia to 218 of water; the glands were immediately blackened, and after 1 hr., there was some inflection, and the protoplasmic contents of the cells became plainly aggregated. This shows that the leaves had not been much injured by their immersion for 25 hrs. in the nitrate.

Potassium Sulphate—Half-minims of a solution of one part to 437 of water placed on the discs of six leaves, after 20 hrs. 30 m. no effect was produced; after an additional 24 hrs. three remained quite unaffected; two seemed injured, and the sixth seemed almost dead, with its tentacles inflected.

Nevertheless, after two additional days, all six leaves recovered. The immersion of three leaves for 24 hrs., each in thirty minims of a solution of one part to 875 of water, produced no apparent effect. They were then treated with the same solution of carbonate of ammonia, with the same result as in the case of the nitrate of potash.

Potassium Phosphate—Half-minims of a solution of one part to 437 of water were placed on the discs of six leaves, which were observed during three days; but no effect was produced. The partial drying up of the fluid on the disc slightly drew together the tentacles on it, as often occurs in experiments of this kind. The leaves on the third day appeared quite healthy.

Potassium Citrate—Half-minims of a solution of one part to 437 of water, left on the discs of six leaves for three days, and the immersion of three leaves for 9 hrs., each in 30 minims of a solution of one part to 875 of water, did not produce the least effect.

Potassium Oxalate—Half-minims were placed on different occasions on the discs of seventeen leaves; and the results perplexed me much, as they still do. Inflection supervened very slowly. After 24 hrs. four leaves out of the seventeen were well inflected, together with the blades of two; six were slightly affected, and seven not at all. Three leaves of one lot were observed for five days, and all dies; but in another lot of six all excepting one looked healthy after four days. Three leaves wee immersed during 9 hrs., each in 30 minims of a solution of one part to 875 of water, and were not in the least affected; but they ought to have been observed for a longer time.

Potassium Chloride—Neither half-minims of a solution of one part to 437 of water, left on the discs of six leaves for three days, nor the immersion of three leaves during 25 hrs., in 30 minims of a solution of one part to 875 of water, produced the least effect. The immersed leaves were then treated with carbonate of ammonia, as described under nitrate of potash, and with the same result.

Potassium Iodide—Half-minims of a solution of one part to 437 of water were placed on the discs of seven leaves. In 30 m. one leaf had the blade inflected; after some hours three leaves had most of their submarginal tentacles moderately inflected; the remaining three being very slightly affected. Hardly any of these leaves had their outer tentacles inflected. After 21 hrs. all re-expanded, excepting two which still had a few submarginal tentacles inflected. Three leaves were next immersed for 8 hrs. 40 m., each in 30 minims of a solution of one part to 875 of water, and were not in least affected. We do not know what to conclude from this conflicting evidence; but it is clear that the iodide of potassium does not generally produce any marked effect.

Potassium Bromide—Half-minims of a solution of one part to 437 of water were placed on the discs of six leaves; after 22 hrs. one had its blade and many tentacles inflected, but we suspect that an insect might have alighted on it and then escaped; the five other leaves were in no way affected. We tested three of these leaves with bits of meat, and after 24 hrs. they became splendidly inflected. Three leaves were also immersed for 21 hrs. in 30 minims of a solution of one part to 875 of water; but they were not at all affected, excepting that the glands looked rather pale.

Other Salts

Lithium Acetate—Four leaves were immersed together in a vessel containing 120 minims of a solution of one part to 437 of water; so that each received, if the leaves absorbed equally, $\frac{1}{16}$ of a grain. After 24 hrs. there was no inflection. We then added, for the sake of testing the leaves, some strong solution (*viz.*, 1 gr. to 20 oz., or one part to 8750 of water) of phosphate of ammonia, and all four became in 309 m. closely inflected.

Lithium Nitrate—Four were immersed, as in the least case, in 120 minims of solution of one part to 437 of water; after 1 hr. 30 m. all four were a little, and after 24 hrs. greatly, inflected, we then diluted the solution with some water, but they still remained somewhat inflected on the third day.

Caesuim Chloride—Four leaves were immersed, as above, in 126 minims of a solution of one part to 437 of water. After 1 hr. 5 m. the glands were darkened; after 4 hrs. 20 m. there was a trace of inflection; after 6 hrs. 40 m., two leaves were greatly, but not closely, and the other two considerably inflected. After 22 hrs. the inflection was extremely great, and two had their blades inflected. We then transferred the leaves into water, and in 46 hrs. from their first immersion they were almost re-expanded.

Rubidium Chloride—Four leaves which were immersed, as above, in 120 minims of a solution of one part to 437 of water, were not acted on in 22 hrs. We then added some of the strong solution (1 gr. to 20 oz.) of phosphate of ammonia, and in 30 m. all were immensely inflected.

Silver Nitrate—Three leaves were immersed in ninety minims of a solution of one part to 437of water; so that each received, as before, $\frac{1}{16}$ of a grain. After 5 m. slight inflection, and after 11 m., very strong inflection, the glands becoming excessively black; after 40 m. all the tentacles were closely inflected. After 6 hrs. the leaves were taken out of the solution, washed, and placed in water; but next morning they were evidently dead.

Calcium Acetate—Four leaves were immersed in 120 minims of a solution of one part to 437 of water; after 24 hrs. none of the tentacles were inflected, excepting a few where the blade joined the petiole; and this may have been caused in a few where the blade joined the petiole; and this may have been caused by the absorption of the salt by the cut-off end of the petiole. We then added some of the solution (1 gr. to 20 oz.) of phosphate of ammonia, but this to my surprise excited only slight inflection, even after 24 hrs. Hence it would appear that the acetate had rendered the leaves torpid.

Calcium Nitrate—Four leaves were immersed in 120 minims of a solution of one part to 437 of water, but were not affected in 24 hrs. We then added some of the solution of phosphate of ammonia (1 gr. to 20 oz.), but this caused only very slight inflection after 24 hrs. A fresh leaf was next put into a mixed solution of the above strengths of the nitrate of calcium and phosphate of ammonia, and it became closely inflected in between 5 m. and 10 m. half-minims of a solution of one part of the nitrate of calcium to 218 of water were dropped on the discs of three leaves, but produced no effect.

Acetate, Nitrate, and Chloride of Magnesium—Four leaves were immersed in 120 minims of solutions, of one part to 437 of water of each of these three salts; after 6 hrs. there was no inflection; but after 22 hrs. one of the leaves in the acetate was rather more inflected then generally occurs from an immersion for this length of time in water. Some of the solution (1 gr. to 20 oz.) of phosphate of ammonia was then added to the three solutions. The leaves in the acetate mixed with the phosphate underwent some inflection; and this was well pronounced after 24 hrs. Those in the mixed nitrate were decidedly inflected in 4 hrs. 30 m., but the degree of inflection did not afterwards much increase; whereas the four leaves in the mixed chloride were greatly inflected in a few minutes, and after 4 hrs. had almost every tentacle closely inflected.

We thus see that the acetate and nitrate of magnesium injure the leaves, or at least prevent the subsequent action of phosphate of ammonia; whereas the chloride has no such tendency.

Magnesium Sulphate—Half-minims of a solution of one part to 218 of water were placed on the discs of ten leaves, and produced no effect.

Barium Acetate—Four leaves were immersed in 120 minims of a solution of one part to 437 of water, and after 22 hrs. there was no inflection, but the glands were blackened. The leaves were then placed in a solution (1 gr. to 20 oz.) of phosphate of ammonia, which caused after 26 hrs. only a little inflection in two of the leaves.

Barium Nitrate—Four leaves were immersed in 120 minims of a solution of one part to 437 of water; and after 22 hrs. there was no more than that slight degree of inflection which often follows from an immersion of this length in pure water. We then added some of the same solution of phosphate of ammonia, and after 30 m. one leaf was greatly inflected, two others moderately, and the fourth not at all. The leaves remained in this state for 24 hrs.

Strontium Acetate—Four leaves, immersed in 120 minims of a solution of one part to 437 of water, were not affected in 32 hrs. They were then placed in some of the same solution of phosphate of ammonia, and in 25 m. two of them were greatly inflected; after 8 hrs. the third leaf was considerably inflected, and the fourth exhibited a trace of inflection. They were in the same state next morning.

Strontium Nitrate—Five leaves were immersed in 120 minims of a solution of one part to 437 of water; after 22 hrs. there was some slight inflection, but not more than sometimes occurs with leaves in water. They were then placed in the same solution of phosphate of ammonia; after 8 hrs. three of them were moderately inflected, as were all five after 24 hrs.; but not one was closely inflected. It appears that the nitrate of strontium renders the leaves half torpid.

Cadmium Chloride—Three leaves were immersed in ninety minims of a solution of one part to 437 of water; after 5 hrs. 20 m. slight inflection occurred, which increased during the next three hours. After 24 hrs. all three leaves had their tentacles well inflected, and remained so for an additional 24 hrs.; glands not discoloured.

Mercury Perchloride—Three leaves were immersed in ninety minims of a solution of one part to 437 of water; after 22 m. there was some slight inflection, which in 18 m. became well pronounced; the glands were now blackened. After 5 hrs. 35 m. all the tentacles closely inflected; after 24 hrs. still inflected and discoloured. The leaves were then removed and left for two days in water; but they never re-expanded, being evidently dead.

Zinc Chloride—Three leaves immersed in ninety minims of a solution of one part to 437 of water were not affected in 25 hrs. 30 m.

Aluminium Chloride—Four leaves were immersed in 120 minims of a solution of one part to 437 of water; after 7 hrs. 45 m. no inflection; after 24 hrs. one leaf rather closely, the second moderately, the third and fourth hardly at all, inflected. The evidence is doubtful, but we think some power in slowly causing inflection must be attributed to this salt. These leaves were then placed in the solution (1 gr. to 20 oz.) of phosphate of ammonia, and after 7 hrs. 30 m., the three, which had been but little affected by the chloride, became rather closely inflected.

Aluminium Nitrate—Four leaves were immersed in 120 minims of a solution of one part to 437 of water; after 7 hrs. 45 m. there was only a trace of inflection; after 24 hrs. one leaf was moderately inflected. The evidence is here again doubtful, as in the case of the chloride of aluminium. The leaves were then transferred to the same solutions as before, of phosphate of ammonia; this produced hardly any effect in 7 hrs. 30 m.; but after 25 hrs. one leaf was pretty closely inflected the three others very slightly, perhaps not more so than from water.

Sulphate of Aluminium and Potassium *(common alum)* – Half-minims of a solution of the usual strength were placed on the discs of nine leaves, but produced no effect.

Chloride of Gold—Seven leaves were immersed in so much of a solution of one part to 437 of water that each received 30 minims, containing $\frac{1}{16}$ of a grain, or 4.048 mg., of the chloride. There was some inflection in 8 m., which became extreme in 45 m. In 3 hrs. the surrounding fluid was coloured purple, and the glands were blackened. After 6 hrs. the leaves were transferred to water; next morning they were found discoloured and evidently killed. The secretion decomposes the chloride very readily; the glands themselves becoming coated with the thinnest layer of metallic gold, and particles float about on the surface of the surrounding fluid.

Chloride of Lead—There leaves were immersed in ninety minims of a solution of one part to 437 of water. After 23 hrs. there was not a trace of inflection; the glands were not blackened, and the leaves did not appear injured. They were then transferred to the solution (1 gr. to 20 oz.) of phosphate of ammonia, and after 24 hrs. two of them were somewhat, the third very little, inflected; and they thus remained for another 24 hrs.

Chloride of Tin—Four leaves were immersed in 120 minims of a solution of about one part (all not being dissolved) to 437 of water. After 4 hrs. 30 m. all four leaves had their sub-marginal tentacles inflected; after 22 hrs. every single tentacle and the blades were closely inflected. The surrounding fluid was now coloured pink. The leaves were washed and transferred to water, but next morning were evidently dead. This chloride is a deadly poison, but acts slowly.

Tartrate of Antimony—Three leaves were immersed in ninety minims of a solution of one part to 437 of water. After 8 hrs. 30 m. there was slight inflection; after 24 hrs. two of the leaves were closely, and the third moderately, inflected; glands not much darkened. The leaves wee washed and placed in water, but they remained in the same state for 48 additional hrs. This salt is probably poisonous, but acts slowly.

Arsenious Acid—A solution of one part to 437 of water; three leaves were immersed in ninety minims; in 25 m. considerable inflection; in 1 hr. great inflection; glands not discoloured. After 6 hrs. the leaves were transferred to water; next morning they looked fresh, but after four days were pale-coloured, had not re-expanded, and were evidently dead.

Chloride of Iron—Three leaves were immersed in ninety minims of a solution of one part to 437 of water; in 8 hrs. no inflection; but after 24 hrs. considerable inflection; glands blackened; fluid coloured yellow, with floating flocculent particles of oxide of iron. The leaves were then placed in water; after 48 hrs. they had re-expanded a very little, but I think were killed; glands excessively black.

Chromic Acid—One part to 437 of water; there leaves were immersed in ninety minims; in 30 m. some, and in 1 hr. considerable, inflection; after 2 hrs. all the tentacles closely inflected; glands discoloured. Placed in water, next day leaves quite discoloured and evidently killed.

Manganese Chloride—Three leaves immersed in ninety minims of a solution of one part to 437 of water; after 22 hrs. no more inflection than often occurs in water; glands not blackened. The leaves were then placed in the usual solution of phosphate of ammonia, but no inflection was caused even after 48 hrs.

Copper Chloride—Three leaves immersed in ninety minims of a solution of one part to 437 of water; after 2 hrs. some inflection; after 3 hrs. 45 m. tentacles closely inflected, with the glands blackened. After 22 hrs., still closely inflected, and the leaves flaccid. Placed in pure water, next day evidently dead. A rapid poison.

Chloride of Nickel—Three leaves immersed in ninety minims of a solution of one part to 437 of water; in 25 m., considerable inflection, and in 3 hrs. all the tentacles closely inflected. After 22 hrs. still closely inflected; most of the glands, but not all, blackened. The leaves were then placed in water; after 24 hrs. remained inflected; were somewhat discoloured, with the glands and tentacles dingy red. Probably killed.

Chloride of Cobalt—Three leaves immersed in ninety minims of a solution of one part to 437 of water; after 23 hrs. there was not a trace of inflection, and the glands were not more blackened then often occurs after an equally long immersion in water.

Platinum Chloride—Three leaves immersed in ninety minims of a solution of one part, to 437 of water; in 6 m. some inflection, which became immense after 48 m. After 3 hrs. the glands were rather pale. After 24 hrs. all the tentacles still closely inflected; glands colourless; remained in same state for four days; leaves evidently killed.

Concluding Remarks on the Action of the Foregoing Salts

Of the fifty-one salts and metallic acids which were tried, twenty-five caused the tentacles to be inflected, and twenty-six had no such effect, two rather doubtful cases occurring in each series. In the table at the head of this discussion, the salts are arranged according to their chemical affinities; but their action on Drosera does not seem to be thus governed. The nature of the base is far more important, as far as can be judged from the few experiments here given, then that of the acid; and this is the conclusion at the which physiologists have arrived with respect to animals. We see this fact illustrated in all the nine salts of soda causing inflection, and in not being poisonous except when given in large doses; whereas seven of the corresponding salts of potash do not cause inflection, and some of them are poisonous. Two of them, however, *viz.*, the oxalate and iodide of potash, slowly induced a slight and rather doubtful amount of inflection. This difference between the two series is interesting, as Dr. Burdon Sanderson informed that sodium salts may be introduced enlarge doses into the circulation of mammals without any injurious effects; whilst small doses of potassium salts cause death by suddenly arresting the movements of the heart. An excellent instance of the different action of the two series is presented by the phosphate of soda quickly causing vigorous inflection, whilst phosphate of potash is quite inefficient. The great power of the former is probably due to the presence of phosphorus, as in the cases of phosphate of lime and of ammonia. Hence we may infer that Drosera cannot obtain phosphorus from the phosphate of potash. This is remarkable, as Dr. Burdon Sanderson the phosphate of potash is certainly decomposed within the bodies of animals. Most of the salts of soda act very rapidly; the iodide acting slowest. The oxalate, nitrate, and citrate seem to have a special tendency to cause the blade of the leaf to be inflected. The glands of the disc, after absorbing the citrate, transmit hardly any motor impulse to the outer tentacles; and in this character the citrate of soda resembles the citrate of ammonia, or a decoction of soda resembles the citrate of ammonia, or a decoction of grass-leaves; these three fluids all acting chiefly on the blade.

It seems opposed to the rule of the preponderant influence of the base that the nitrate of lithium causes moderately rapid inflection, whereas the acetate causes none; but this metal is closely allied to sodium and potassium, which act so differently; therefore, we might except that its action would be intermediate. We see also, that Caesium causes inflection, and rubidium does not; and these two metals allied to sodium and potassium. Most of the earthy salts are inoperative. Two salts of calcium, four of magnesium, two of barium., and two of strontium did not cause any inflection, and thus follow the rule of the preponderant power of the base. Of three salts of aluminium, one did not act, a second showed a trace of action, and third acted slowly and doubtfully, so that there effects are nearly alike.

Of the salts and acids of ordinary metals, seventeen were tried, and only four, namely, those of zinc, lead, manganese, and cobalt, failed to cause inflection. The salts of cadmium, tin, antimony, and iron act slowly; and the three latter seem more or less poisonous. The salts of silver, mercury, gold, copper, nickel, and platinum, chromic and arsenious acids, cause great inflection with extreme quickness, and are deadly poisons. It is surprising, judging from animals, that leaf and barium should not be poisonous. Most of the poisonous salts make the gland black, but chloride of platinum made them very pale. We shall have occasion, in the next chapter, to add a few remarks on the different effects of phosphate of ammonia on leaves previously immersed in various solutions.

Acids

We will first give, as in the case of the salts, a list of the twenty-four acids which were tried, divided into two series, according as they cause or do not cause inflection. After describing the experiments, a few concluding remarks will be added.

Acids, Much Diluted, which Cause Inflection	Acids, Diluted to the Same Degree, which do not Cause Inflection
1. Nitric, strong inflection; poisonous.	1. Gallic; not poisonous.
2. Hydrochloric, moderate and slow inflection; not poisonous.	2. Tannic; not poisonous.
3. Hydriodic, strong inflection; poisonous.	3. Tartaric; not poisonous.
4. Iodic, strong inflection; poisonous.	4. Citric; not poisonous.
5. Sulphuric, strong inflection; somewhat poisonous.	5. Uric; (?) not poisonous.
6. Phosphoric, strong inflection; poisonous.	
7. Boracic, moderate and rather slow inflection; not poisonous.	
8. Formic, very slight inflection; not poisonous.	
9. Acetic, strong and rapid inflection; poisonous.	
10. Propionic, strong but not very rapid inflection; poisonous.	
11. Oleic, quick inflection; very poisonous.	
12. Carbolic, very slow inflection; poisonous.	
13. Lactic, slow and moderate inflection; poisonous.	
14. Oxalic, moderately quick inflection; very poisonous.	
15. Malic, very slow but considerable inflection; not poisonous.	
16. Benzoic, rapid inflection; very poisonous.	
17. Succinic, moderately quick inflection; moderately poisonous.	
18. Hippuric, rather slow inflection; poisonous.	
19. Hydrocyanic, rather rapid inflection; very poisonous.	

Nitric Acid—Four leaves were placed, each in thirty minims of one part by weight of the acid to 437 of water, so that each received $\frac{1}{16}$ of a grain, of 4.048 mg. This strength was chosen for this and most of the following experiments, as it is the same as that of most of the foregoing saline solutions. In 2 hrs. 30 m. some of the leaves were considerably, and in 6 hrs. 30 m. all were immensely, inflected, as were their blades. The surrounding fluid was slightly coloured pink, which always shows that the leaves have been injured. They were then left in water for three days; but they remained inflected and were evidently killed. Most of the glands had become colourless. Two leaves were then immersed, each in thirty minims of one part to 1000 of water; in a few hours there was some inflection; and after 24 hrs. both leaves had almost all their tentacles and blades inflected. They were left in water for three days, and one partially re-expanded and recovered. Two leaves were next immersed, each in thirty minims of one part to 2000 of water; this produced very little effect, except that most of the tentacles close to the summit of the petiole were inflected, as if the acid had been absorbed by the cut-off end.

Hydrochloric Acid—One part to 437 of water; four leaves were immersed as before, each in thirty minims. After 6 hrs. only one leaf was considerably inflected. After 8 hrs. 15 m., one had its tentacles and blade well inflected; the other three were moderately inflected, and the blade of one slightly. The surrounding fluid was not coloured at all pink. After 25 hrs. three of these four leaves began to re-expand, but their glands were of a pink instead of a red colour; after two more days they fully re-expanded; but the fourth leaf remained inflected, and seemed much injured or killed, with its glands white. Four leaves were then treated, each with thirty minims of one part to 875 of water; after 21 hrs. they were moderately inflected; and, on being transferred to water, fully re-expanded in two days, and seemed quite healthy.

Hydriodic Acid—One to 437 of water; three leaves were immersed as before, each in thirty minims. After 45 m. the glands were discoloured, and the surrounding fluid became pinkish, but there was no inflection. After 5 hrs. all the tentacles were closely inflected; and an immense amount of mucus was secreted, so that the fluid cold be drawn out into long ropes. The leaves were then placed in water, but never re-expanded, and were evidently killed. Four leaves were next immersed in one part to 875 of water; the action was not slower, but after 22 hrs. all four leaves were closely inflected, and were affected in other respects as above described. These leaves did not re-expand, though left for four days in water. This acid acts far more powerfully than hydrochloric, and is poisonous.

Ionic Acid—One to 437 of water; three leaves were immersed, each in thirty minims; after 3 hrs. strong inflection; after 4 hrs. glands dark brown; after 8 hrs. 30 m. close inflection, and the leaves had become flaccid; surrounding fluid not coloured pink. These leaves were then placed in water, and next day were evidently dead.

Sulphuric Acid—One to 437 of water; four leaves were immersed each in thirty minims; after 4 hr. great inflection; after 6 hrs. surrounding fluid just tinged pink; they were then placed in water, and after 46 hrs. two of them were still closely inflected, two beginning to re-expand; many of the glands colourless. This acid is not so poisonous as hydriodic or iodic acids.

Phosphoric Acid—One to 437 of water; three leaves were immersed together in ninety minims; after 5 hrs. 30 m. some inflection, and some glands colourless; after 8 hrs. all the tentacles closely inflected, and many glands colourless: surrounding fluid pink. Left in water for two days and a half, remained in the same state and appeared dead.

Boracic Acid—One to 437 of water; four leaves were immersed together in 120 minims; after 6 hrs. very slight inflection; after 8 hrs. 15 m. two were considerably inflected, the other two slightly. After 24 hrs. one leaf was rather closely inflected, the second less closely, the third and fourth moderately. The leaves were washed and put into water; after 24 hrs. they were almost fully re-expanded and looked healthy. This acid agrees closely with hydrochloric acid of the same strength in its power of causing inflection, and in not being poisonous.

Formic Acid—Four leaves were immersed together in 120 minims of one part to 437 of water; after 40 m. slight, and after 6 hrs. 30 m., very moderate inflection; after 22 hrs. only a little more inflection than often occurs in water. Two of the leaves were then washed and placed in a solution (1 gr. to 20 oz.) of phosphate of ammonia; after 24 hrs. they were considerably inflected, with the contents of their cells aggregated, showing that the phosphate had acted, though not to the full and ordinary degree.

Acetic Acid—Four leaves were immersed together in 120 minims of one part to 437 of water . In 1 hr. 20 m. the tentacles of all four and the blades of two were greatly inflected. After 8 hrs. the leaves had become flaccid, but still remained closely inflected, the surrounding fluid being coloured pink. They were then washed and placed in water; next morning they were still inflected and of a very dark red colour, but with their glands colourless. After another day they were dingy coloured, and evidently dead. This acid is far more powerful then formic, and is highly poisonous. Half-minim drops of a stronger mixture (*viz.*, one part by measure to 320 for water) were placed on the discs of five leaves; none of the exterior tentacle, only those on the borders of the disc which actually absorbed the acid, became inflected. Probably the dose was too strong and paralyzed the leaves, for drops of a weaker mixture caused much inflected; nevertheless, the leaves all died after two days.

Propionic Acid—Three leaves were immersed in ninety minims of a mixture of one part to 437 of water; in 1 hr. 50 m. there was no inflection; but after 3 hrs. 40 m. one leaf was greatly inflected, and the other two slightly. The inflection continued to increase, so that in 8 hrs. all leaves were closely inflected. Next morning, after 20 hrs. most of the glands were very pale, but some few were almost black. No mucus had been secreted, and the surrounding fluid was only just perceptibly tinted of a pale pink. After 46 hrs. the leaves became slightly flaccid and were evidently killed, as was afterwards proved to be the case by keeping them in water. The protoplasm in the closely inflected tentacles was not in the least aggregated, but towards their bases it was collected in little brownish masses at the bottoms of the cells. This protoplasm was dead, for, on leaving the leaf in a solution of carbonate of ammonia, no aggregation ensued. Propionic acid is highly poisonous to Drosera, like its ally acetic acid, but induces inflection at a much slower rate.

Oleic Acid—Three leaves were immersed in this acid; some inflection was almost immediately caused, which increases slightly, but then ceased, and the leaves seemed killed. Next morning they were rather shrivelled, and many of the glands had fallen off the tentacles. Drops of this acid were placed on the discs of four leaves; in 40 m. all the tentacles were greatly inflected, excepting the extreme marginal ones; and many of these after 3 hrs. became inflected. We were led to try this acid from supposing that it was present (which does not seem to be the case) in olive oil, the action of which is anomalous. Thus drops of this oil placed on the disc do not cause the outer tentacles to be inflected; yet when minute drops were added to the secretion surrounding the glands of the outer tentacle, these were occasionally, but by no means always inflected. Two leaves were also immersed in this oil, and there was no inflection for about 12 hrs.; but after 23 hrs. almost all the tentacles were inflected. Three leaves were likewise immersed in unboiled linseed oil, and soon became somewhat, and in 32 hrs. greatly inflected. After 1 hr. the secretion round the glands was coloured pink. We infer from this latter fact that the power of linseed oil to cause inflection cannot be attributed to the albumin which it is said to contain.

Carbolic Acid—Two leaves were immersed in sixty minims of a solution of 1 gr. to 4537 of water; in 7 hrs. one was slightly, and in 24 hrs. both were closely inflected, with a surprising amount of mucus secreted. These leaves were washed and left for two days in water; they remained inflected; most of their glands became pale, and they seemed dead. This acid is poisonous, but does not act nearly so rapidly or powerfully as might have been expected from its known destructive power on the lowest organisms. Half-minims of the same solutions were placed on the discs of three leaves; after 24 hrs. no inflection of the outer tentacles caused, and when bits of meat were given them they became fairly will inflected. Again half-minims of a stronger solution, of one part to 218 of water, were placed on the discs of three leaves; no inflection of the outer tentacles ensued; bits of meat were then given as before; one leaf alone became well inflected, the discal glands of the other two appearing much injured and dry. We thus see that the glands of the discs, after absorbing this acid, rarely transmit any motor impulse to the outer tentacles; though these, when their own glands absorb the acid, are strongly acted on.

Lactic Acid—Three leaves were immersed in ninety minims of one part to 437 of water. After 48 m. there was no inflection, but the surrounding fluid was coloured pink; after 8 hrs. 30 m. one leaf alone was a little inflected, and almost all the glands on all three leaves were of a very pale colour. The leaves were then wasted and placed in a solution (1 gr. to 20 oz.) of phosphate of ammonia; after about 16 hrs. there was only a trace of inflection. They were left in the phosphate for 48 hrs., and remained in the same state, with almost all their glands discoloured. The protoplasm within the cells was not aggregated, except in a very few tentacles, the glands of which were not much discoloured. We believe, therefore, that almost all the glands and tentacles had been killed by the acid so suddenly that hardly any inflection was caused. Four leaves were next immersed in 120 minims of a weaker solution, of one part to 875 of water; after 2 hrs. 30 m. the surrounding fluid was quite pink; the glands were pale, but there was no inflection; after 7 hrs. 30 m. two of the leaves showed some inflection, and the glands were almost white; after 21 hrs. two of the leaves were considerably inflected, and a third slightly; most of the glands were white, the others dark red. After 45 hrs. one leaf had almost every tentacles inflected; a second a large number; the third and fourth very few; almost all the glands were white., excepting those on the discs of two of the leaves, and many of these were very dark red. The leaves appeared dead. Hence lactic acid acts in a very peculiar manner, causing inflection at an extraordinarily slow rate, and being highly poisonous. Immersion even weaker solution, *viz.*, of one part to 1312 and 1750 of water, apparently killed the leaves (the tentacles after a time being bowed backwards), and rendered the glands white, but caused no inflection.

Gallic, Tannic,Tartaric, and Citric Acids—One part to 437 of water. Three or four leaves were immersed, each in 30 minims of these four solution, so that each leaf received $\frac{1}{16}$ of a grain, or 4.048 mg. No inflection was caused in 24 hrs., and the leaves did not appear at all injured. Those which had been in the tannic and tartaric acids were placed in a solution (1 gr. to 20 oz.) of phosphate of ammonia, but no inflection ensued in 24 hrs. On the other hand, the four leaves which had been in the citric acid, when treated with the phosphate, became decidedly inflected 50 m., and strongly inflected after 5 hrs., and so remained for the next 24 hrs.

Malic Acid—Three leaves were immersed in ninety minims of a solution of one part to 437 of water; no inflection was caused in 8 hrs. 20 m., but after 24 hrs. two of them were considerably, and the third slightly, inflected—more so than could be accounted for by the action of water. No great amount of mucus was secreted. They were then placed in water, and after two days partially re-expanded. Hence this acid is not poisonous.

Oxalic Acid—Three leaves were immersed in ninety minims of a solution of 1 gr. to 437 of water; after 2 hrs. 10 m. there was much inflection; glands pale; the surrounding fluid of dark pink colour; after 8 hrs. excessive inflection. The leaves were then placed in water; after about 16 hrs. the tentacles were of a very dark red colour, like those of the leaves in acetic acid. After 24 additional hours, the three leaves were dead and their glands colourless.

Benzoic Acid—Five leaves were immersed, each in thirty minims of a solution of 1 gr. to 437 of water. This solution was so weak that it only just tasted acid, yet, as we shall see, was highly poisonous to Drosera. After 52 m. the submarginal tentacles were somewhat inflected, and all the glands very pale-coloured; the surrounding fluid was coloured pink. On one occasion the fluid became pink in the course of only 12 m. and the glands as white as if the leaf had been dipped in boiling water. After 4 hrs. much inflection; but none of the tentacles were closely inflected, owing to their having been paralysed before they had time to complete their movement. An extraordinary quantity of mucus was secreted. Some of the leaves were left in the solution; others, after an immersion of 6 hrs. 30 m., were placed in water. Next morning both lots were quite dead; the leaves in the solution being flaccid, those in the water (now coloured yellow) of a pale brown tint, and glands white.

Succinic Acid—Three leaves were immersed in ninety minims of a solution of one gr. to 437 of water; after 4 hrs. 15 m. considerable, and after 23 hrs. great inflection; many of the glands pale; fluid coloured pink. The leaves were then washed and placed in water; after two days there was some re-expansion, but many of the glands were still white. This acid is not nearly so poisonous as oxalic or benzoic.

Uric Acid—Three leaves were immersed in 180 minims of a solution of 1 gr. to 875 of warm water, but all the acid was not dissolved; so that each received nearly $\frac{1}{16}$ of a grain. After 25 m. there was some slight inflection; but this never increases; after 9 hrs. the glands were not discoloured, nor was the solution coloured pink; nevertheless, much mucus was secreted. The leaves were then placed in water, and by next morning fully re-expanded. We doubt whether this acid really caused inflection, for the slight movement which at first occurred may have been due to the presence of a trace of albuminous matter. But it produces some effect, as shown by the secretion of so much mucus.

Hippuric Acid—Four leaves were immersed in 120 minims of a solution of 1 gr. to 437 of water. After 2 hrs. the fluid was coloured pink; glands pale, but no inflection. After 6 hrs. some inflection; after 9 hrs. all four leaves greatly inflected; much mucus secreted; all the glands very pale. The leaves were then left in water for two days; they remained closely inflected, with their glands colourless, and we do not doubt were killed.

Hydrocyanic Acid—Four leaves were immersed, each in thirty minims of one part to 437 of water; in 2 hrs. 45 m. all the tentacles were considerably inflected, with many of the glands pale; after 3 hrs. 45 m. all strongly inflected, and the surrounding fluid coloured pink; after 6 hrs. all closely inflected. After an immersion of 8 hrs. 20 m. the leaves were washed and placed in water; next morning, after about 16 hrs., they were evidently dead. Two leaves were immersed in a stronger mixture of one part to fifty of water; in 1 hr. 15 m. the glands became as white as porcelain, as if they had been dipped in boiling water; very few of the tentacles were inflected; but after 4 hrs. almost all were inflected. These leaves were then placed in water and next morning were evidently dead. Half-minim drops of the same strength (*viz.*, one part to fifty of water) were next placed on the discs of five leaves; after 21 hrs. all the outer tentacles were inflected, and the leaves appeared much injured. We like-wise touched the secretion round a large number of glands with minute drops (about 1/20 of a minim, or .00296 c.c.) of cheele's mixture (containing 4 per cent of anhydrous acid); the gland first became bright red, and after 3 hrs. 15 m. about two-thirds of the tentacles bearing these glands were inflected, and remained so for the two succeeding days, when they appeared dead.

Concluding Remarks on the Action of Acids

It is evident that acids have a strong tendency to cause the inflection of the tentacles; for out of twenty-four acids tried, nineteen thus acted, either rapidly and energetically, or slowly and slightly. This fact is remarkable, as the juices of many plants contain more acid, judging by the taste, then the solutions employed in my experiments. From the powerful effects of so many acids on Drosera, we are led to infer that those naturally contained in the tissues of this plant, as well as of others, much play some important part in their economy. Of the five cases in which acids did not cause the tentacles to be inflected, one is doubtful; for uric acid did act slightly, and caused a copious secretion of mucus. Mere sourness to the taste is no criterion of the power of an acid on Drosera. As citric and tartaric acids are very sour, yet do not excite inflection, it is remarkable how acids differ in their power. Thus, hydrochloric acid far less powerful then hydrodic and many other acids of the same strength, and is not poisonous. This is an interesting fact, as hydrochloric acid plays so important a part in the digestive process of animals. Formic acid induces very slight inflection, and is not poisonous; whereas its ally, acetic acid, acts rapidly and powerfully, and is poisonous. Malic acid acts slightly, whereas citric and tartaric acids produce no effect. Lactic acid is poisonous and is remarkable from inducing inflection only after a considerable interval of time. Nothing surprised more than that a solution of benzoic acid, so weak as to be hardly acidulous to the taste, should act with great rapidity and be highly poisonous; for we were informed that it procuress no marked effect on the animal economy. It may be seen, by looking down the list at the head of this discussion, that most of the acids are poisonous, often highly so. Diluted acids are known to induce negative osmose, and the poisonous action of so many acids on Drosera is, perhaps, connected with this power, for we have seen that the fluid in which they were immersed often became pink, and the glands pale-coloured or white. Many of the poisonous acids, such as hydriodic, benzoic, Hippuric, and carbolic, caused the secretion of an extraordinary amount of mucus, so that long ropes of this matter hung from the leaves when they were lifted out of the solutions. Other acids, such as hydrochloric and malic, have no such tendency; in these two latter cases the surrounding fluid was not coloured pink, and the leaves were not poisoned. On the other hand, propionic acid, which is poisonous, does not cause much mucus to be secreted, yet the surrounding fluid became slightly pink. Lastly, as in the case of saline solutions, leaves after being immersed in certain acids, were soon acted on by phosphate of ammonia; on the other hand, they were not thus affected after immersion in certain other acids.

❑❑❑

9

The Effects of Certain Alkaloid Poisons, Other Substances and Vapours

Strychnine, Salts of—Quinine, Sulphate of, does not soon arrest the movement of the protoplasm—Other salts of quinine—Digitaline—Nicotine—Atropine—Veratrine—Colchicine—Theine—Curare—Morphia—Hyoscyamus—Poison of the Cobra, apparently accelerates the movements of the protoplasm—Camphor, a powerful stimulant, its vapour narcotic—Certain essential oils excite movement—Glycerine—Water and certain solutions retard or prevent the subsequent action of phosphate of ammonia—Alcohol innocuous, its vapour narcotic and poisonous—Chloroform, sulphuric and nitric ether, their stimulant, poisonous, and narcotic power—Carbonic acid narcotic, not quickly poisonous—Concluding remarks.

Effect of Alkaloid Poisons

Acetate of Strychnine—Half-minims of a solution of one part to 437 of water were placed on the discs of six leaves; so that each received $\frac{1}{960}$ of a grain, or .0675 mg. In 2 hrs. 30 m. the outer tentacles on some of them were inflected, but in an irregular manner, some times only on one side of the leaf. The next morning, after 22 hrs. 30 m., the inflection had not increased. The glands on the central disc were blackened, and had ceased secreting. After an additional 24 hrs. all the central glands seemed dead, but the inflected tentacles had re-expanned and appeared quite healthy. Hence the poisonous action of strychnine seems confined to the glands which have observed it; nevertheless, these glands transmit a motor impulse to the exterior tentacles. Minute drops (about $\frac{1}{20}$ of a minim) of the same solution applied to the glands of the outer tentacles occasionally caused them to bend. The poison does not seem to act quickly, for having applied to several glands similar drops of a rather stronger solution, of one part to 292 of water, this did not prevent the tentacles bending, when their glands were excited, after an interval of a quarter to three quarters of an hour, by being rubbed or given bits of meat. Similar drops of a solution of one part to 218 of water (2 grs. to 1 oz.) quickly blackened the glands; some few tentacle thus treated moved, whilst others did not. The latter, however on being subsequently moistened with saliva or given bits of meat, become incurved, though with extreme slowness; and this shows that they had been injured. Stronger solutions (but the strength was not ascertained) sometimes arrested all power of movement very quickly; thus bits of meat were placed on the glands of several exterior tentacles, and as soon as they began to move, minute drops of the strong solutions were added. They continued for a short time to go on bending, and then suddenly stood still; other tentacles on the same leaves, with meat on their glands, but not wetted with the strychnine, continued to bend and soon reached the centre of the leaf.

Citrate of Strychnine—Half-minims of a solution of one part to 437 of water placed on the discs of six leaves; after 24 hrs. the outer tentacles showed only a trace of inflection. Bits of meat were then placed on three of these leaves, but in 24 hrs. only slight and irregular inflection occurred, proving that the leaves had been greatly injured. Two of the leaves to which meat had not been given had their discal glands dry and much injured. Minute drops of a strong solution of one part to 109 of water (4 grs. to 1 oz.) were added to the secretion round several glands, but did not produce nearly so plain an effect as the drops of a much weaker solution of the acetate. Particles of the dry citrate were placed on six glands; two of these moved some way towards the centre, and then stood still being no doubt killed; 1 alone reached the centre. Five leaves were immersed, each in thirty minims of solution of one part to 437 of water; so that each received $\frac{1}{16}$ of a grain; after about 1 hr. some of the outer tentacles became inflected, and the glands were oddly mottled with black and white. These glands, in from 4 hrs. to 5 hrs., became whitish and opaque, and the protoplasm in the cells of the tentacles was well-aggregated. By this time two of the leaves were greatly inflected, but the three others not much more inflected than they were before. Nevertheless two of fresh leaves, after an immersion respectively for 2 hrs. and 4 hrs. in the solution. were not killed; for on being left for 1 hr. 30 m. in a solution of one part of carbonate of ammonia to 218 of water, their tentacles became more inflected, and there was much aggregation. The glands of two other leaves, after an immersion for 2 hrs. in a stronger solution, of one part of the citrate to 218 of water, became of an opaque, pale pink colour, which before long disappeared, leaving them white. One of these two leaves had its blade and tentacles greatly inflected; the other hardly at all; but the protoplasm in the cells of both was aggregated down to the bases of the tentacles, with the spherical masses in the cells close beneath the glands blackened. After 24 hrs. one of these leaves was colourless, and evidently dead.

Sulphate of Quinine—Some of this salt was added to water, which is said to dissolve $\frac{1}{1000}$ part of its weight. Five leaves were immersed each in thirty minims of this solutions, which tasted bitter. In less than 1 hr. some of them had a few tentacles inflected. In 3 hrs. most of the glands became whitish, others dark-coloured, and many oddly mottled. After 6 hrs. two of the leaves had a good many tentacles inflected, but this very moderate degree of inflection never increased. One of the levers was taken out of the solution after 4 hrs., and placed in water; by the next morning some few of the inflected tentacles had re-expanded, showing that they were not dead; but the glands were still much discoloured. Another leaf, not included in the above lot, after an immersion of 3 hrs. 15 m., was carefully examined; the protoplasm in the cells of the outer tentacles, and of the short green ones on the discs had become strongly aggregated down to their bases; and we distinctly saw that the little masses changed their positions and shapes rather rapidly; some coalescing and again separating. We were surprised at the fact, because quinine is said to arrest all movement in the whist corpuscles of the blood; but as, according to Binz, this is due to their being no longer supplied with oxygen by the red corpuscles, any such arrestment of movements could not be expected in Drosera. That the glands had absorbed some of the salt was evident from their change of colour; but we at first thought that the solution might not have travelled down the cells of the tentacles, where the protoplasm was seen in active movement. This view, however, was erroneous, for a leaf which had been immersed for 3 hrs. in the quinine solutions was then placed in a little solution of one part of carbonate of ammonia to 218 of water; and in 30 m. the glands and the upper cells of the tentacles became intensely black, with the protoplasm presenting a very unusual appearance; for it had become aggregated into reticulated dingy-coloured masses, having rounded and angular interspaces. As we have never seen this effect produced by the carbonate of ammonia alone. It must be attributed to the previous action of the quinine. These reticulated masses were watched for some time, but not change their forms; so that the protoplasm no doubt had been killed by the combined action of the two salts, though exposed to them for only a short time.

Another leaf, after an immersion for 24 hrs. in the quinine solution, became somewhat flaccid, and the protoplasm in all the cells was aggregated. Many of the aggregated masses were discoloured, and presented a granular appearance: they were spherical, or elongated, or still more commonly consisted of little curved chains of small globules. None of these masses exhibited the least movement, and no doubt were all dead.

Half-minims of the solution were placed on the discs of six leaves; after 23 one had all its tentacles, two had a few, and the others none inflected; so that the discal glands, when irritated by this salt, do not transmit any strong motor impulse to the outer tentacles. After 48 hrs. the glands, when irritated by this salt, do not transmit any strong motor impulse to the outer tentacles. After 48 hrs. the glands on the discs of all six leaves were evidently much injured or quite killed. It is clear that this salt is highly poisonous.

Acetate of Quinine—Four leaves were immersed, each in thirty minims of a solution of one part to 437 of water. The solution was tested with litmus paper, and was not acid. After only 10 m. all four leaves were greatly, and after 6 hrs. immensely, inflected. They were then left in water for 60 hrs., but never re-expanded; the glands were white, and the leaves evidently dead. This salt is far more efficient then the Sulphate in causing inflection, and, like that salt, is highly poisonous.

Nitrate of Quinine—Four leaves were immersed, each in thirty minims of a solution of one part to 437 of water. After 6 hrs. there was hardly a trace of inflection; after 22 hrs. three of the leaves were moderately, and the fourth slightly inflected; so that this induces, though rather slowly, well-marked inflection. These leaves, on being left in water for 48 hrs., almost completely re-expanded, but the glands were much discoloured. Hence this salt is not poisonous in any high degree. The different action of the three foregoing salts of quinine is singular.

Digitaline—Half-minims of a solution of one part to 437 of water were placed on the discs of five leaves. In 3 hrs. 45 m. some of them had their tentacles, and one had its blade, moderately inflected. After 8 hrs. three of them were well inflected; the fourth had only a few tentacles inflected, and the fifth (an old leaf) was not at all glands on their discs became pale. On the third day the leaves appeared much injured. Nevertheless, when bits of meat were placed on two of them, the outer tentacles became inflected. A minute drop (about $\frac{1}{20}$ of a minim) of the solution was applied to three glands, and after 6 hrs. all three tentacles were inflected, but next day had nearly re-expanded; so that this very small dose of $\frac{1}{28800}$ of a grain (.00225 mg.) acts on a tentacle, but is not poisonous. It appears from these several facts that digitaline causes inflection, and poisons the glands which adsorb a moderately large amount.

Nicotine—The secretion round several glands was touched with a minute drop of the pure fluid, and the glands were instantly blackened; the tentacles becoming inflected in a few minutes. Two leaves were immersed in a weak solution of two drops to 1 oz., or 437 grains, of water. When examined after 3 hrs. 20 m., only twenty-one tentacles on one leaf were closely inflected, and six on the other slightly so; but all the glands were blackened, or very dark coloured, with the protoplasm in all the cells of all the tentacles much aggregated and dark coloured. The leaves were not quite killed, for on being placed in a little solution of carbonate of ammonia (2 grs. to oz.) a few more tentacles became inflected, the remainder not being acted on during the next 24 hrs.

Half-minims of a stronger solution (two drops to ½ oz. of water) were placed on the discs of six leaves, and in 30 m. all those tentacles became inflected; the glands of which had actually touched the solutions, as shown by their blackness; but hardly any motor influence was transmitted to the outer tentacles. After 22 hrs. most of the glands on the discs appeared dead; but this could not have been the case, as, when bits of meat were placed on three of them, some few of the outer tentacles were inflected in 24 hrs. hence nicotine has great tendency to blacken the glands and to induce aggregation of the protoplasm, but, except when pure, has very moderate power of inducing inflection, and still less power of causing a motor influence to be transmitted from the discal glands to the outer tentacles. It is moderately poisonous.

Atropine—A grain was added to 437 grains of water, but was not all dissolved; another grain was added to 437 grains of a mixture of one part of alcohol to seven parts of water; and a third solutions was made by adding one part of valerianate of atropine to 437 of water. Half-minims of these three solutions were placed, in each case, on the discs of six leaves; but no effect whatever was produced, excepting that the glands on the discs to which the valerianate was given were slightly discoloured. The six leaves on which drops of the solutions of atropine is diluted alcohol had been left for 21 hrs. were given bits of meat, and all became in 24 hrs. fairly well inflected; so that atropine does not excite movement, and is not poisonous. We also tried in the same manner the alkaloid sold as daturine, which is believed not to differ from atropine, and it produced no effect. Three of the leaves on which drops of this latter solutions had been left for 24 hrs. were likewise given bits of meat, and they had in the course of 24 hrs. a good many of their sumarginal tentacles inflected.

Veratrine, Colchicine, Theine—Solutions were made of these three alkaloids by adding one part to 437 of water. Half-minims were placed, in each case, on the discs of at least six leaves, but no inflections was caused, except perhaps a very slight amount by the theine. Half-minims of a strong infusion of tea likewise produced, as formally stated, no effect. We also tried similar drops of an infusion of one part of the extract of colchicum, sold by druggists, to 218 of water; and the leaves were observed for 48 hrs., without any effect being produced. The seven leaves on which drops of Veratrine had been left for 26 hrs. were given bits of meat, and after 21 hrs., were well inflected. These three alkaloids are therefore quite innocuous.

Curare—One part of this famous poison was added to 218 of water, and three leaves were immersed in ninety minims of the filtered solution. In 3 hrs. 30. some of the tentacles were a little inflected; as was the blade of one, after 4 hrs. After 7 hrs. the glands were wonderfully blackened, show in that matter of some kind had been absorbed. In 9 hrs. two of the leaves had most of their tentacles sub-inflected, but the inflection did not increase in the course of 24 hrs. One of these leaves, after being immersed for 9 hrs. in the solutions, was placed in water, and by next morning had largely re-expanded; the other two, after their immersion for 24 hrs., were likewise placed in water, and in 24 hrs. were considerably re-expanded, though their glands were as black as ever. Half-minims were placed on the discs of six leaves, and no inflection ensued; but after three days the glands in the discs appeared rather dry, yet to my surprise were not blackened. On another occasion drops were placed on the disc's of six leaves, and a considerable amount of inflection was soon caused; but as we had not filtered the solutions, floating particles may have acted on the glands. After 24 hrs. bits of meat were placed on the discs of three of these leaves, and next day they became strongly inflected. As we at first thought that the poison might not have been dissolved in pure water, one grain was added to 437 grains of a mixture of one at of alcohol to seven of water, and half-minims were placed on the discs of six leaves. These were not at all affected, and when after a day bits of meat were given then, they were slightly inflected in 5 hrs. and closely after 24 hrs. It follows from these several facts that a solution of curare induces a very moderate degree of inflection, and this may perhaps be due to the presence of a minute quantity of albumen. It certainly is not poisonous. The protoplasm in one of the leaves, which had been immersed for 24 hrs., and which had become slightly inflected, had undergone a very slight amount of aggregation—not more than often ensues from an immersion of this length of time in water.

Acetate of Morphia—We tried a great number of experiments with this substance, but with no certain result. A considerable number of leaves were immersed from between 2 hrs. and 6 hrs. in a solution of one part to 218 of water, and did not become inflected. Nor were they poisoned; for when they were washed and placed in weak solutions of phosphate and carbonate of ammonia, they soon became strongly inflected, with the protoplasm of the cells well aggregated. If, however, whilst the leaves were immersed in the morphia, phosphate of ammonia was added, inflection did not rapidly ensue. Minute drops of the solution were applied in the usual manner to the secretion round between thirty and forty glands; and when, after an interval of 6m., bits of meat, a little saliva, or particles of glass, were placed on them, the movement of the tentacles was greatly retarded. But on other occasions no such retardation occurred. Drops of water similarly applied never have any retarding power. Minute drops of a solution of sugar of the same strength (one part to 218 of water) sometimes retarded the subsequent action of meat and of particles of glass, and sometimes did not do so. At one time it was felt that morphia acted as narcotic on Drosera, but after having found in what a singular manner immersion in certain non-poisonous salts and acids prevents the subsequent action of phosphate of ammonia, whereas other solutions have no such power, first conviction seems very doubtful.

Extract of Hyoscyamus—Several leaves were placed, each in thirty minims of an infusion of 3 hrs. of the extract sold by druggists to 1 oz. of water. One of them, after being immersed for 5 hrs. 15 m., was not inflected, and was then put into a solution (1 gr. to oz.) of carbonate of ammonia; after 2 hrs. 40 m. It was found considerably inflected, and the glands much blackened. Four of the leaves, after being immersed for 2 hrs. 14 m., were placed in 120 minims of a solution slightly inflected from the hyoscyamus, probably owing to the presence of some albuminous matter, as formerly explained, but the inflection immediately increased, and after 1 hr. was strongly pronounced; so that hyoscyamus does not act as a narcotic or poison.

Effect of Other Poisons and Substances

Poison from the Fang of Living Adder—Minute drops were placed on the glands of many tentacles; these were quickly inflected, just as if saliva had been given them. Next morning, after 17 hrs. 30 m., all were beginning to re-expand, and they appeared uninjured.

Poison from the Cobra—It is an albuminous substance, and is believed to replace the ptyaline of saliva. A minute drop (about $\frac{1}{20}$ of a minim) of a solution of one part to 437 of water was applied to the secretion round four glands; so that each received only about $\frac{1}{38400}$ of a grain (.0016 mg.). The operation was repeated on four other glands; and in 15 m. several of the eight tentacles became well inflected, and all of them in 2 hrs. Next morning, after 24 hrs., they were still inflected, and the glands of a very pale pink colour. After an additional 24 hrs. they were nearly re-expanded, and completely so on the succeeding day; but most of the glands remained almost white.

Half-minims of the same solution were placed on the discs of three leaves, so that each received $\frac{1}{900}$ of a grain (.0675 mg.); in 4 hrs. 15 m. the outer tentacles were much inflected; and after 6 hrs. 30 m. those on two of the leaves were closely inflected and the blade of one; the third leaf was only moderately affected. The leaves remained in the same state during the next day, but after 48 hrs. re-expanded.

Three leaves were now immersed, each in thirty minims of the solutions, so that each received $\frac{1}{16}$ of a grain, or 4.048 mg. In 6 m. there was some inflection, which steadily increased, so that after 2 hrs. 30 m. all three leaves were closely inflected; the glands were at first somewhat darkened, then rendered pale; and the protoplasm within the cells of the tentacles was partially aggregated. The little masses of protoplasm were examined after 3 hrs., and again after 7 hrs., and on no other occasion have we seen them undergoing such rapid changes of form. After 8 hrs. 30 m. the glands had become quite white; they had not secreted any great quantify of mucus. The leaves were now placed in water, and after 40 hrs. re-expanded, showing that they were not much or at all injured. During their immersion in water the protoplasm within the cells of the tentacles was occasionally examined, and always found in strong movement.

Two leaves were next immersed, each in thirty minims of a much stronger solution, of one part to 109 of water; so that each received ¼ of a grain, or 16.2 mg. After 1hr. 45 m. the submarginal tentacles were strongly inflected, with the glands somewhat pale; after 3 hrs. 30m. both leaves had all their tentacles closely inflected and the glands white. Hence the weaker solutions, as in so many other cases, induced more rapid inflection than the stronger one; but the glands were sooner rendered white by the latter. After an immersion of 24 hrs. some of the tentacles were examined, and the protoplasm, still of a fine purple colour, was found aggregated into chains of small globular masses. These changed their shapes with remarkable quickness. After an immersion of 48 hrs. they were again examined, and their movements were so plain that they could easily be seen under a weak power. The leaves were now placed in water, and after 24 hrs. (*i.e.,* 72 hrs. from their first immersion) the little masses of protoplasm, which had become of a dingy purple, were still in strong movement, changing their shapes, coalescing, and again separating.

In 8 hrs. after these two leaves had been placed in water (*i.e.,* in 56 hrs. after their immersion in the solution) they began to re-expand, and by the next morning were more expanded. After an additional day (*i.e.,* on the fourth day after their immersion in the solution) they were largely, but not quite fully, expanded. The tentacles were now examined, and the aggregated masses were almost wholly re-dissolved; the cells being filled with homogeneous purple fluid, with the exception here and there of a single globular mass. We thus see how completely the protoplasm had escaped all injury from the poison. As the glands were soon rendered quite white, it occurred that their texture might have been modified in such a manner as to prevent the poison passing into the cells beneath, and consequently that the protoplasm within these cells has not been at all affected. Accordingly we placed another leaf, which had been immersed for 48 hrs. in the poison and afterwards for 24 hrs. in water, in a little solution of one part of carbonate of ammonia to 218 of water; in 30 m. the protoplasm in the cells beneath the glands became darker, and in the course of 24 hrs. the tentacles were filled down to their bases with dark-coloured spherical masses. Hence the glands had not lost their power of absorption, as far as the carbonate of ammonia is concerned.

From these facts it is manifest that the poison of the cobra, though so deadly to animals, is not at all poisonous to Drosera; yet it causes strong and rapid inflection of the tentacles, and soon discharges all colour from the glands. It seems even to act as a stimulant to the protoplasm, for after considerable experience in observing the movements of this substance in Drosera, we have never seen it on any other occasion in so active a state. Ciliated epithelium from the mouth of a frog was placed in a solution same time in pure water for comparison. The movements of the cilia in the solution seemed at first increased, but soon languished, and after between 15 and 20 minutes ceased; whilst those in the water were still acting vigorously. The white corpuscles of the blood of a frog, and the cillia on two infusorial animals, a Paramaecium and Volvox, were similarly affected by the poison. Dr. Fayrer found that the muscle of a frog lost its irritability after an immersion of 20 m. in the solutions not them responding to a strong electrical current. On the other hand, the movements of the cilia on the mantle of an Unio were not always arrested, even when left for a considerable time in a very strong solution. On the whole, it seems that the poison of the cobra acts far more injuriously on the protoplasm of the higher animals than on that of Drosera.

There is one point which may be noticed. We have occasionally observed that the drops of secretion round the glands were rendered somewhat turbid by certain solutions, and more especially by some acids, a film being formed on the surfaces of the drops; but we never saw this effect produced in so conspicuous a manner as by the cobra poison. When the stronger solution was employed, the drops appeared in 10 m. like little white rounded clouds. After 48 hrs. the secretion was changed into threads and sheets of a membranous substance, including minute granules of various sizes.

Camphor—Some scraped camphor was left for a day in a bottle with distilled water, and then filtered. A solution thus made is said to contain $\frac{1}{1000}$ of its weight of camphor; it smelt and tasted of this substance. Ten leaves were immersed in this solution; after 15 m. five of them were well inflected, two showing a first trace of movement in 11 m. and 12 m.; the sixth leaf did not begin to move until 15 m. had elapsed, but was fairly well inflected in 17 m. and quite closed in 24 m.; the seventh began to move in 17 m., and was completely shut in 26 m. The eight, ninth, and tenth leaves were old and of a very dark red colour, and these were not inflected after an immersion of 24 hrs.; so that in making experiments with camphor it is necessary to avoid such leaves. Some of these leaves, on being left in the solution for 4 hrs., became of a rather dingy pink colour, and secreted much mucus; although their tentacles were closely inflected, the protoplasm within the cells was not at all aggregated. On another occasion, however, after a longer immersion of 24 hrs., there was well-marked aggregation. A solution made by adding two drops of camphorated spirits to an ounce of water did not act on one leaf; whereas thirty minims added to an ounce of water acted on two leaves immersed together.

Number of Leaves	Length of Immersion in the Solution of Camphor	Length of Times between the Act of Brushing and the Inflection of the Tentacles	Length of Time between the Immersion of the Leaves in the Solution and the First Sign of the Inflection of the Tentacles
1.	5 m.	3 m. considerable inflection; 4 m. all the tentacles except 3 or 4 inflected.	8 m.
2.	5 m.	6 m. first sign of inflection.	11 m.
3.	5 m.	6 m. 30 s. slight inflection; 7 m. 30 s. plain inflection.	11 m. 30 s.
4.	4 m. 30 s.	2 m. 30 s. a trace of inflection; 3 m. plain; 4 m. strongly marked.	7 m.
5.	4 m.	2 m. 30 s. a trace of inflection; 3 m. plain inflection.	6 m. 30 s.
6.	4 m.	2 m. 30 s. decided inflection; 3 m. 30 s. strongly marked.	6 m. 30 s.
7.	4 m.	2 m. 30 s. slight inflection; 3 m. plain; 4 m. well-marked.	6 m. 30 s.
8.	3 m.	2 m. trace of inflection; 3 m. considerable, 6 m. strong inflection.	5 m.
9.	3 m.	2 m. trace of inflection; 3 m. considerable, 6 m. strong inflection.	5 m.

M. Vogel has shown that the flowers of various plants do not wither so soon when their stems are placed in a solution of camphor as when in water; and that if already slightly withered, they recover more quickly. The germination of certain seeds also accelerated by the solution. So that camphor acts as a stimulant, and it is the only known stimulant for plants. We wished, therefore, to ascertain whether camphor would render the leaves of Drosera more sensitive to mechanical irritation than they naturally are. Six leaves were left in distilled water for 5 m. or 6 m., and then gently brushed twice or thrice, whilst still under water, with a soft camel-hair brush; but no movement ensued. Nine leaves, which had been immersed in the above solution of camphor for the times stated in the above table, were next brushed only once with the same brush and in the same manner as before; the result are given in the table. Our first trials were made by brushing the leaves whilst still immersed in the solution; but it occurred to me that the viscid secretion round the glands would thus be removed, and the camphor might act more effectually on them. In all the above trials, therefore, each leaf was taken out of the solution, waved for about 15 s. in water, then placed in fresh water and brushed, so that the brushing would not allow the freer access of the camphor; but this treatment made no difference in the results.

Other leaves were left in the solutions without being brushed; one of these first showed a trace of inflection after 11 m.; a second after 12 m.; five were not inflected until 15 m. had elapsed, and two not until a few minutes later. On the other hand, it will be seen in the right-hand column of the table that most of leaves subjected to the solution, and which were brushed, became inflected in a much shorter time. The movement of the tentacles of some of these leaves was so rapid that it could be plainly seen through a very weak lens.

Two or three other experiments are worth giving. A large old leaf, after being immersed for 10 m. in the solution, did not appear likely to be soon inflected; so we brushed it, and in 2 m. it began to move, and in 3 m. was completely shut. Another leaf, after an immersion of 15 m., showed no signs of inflection, so was brushed, and in 4m, was grandly inflected. A third leaf after an immersion of 17 m., likewise showed no sings of inflection; it was then brushed, but did not move for 1 hr.; so that here was a failure. It was again brushed, and now in 9 m. A few tentacles became inflected; the failure therefore was not complete.

We may conclude that a small dose of camphor in solution is a powerful stimulant to Drosera. It not only soon excites the tentacles to bend, but apparently renders the glands sensitive to a touch, which by itself does not cause any movement. Or it may be that a slight mechanical irritation not enough to cause any inflection yet gives some tendency to movement, and thus reinforces the action of the camphor. This latter view would have appeared to me the more probable one, had it not been shown by M. Vogel that camphor is a stimulant in other ways to various plants and seeds.

Two plants bearing four or five leaves, and with their root in a little cup of water, were exposed to the vapour of some bits of camphor (about as large as a filbert nut), under a vessel holding ten fluid ounces. After 10 hrs. no inflection ensued; but the glands appeared to be secreting more copiously. The leaves were in a narcotised condition, for on bits of meat being placed on two of them, there was no inflection in 3 hrs. 15 m., and even after 13 hrs. 15 m. only a few of the outer tentacles were slightly inflected; but this degree of movement shows that the leaves had not been killed by an exposure during 10 hrs. to the vapour of camphor.

Oil of Caraway—Water is said to dissolve about a thousands part of its weight of this oil. A drop was added to an ounce of water and the bottle occasionally shaken during a day; but many minute globules remained undissolved. Five leaves were immersed in this mixture; in from 4 m. to 5 m. there was some inflection, which became moderately pronounced in two or three additional minutes. After 14 m. all five leaves were well, and some of them closely, inflected. After 6 hrs. the glands were white, and much mucus had been secreted. The leaves were now flaccid, of a peculiar dull-red colour, and evidently dead. One of the leaves, after an immersion of 4 m. was brushed, like the leaves in the camphor, but this produced no effect. A plant with its roots in water was exposed under a 10-oz. vessel to the vapour of this oil, and in 1 hr. 20 m. one leaf showed a trace of inflection. After 5 hrs. 20 m. the cover was taken off and the leaves examined; one had all its tentacles closely inflected, the second about half in the same state; and the third all sub-inflected. The plant was left in the open air for 42 hrs., but not a single tentacle expanded; all the glands appeared dead, except here and there one, which was still secreting. It is evident that this oil is highly exciting and poisonous to Drosera.

Oil of Cloves—A mixture was made in the same manner as in the last case, and three leaves were immersed in it. After 30 m. there was only a trace of inflection which never increased. After 1 hr. 30 m. the glands were pale, and after 6 hrs. white. No doubt the leaves were much injured or killed.

Turpentine—Small drops placed on the discs of some leaves killed them, as did likewise drops of creosote. A plant was left for 15 m. under a 12-oz. vessel, with its inner surface wetted with twelve drops of turpentine; but no movement of the tentacles ensued. After 24 hrs. the plant was dead.

Glycerine—Half-minims were placed on the discs of three leaves; in 2 hrs. some of the outer tentacles were irregularly inflected; and in 19 hrs. the leaves were flaccid and apparently dead; the glands which had touched the glycerine were colourless. Minute drops (about $\frac{1}{20}$ of a minim) were applied to the glands of several tentacles, and in a few minutes these moved and soon reached the centre. Similar drops of a mixture of four dropped drops to 1 oz. of water were likewise applied to several glands; but only a few of the tentacles moved, and these very slowly and slightly. Half minims of this same mixture placed on the discs of some leaves caused, to my surprise, no inflection in the course of 48 hrs. Bits of meat were then given them, and next day they were well inflected; notwithstanding that some of the discal glands had been rendered almost colourness. Two leaves were immersed in the same mixture, but only for 4 hrs.; they were not inflected, and on being afterwards left for 2 hrs. 30 m. in a solution (1 gr. to 1 oz.) of carbonate of ammonia, their glands were blackened, their tentacles inflected and the protoplasm within their cells aggregated. It appears from these facts that a mixture of four drops of glycerine to an ounce of water is not poisonous, and excites very little inflection; but that pure glycerine is poisonous, and if applied in very minute quantities to the glands of the outer tentacles causes their inflection.

The Effects of Immersion in Water and in Various Solutions on the Subsequent Action of Phosphate and Carbonate of Ammonia

We have seen in the third and seventh chapters that immersions in distilled water causes after a time some degree of aggregation of the protoplasm, and a moderate amount of inflection, especially in the case of plants which have been kept at a rather high temperature. Water does not excite a copious secretion of mucus. We have here to consider the effects of immersion in various fluids on the subsequent action of salts of ammonia and other stimulants. Four leaves which had been left for 24 hrs. in water were given bits of meat, but did not clasp them. Ten leaves, after a similar immersion, were left for 24 hrs. in a powerful solution (1 gr. to 20 oz.) of phosphate of ammonia, and only one showed even a trace of inflection. Three of these leaves, on being left for an additional day in the solution, still remained quite unaffected. When, however, some of these leaves, which had been first immersed in water for 24 hrs., and then in the phosphate for 24 hrs. were placed in a solution of carbonate of ammonia (one part to 218 of water), the protoplasm in the cells of the tentacles became in a few hours strongly aggregated, showing that this salt had been absorbed and taken effect.

A short immersion in water for 20 m. did not retard the subsequent action of the phosphate, or of splinters of glass placed on the glands; but in two instances an immersion for 50 m. prevented any effect from a solution of camphor. Several leaves which had been left for 20 m. in a solution of one part of white sugar to 218 of water were placed in the phosphate solution, the action of which was delayed; whereas a mixed solution of sugar and the phosphate did not in the least interfere with the effects of the latter. Three leaves, after being immersed for 20 m. in the sugar solutions, were placed in a solution of carbonate of ammonia (one part to 218 of water); in 2 m. or 3 m. other glands were blackened, and after 7 m. the tentacles were considerably inflected, so that the solutions of sugar, though it delayed the action of the phosphate, did not delay that of the carbonate. Immersion in a similar solution of gum arabic for 20 m. had no retarding action on the phosphate. Three leaves were left for 20 m. in a mixture of one part of alcohol to seven parts of water, and then placed in the phosphate solution: in 2 hrs. 15 m. there was a trace of inflection in one leaf, and in 5 hrs. 30 m. a second was slightly affected; the inflection subsequently increased, though slowly. Hence diluted alcohol, which, as we shall see, is hardly at all poisonous, plainly retards the subsequent action of the phosphate.

It was shown in the last chapter that leaves which did not become inflected by nearly a day's immersion in solution of various salts and acids behaved very differently from one another when subsequently placed in the phosphate solution. We here give a table summing up the result.

Name of the Salts and Acids in Solution	Period of Immersion of the Leaves in Solutions of One Part to 437 of Water	Effects Produced on the Leaves by their Subsequent Immersion for Stated Periods in a Solution of One Part of Phosphate of Ammonia to 8750 of Water, or 1 gr. to 20 oz.
Rubidium chloride	22 hrs.	After 30 m. strong inflection of the tentacles.
Potassium carbonate	20 m.	Scarcely any inflection until 5 hrs. had elapsed.
Calcium acetate	24 hrs.	After 24 hrs. very slight inflection.
Calcium nitrate	24 hrs.	After 24 hrs. very slight inflection.
Magnesium acetate	22 hrs.	Some slight inflection, which became well-pronounced in 24 hrs.
Magnesium nitrate	22 hrs.	After 4 hrs. 30 m. a fair amount of inflection, which never increased.
Magnesium chloride	22 hrs.	After a few minutes great inflection; after 4 hrs. all four leaves with almost every tentacle closely inflected.
Barium acetate	22 hrs.	After 24 hrs. two leaves out of four slightly inflected.
Barium nitrate	22 hrs.	After 30 m. one leaf greatly, and two others moderately, inflected; they remained thus for 24 hrs.
Strontium acetate	22 hrs.	After 25 m. two leaves greatly inflected; after 8 hrs. a third leaf moderately, and the fourth very slightly, inflected. All four thus remained for 24 hrs.
Strontium nitrate	22 hrs.	After 8 hrs. three leaves out of five moderately inflected; after 24 hrs. all five in this state; but not one closely inflected.
Aluminium chloride	24 hrs.	Three leaves which had either been slightly or not at all affected by the chloride became after 7 hrs. 30 m. rather closely inflected.
Aluminium nitrate	24 hrs.	After 25 hrs. slight and doubtful effect.
Lead chloride	23 hrs.	After 24 hrs. two leaves somewhat inflected, the third very little; and thus remained.
Manganese chloride	22 hrs.	After 48 hrs. not the least inflection.
Lactic acid	48 hrs.	After 24 hrs. a trace of inflection in a few tentacles, the glands of which had not been killed by the acid.
Tannic acid	24 hrs.	After 24 hrs. no inflection.
Tartaric acid	24 hrs.	After 24 hrs. no inflection.
Citric acid	24 hrs.	After 50 m. tentacles decidedly inflected, and after 5 hrs. strongly inflected; so remained for the next 24 hrs.
Formic acid	22 hrs.	Not observed until 24 hrs. had elapsed; tentacles considerably inflected, and protoplasm aggregated.

In a large majority of these twenty cases, a varying degree of inflection was slowly caused by the phosphate. In four cases, however, the inflection was rapid, occurring in less than half an hour or at most in 50 m. In three cases the phosphate did not produce the least effect. Now what are we to infer from these facts? We know from ten trials that immersion in distilled water for 24 hrs. prevents the subsequent action of the phosphate solution. It would therefore appear as if these solutions of chloride of manganese, tannic and tartaric acids, which are not poisonous, acted exactly like water, for the phosphate produced no effect on the leaves which had been previously immersed in these three solutions. The majority of the other solutions behaved to a certain extent like water, for the phosphate produced, after a considerable interval of time, only a slight effect. On the other hand, the leaves, which had been immersed in the solutions of the chloride of rubidium and magnesium, of acetate of strontium, nitrate of barium, and citric acid, were quickly acted on by the phosphate. Now, was water absorbed from these five weak solutions, and yet owing to the presence of the salts, did not prevent the subsequent action of the phosphate? Or may we not suppose that the interstices of the walls of the glands were blocked up with the molecules of these five substances, so that they were rendered impermeable to water; for had water entered, we know from the ten trials that the phosphate would not afterwards have produced any effect? It further appears that the molecules of the carbonate of ammonia can quickly pass into glands which, for having been immersed for 20 m. in weak solution of sugar, either absorbed the phosphate very slowly or are acted on by it very slowly. On the other hand, glands, however they may have been treated, seem easily or permit the subsequent entrance of the molecules of carbonate of ammonia. Thus leaves which had been immersed in a solution (of one part to 437 of water) of nitrate of potassium for 48 hrs.—of Sulphate of potassium for 24 hrs.—and of the chloride of potassium for 25 hrs.—of being placed in a solution of one part to carbonate of ammonia to 218 of water, had their glands immediately blackened, and after 1 hr. their tentacles somewhat inflected, and the protoplasm aggregated. But it would be an endless task to endeavour to ascertain the wonderfully diversified effects of various solutions of Drosera.

Alcohol *(one part to seven of water)*—It has already been shown that half-minims of this strength placed on the discs of leave do not cause any inflection;

and that when two days afterwards the leaves were given bits of meat, they became strongly inflected. Four leaves were immersed in this mixture and two of them after 30 m. were brushed with a camel-hair brush, like leaves in a solution of camphor, but this produced no effect. Nor did these four leaves, on being left for 24 hrs. in the diluted alcohol, undergo any inflection. They were then removed; one being placed in infusion of raw meat, and bits of meat on the discs of the other three, with their stalks in water. Next day one seemed a little injured, whilst two others showed merely a trace of inflection. We must, however, bear in mind that immersion for 24 hrs. in water prevents leaves from clasping meat. Hence alcohol of the above strength is not poisonous, nor does it stimulate the leaves like camphor does.

The vapour of alcohol acts differently. A plant having three good leaves was left for 25 m. under a receiver holding 19-oz. with sixty minims of alcohol in a watch-glass. No movement ensued, but some few of the glands were blackened and shrivelled, whilst many became quite pale. These were scattered over all the leaves in the most irregular manner, reminding me of the manner in which the glands were affected by the vapour of carbonate of ammonia. Immediately on the removal of the receiver parcels of raw meat were placed on many of the glands, those which retained their proper colour being chiefly selected. But not a single tentacle was inflected during the next 4 hrs. After the first 2 hrs. the glands of all the tentacles began to dry; and next morning, after 22 hrs., all three leaves appeared almost dead, with their glands dry; the tentacles on one leaf alone being partially inflected.

A second plant was left for only 5 m. with some alcohol in a watch glass, under a 12-oz. receiver, and particles of merit were then placed on the glands of several tentacles. After 10 m. some of them began to curve inwards, and after 55 m., nearly all were considerably inflected; but a few did not move. Some anæsthetic effect is here probable, but by no means certain. A third plant was also left for 5 m., under the same small vessels, with its whole inner surface wetted with about a dozen drops of alcohol. Particles of meat were now placed on the glands of several tentacles, some of which first began to move in 25 m.; after 40 m. most of them were somewhat inflected, and after 1 hr. 10 m. almost all were somewhat inflected. From their slow rate of movement there can be no doubt that the glands of these tentacles had been rendered insensible for a time by exposure during 5 m. to the vapour of alcohol.

Vapour of Chloroform—The action of this vapour on Drosera is very variable, depending on the constitution or age of the plant, or on some unknown condition. It sometimes causes the tentacles to move with extraordinary rapidity, and sometimes produces no such effect. The glands are sometimes rendered for a time insensible to the action of raw meat, but sometimes are not thus affected, or in a very slight degree. A plant recovers from a small dose, but is easily killed by a larger one.

A plant was left for 30 m. under a bell-glass holding 19 fluid oz. (539.9 c.c.) with eight drops of chloroform, and before the cover was removed, most of the tentacles became much inflected, though they did not reach the centre. After the cover was removed, bits of meat were placed on the glands of several of the somewhat incurved tentacles; these glands were found much blackened after 6 hrs. 30 m., but no further movement ensued. After 24 hrs. the leave appeared almost dead.

A smaller bell-glass, holding 12 fluid oz. (340.8 c.c.), was now employed, and a plant was left for 90 s. under it, with only two drops of chloroform. Immediately on the removal of the glass all the tentacles curved inwards so as to stand perpendicularly up; and some of them could actually be seen moving with extraordinary quickness by little starts, and therefore in an unnatural manner; but they never reached the centre. After 22 hrs. they fully re-expanded, and on meat being place on their glands, or when roughly touched by a needle, they promptly became inflected; so that these leaves had not been in the least injured.

Another plant was placed under the same small bell-glass with three drops of chloroform, and before two minutes had elapsed, the tentacles began to curl inwards with rapid little jerks. The glass was then removed, and in the course of two or three additional minutes almost every tentacle reached the centre. On several other occasions the vapour did not excite any movement of this kind.

There seems also to be great variability in the degree and manner in which chloroform renders the glands insensible to the subsequent action of meat. In the plant last referred to, which had been exposed for 2 m. to three drops of chloroform, some few tentacles curved up only to a perpendicular position, and particles of meat were placed on their glands; this caused them in 5 m. to begin moving, but they moved so slowly that they did not reach the centre until 1 hr. 30 m. had elapsed. Another plant was similarly exposed, that is, for 2 m., to three drops of chloroform, and on particles of meat being placed on the glands of several tentacles, which had curved up into a perpendicular position, one of these began to bend in 8 m., but afterwards moved very slowly; whilst none of the other tentacles moved for the next 40 m. Nevertheless, in 1 hr. 45 m. from the time when the bits of meat had been given, all the tentacles reached the centre. In this case some slight anaesthetic effect apparently had been produced. On the following day the plant had perfectly recovered.

Another plant bearing two leaves was exposed for 2 m. under the 19 oz. vessel to two drops of chloroform; it was then taken out and examined; again exposed for 2 m. to two drops; taken out, and re-exposed for 3 m. to three drops; so that altogether it was exposed alternately to the air and during 7 m. to the vapour of seven drops of chloroform. Bits of meat were now placed on thirteen glands on the two leaves. On one of these leaves, a single tentacles first began moving in 40 m., and two others in 54 m. On the second leaf some tentacles first moved in 1 hr. 11 m. After 2 hrs. many tentacles on both leaves were inflected; but none had reached the centre within this time. In this case there could not be the least doubt that the chloroform had exerted an anaesthetic influence on the leaves.

On the other hand, another plant was exposed under the same vessel for a much longer time, *viz.*, 20 m., to twice as much chloroform. Bits of meat were then placed on the glands of many tentacles, and all of them, with a single exception, reached the centre in from 13 m. to 14 m. In this case, little or no anaesthetic the centre in from 13 m. to 14 m. In this case, little or no anaesthetic effect had been produced.

Vapour of Sulphuric Ether—A plant was exposed for 30 m. to thirty minims of this ether in a vessel holding 19 oz.; and bits of raw meat were afterwards placed on many glands which had become pale-coloured; but none of the tentacles moved. After 6 hrs. 30 m. the leaves appeared sickly, and the discal glands were almost dry. By the next morning many of the tentacles were dead, as were all those on which meat had been placed; showing that matter had been absorbed from the meat which had increased the evil effects of the vapour. After four days the plant itself died. Another plant was exposed in the same vessel for 15 m. to forty minims. One young, small and tender leaf had all its tentacles inflected, and seemed much injured. Bits of raw meat were placed on several glands on two other and older leaves. These glands became dry after 6 hrs., and seemed injured; the tentacles never moved, excepting one was ultimately a little inflected. The uninjured, but the whole plant after three days became very sickly.

In the two foregoing experiments the doses were evidently too large and poisonous. With weaker doses, the anaesthetic effect was variable, as in the case of chloroform. A plant was exposed for 5 m. to ten drops under a 12-oz. vessel, and bits of meat were then placed on many glands. None of the tentacles thus treated began to move in a decided manner until 40 m. had elapsed; but then some of them moved very quickly, so that two reached the centre after an additional interval of only 10 m. In 2 hrs. 12 m. from the time when the meat was given, all the tentacles reached the centre. Another plant, with two leaves, was exposed in the same vessel for 5 m. to rather large dose of ether, and bits of meat were placed on several glands. In this case one tentacle on each leaf began to bend in 5 m; and after 12 m. two tentacles on one leaf, and one on the second leaf, reached the centre. In 30 m. after the meat had been given, all the tentacles, both those with and without meat, were closely inflected; so that the ether apparently had stimulated these leaves, causing all the tentacles to bend.

Vapour of Nitric Ether—This vapour seems more injurious than of sulphuric ether. A plant was exposed for 5 m. in a 12-oz. vessel to eight drops in a watch-glass, and we distinctly saw a few tentacles curling inwards before the glass was removed. Immediately afterwards bits of meat were placed on three glands, but no movements ensued in the course of 18 m. The same plant was placed again under the same vessel for 16 m. with ten drops of the ether. None of the tentacles moved, and next morning those with the meat were still in the same position. After 48 hrs. one leaf seemed healthy, but the others were much injured.

Another plant, having two good leaves was exposed for 6 m. under a 19-oz. vessel to the vapour from ten minims of the ether, and bits of meat were then placed on the glands of many tentacles on both leaves. After 36 m. several of them on one leaf became inflected, and after 1 hr. almost all the tentacles, those with and without meat, nearly reached the centre. On the other leaf the glands began to dry in 1 hr. 40 m., and after several hours not a single tentacle was inflected; but by the next morning, after 21 hrs., many were inflected, though they seemed much injured. In this and the previous experiment, it is doubtful, owing to the injury which the leaves has suffered, whether any anaesthetic effect had been produced.

A third plant, having two good leaves, was exposed for only 4 m. in the 19-oz. vessel to the vapour from six drops. Bits of meat were then placed on the glands of seven tentacles on the same leaf. A single tentacle moved after 1 hr. 23 m.; after 21 hrs. 3 m. several were inflected; and after 3 hrs. 3 m. all the seven tentacles with meat were well inflected. From the slowness of these movements it is clear that this leaf had been rendered insensible for a time to the action of the meat. A second leaf was rather differently affected; bits of meat were placed on the glands of five tentacles, three of which were slightly inflected in 28 m.; after 1 hr. 21 m. one reached the centre, but the other two were still only slightly inflected; after 3 hrs. they were much more inflected; but even after 5 hrs. 16 m. all five had not reached the centre. Although some of the tentacles began to move moderately soon, they afterwards moved with extreme slowness. By next morning, after 20 hrs., most of the tentacles on both leaves were closely inflected, but not quite regularly. After 48 hrs. neither leaf appeared injured, though the tentacles were still inflected; after 72 hrs. one was almost dead, whilst the other was re-expanding and recovering.

Carbonic Acid—A plant was placed under a 122-oz. bell-glass filled with this gas and standing over water; but we did not make sufficient allowance for the absorption of the gas by the water, so that towards the latter part of the experiment some air was drawn in. After an exposure of 2 hrs. the plant was removed, and bits of raw meat placed on the glands of three leaves. One of these leaves hung a little down, and was at first partly and soon afterwards completely covered by the water, which rose within the vessel as the gas was absorbed. On this latter leaf the tentacles, to which meat had been given, became well inflected in 2 m. 30 s., that is, at about the normal rate; so that until we remembered that the leaf had been protected from the gas, and might perhaps have absorbed oxygen from the water which was continually drawn inwards, we falsely concluded that the carbonic acid has produced no effect. On the other two leaves, the tentacles with meat behaved very differently from those on the first leaf; two of them first began to move slightly in 1 hr. 50 m., always reckoning from the time when the meat was placed on the glands—were plainly inflected in 2 hrs. 22 m.—and in 3 hrs. 22 m. reached the centre.

Three other tentacles did not begin to move until 2 hrs. 20 m. had elapsed, but reached the centre at about the same time with the others, *viz.*, in 3 hrs. 22 m.

This experiment was repeated several times with nearly the same results, excepting that the interval before the tentacles began to move varied a little. We will give only one other case. A plant was exposed in the same vessel to the gas for 45 m., and bits of meat were then placed on four glands. But the tentacles did not move for 1 hr. 40 m.; after 2 hrs. 30 m. all four were well inflected, and after 3 hrs. reached the centre.

The following singular phenomenon sometimes, but by no means always, occurred. A plant was immersed for 2 hrs., and bits of meat were then placed on several glands. In the course of 13 m. all the submarginal tentacles on one leaf became considerably inflected; those with the meat not in the least degree more than the others. On a second leaf, which was rather old, the tentacles with meat, as well as a few others, were moderately inflected. On a third leaf all the tentacles were closely inflected, though meat had not been placed on any of the glands. This movement may be attributed to excitement from the absorption of oxygen. The last-mentioned leaf, to which no meat had been given, was fully re-expanded after 24 hrs.; whereas the two other leaves had all their tentacles closely inflected over the bits of meat which by this time had been carried to their centres. Thus these three leaves had perfectly recovered from the effects of the gas in the course of 24 hrs.

On another occasion some fine plants, after having been left for 2 hrs. in the gas, were immediately given bits of meat in the usual manner, and on their expousre to the air most of their tentacles became in 12 m. curved into a vertical or sub-vertical position, but in an extremely irregular manner; some only on one side of the leaf and some on the other. They remained in this position for some tie; the tentacles with the bits of meat not having at first moved more quickly or farther inwards than the others without meat. But after 2 hrs. 20 m. the former began to move, and steadily went on bending until they reached the centre. Next morning, after 22 hrs., all the tentacles on these leaves were closely clasped over the meat which had been carried to their centres; whilst the vertical and sub-vertical tentacles on the other leaves to which no meat had been given had fully re-expanded. Judging, however, from the subsequent action of a weak solution of carbonate of ammonia on one of these latter leaves, it had not perfectly recovered its excitability and power of movement in 22 hrs.; but another leaf, after an additional 24 hrs., had completely recovered, judging from the manner in which it clasped a fly placed on its disc.

After the exposure of a plant for 2 hrs. to the gas, one of its leaves was immersed in a rather strong solution of carbonate of ammonia, together with a fresh leaf from another plant. The latter had most of its tentacles strongly inflected within 30 m.; whereas the leaf which had been exposed to the carbonic acid remained for 24 hrs. in the solution without undergoing any inflection, with the exception of two tentacles. This leaf had been almost completely paralysed, and was not able to recover its sensibility whilst still in the solution, which from having been made with distilled water probably contained little oxygen.

Concluding Remarks

As the glands, when excited, transmit some influence to the surrounding tentacles, causing them to bend and their glands to pour forth an increased amount of modified secretion, we were anxious to ascertain whether the leaves included any element having the nature of nerve-tissue, which, though not continuos, served as the channel of transmission. This led us to try the several alkaloids and other substances which are known to exert a powerful influence on the nervous system of animals. We were at first encouraged in my trials by finding that strychnine, digitaline, and nicotine, which all act on the nervous system, were poisonous to Drosera, and caused a certain amount of inflection. Hydrocyanic acid, again, which is so deadly a poison to animals, caused rapid movement of the tentacles. But as several innocuous acids, though much diluted, such as benzoic, acetic, &c., as well as some essential oils, are extremely poisonous to Drosera, and quickly cause strong inflection, it seems probable that strychnine, nicotine, digitaline, and hydrocyanic acid, excite inflection by acting on elements in no way analogous to the nerve-cells of animals. If elements of this latter nature had been present in the leaves, it might have been expected that morphia, hyoscyamus, atropine, veratrine, colchicine, curare, and diluted alcohol would have produced some marked effect; whereas these substances are not poisonous and have no power, or only a very slight one, of inducing inflection. It should, however, be observed that curare, colchicine, and veratrine are muscle-poisons—that is, act on nerves having some special relation with the muscles, and therefore, could not be expected to act on Drosera. The poison of the cobra is most deadly to animals, by paralyzing their nerve-centres, yet is not in the least so to Drosera, though quickly causing strong inflection.

Notwithstanding the foregoing facts, which show how widely different is the effect of certain substances on the health or life of animals and of Drosera; yet there exists a certain degree of parallelism in the action of certain other substances. We have seen that this holds good in a striking manner with the salts of sodium and potassium. Again various metallic salts and acids, namely, those of silver, mercury, gold, tin, arsenic, chromium, copper, and platina, most or all of which are highly poisonous to animals, are equally so to Drosera. But it is a singular fact that the chloride of lead and two salts of barium were not poisonous to this plant. It is an equally strange fact that though acetic and propionic acids are highly poisonous, their ally, formic acid, is not so; and that, whilst certain vegetable acids, namely oxalic, benzoic, &c., are poisonous in a high degree, gallic, tannic, tartaric, and malic (all diluted to an equal degree) are not so. Malic acid induces inflection, whilst the three other just named vegetable acids have no such power. But a pharmacopoeia would be requisite to describe the diversified effects of various substances on Drosera.

Of the alkaloids and their salts which were tried, several had not the least power of inducing inflection: others, which were certainly absorbed, as shown by the changed colour of the glands, had but a very moderate power of this kind; other, again, such as the acetate of quinine and digitaline, caused strong inflection.

The several substances mentioned in this chapter affect the colour of the glands very differently. These often become dark at first, and then very pale or white, as was conspicuously the case with glands subjected to the poison of the cobra and citrate of strychnine. In other cases they are from the first rendered white, as with leaves placed in hot water and several acids; and this is the result of the coagulation of the albumen. On the same leaf some glands become white and others dark-coloured, as occurred with leaves in a solution of the Sulphate of quinine, and in the vapour of alcohol. Prolonged immersion in nicotine, curare, and even water, blackness the glands; and this is due to the aggregation of the protoplasm within their cells. Yet curare caused very little aggregation in the cells of the tentacles, whereas nicotine and sulphate of quinine induced strongly marked aggregation down their bases. The aggregated masses leaves which had been immersed for 3 hrs. 15 m. in a saturated solution of sulphate of quinine exhibited incessant changes of form, but after 24 hrs. were motionless; the leaf being flaccid and apparently dead. On the other hand, with leaves subjected for 48 hrs. to a strong solution of the poison of the cobra, the protoplasmic masses were unusually active, whilst with the higher animals the vibratile cilia and white corpuscles of the blood seem to be quickly paralysed by this substance.

With the salts of alkalies and earths, the nature of the base, and not that of the acid, determines their physiological action on Drosera, as is likewise the case with animals; but this rule hardly applies to the salts of quinine and strychnine, for the acetate of quinine causes much more inflection than the sulphate, and both are poisonous, whereas the nitrate of quinine is not poisonous, and induces inflection at a much slower rate than the acetate. The action of the citrate of strychnine is also somewhat different from that of the sulphate.

Leaves which have been immersed for 24 hrs. in water, and for only 20 m. in diluted alcohol, or in a weak solution of sugar, are afterwards acted on very slowly, or not at all, by the phosphate of ammonia, though they are quickly acted on by the carbonate. Immersion for 20 m. in a solution of gum Arabic had no such inhibitory power. The solution of certain salts and acids affect the leaves, with respect to the subsequent action of the phosphate, exactly like water, whilst others allow the phosphate afterwards to act quickly and energetically. In this latter case, the interstices of the cell walls may have been blocked up by the molecules of the salts first enter, though the molecules of the phosphate could do so, and those of the carbonate still more easily.

The action of camphor dissolved in water is remarkable, for it not only soon induces inflection, but apparently renders the glands extremely sensitive to mechanical irritation; for if they are brushed with a soft brush, after being immersed in the solution for a short time, the tentacles begin to bend in about 2 m. It may, however, be that the brushing, thought not a sufficient stimulus by itself, tends to excite movement merely by reinforcing the direct action of the camphor. The vapour of camphor, on the other hand, serves as a narcotic.

Some essential oils, both in solution and in vapour, cause rapid inflection, other have no such power; those which we tried were all poisonous.

Diluted alcohol (one part to seven of water) is not poisonous, does not induce inflection, nor increase the sensitiveness of the glands to mechanical irritation. The vapour acts as a narcotic or anaesthetic, and long exposure to it kills the leaves.

The vapours of chloroform, sulphuric and nitric ether, act in a singularly variable manner on different leaves, and on the several tentacles of the same leaf. This is owing to differences in the age or constitution of the leaves, and to whether certain tentacles have lately been in action. That these vapours are absorbed by the glands is shown by their changed colour; but as other plants not furnished with glands are affected by these vapours, it is probable that they are likewise absorbed by the stomata of Drosera. They sometimes excite extraordinarily rapid inflection, but this is not an invariable result. If allowed to act for even a moderately long time, they kill the leaves; whilst a small dose acting for only a short time serves as a narcotic or anaesthetic. In this case the tentacles, whether or not they have become inflected, are not excited to further movement by bits of meat placed on the glands, until some considerable time has elapsed. It is generally believed that with animals and plants these vapours act by arresting oxidation.

Exposure to carbonic acid for 2 hrs., and in one case for only 45 m., likewise rendered the glands insensible for a time to the powerful stimulus of raw meat. The leaves, however, recovered their full powers, and did not seem in the least injured, on being left in the air for 24 or 48 hrs. We have seen in the third chapter that the process of aggregation in leaves subject for two hours to this gas and then immersed in a solution of the carbonate of ammonia is much retared, so that a considerable time elapses before the protoplasm in the lower cells of the tentacles becomes aggregated. In some cases, soon after the leaves were removed from the gas and brought into the air, the tentacles moved spontaneoulsy; this being due to the excitement from the access of oxygen. These inflected tentacles, however, could not be excited for some time afterwards to any further movement by their glands being stimulated. With other irritable plants it is known that the exclusion of oxygen prevents their moving, and arrests the movements of the protoplasm within their cells, but this arrest is a different phenomenon from the retardation of the process of aggregation just alluded to.

❑❑❑

10

The Sensitiveness of the Leaves, and the Lines of Transmission of the Motor Impulse

Glands and summits of the tentacles alone sensitive—Transmission of the motor impulse down the pedicels of the tentacles, and across the blade of the leaf—Aggregation of the protoplasm, a reflex action—First discharge of the motor impulse sudden—Direction of the movements of the tentacles—Motor impulse transmitted through the cellular tissue—Mechanism of the movements—Nature of the motor impulse—Re-expansion of the tentacles.

Glands and Summits of Tentacles

We have seen in the previous chapters that many widely different stimulants, mechanical and chemical, excite the movement of the tentacles, as well as of the blade of the leaf; and we must now consider, firstly, what are the points which are irritable or sensitive, and secondly how the motor impulse is transmitted from one point to another. The glands are almost exclusively the seat of irritability, yet this irritability must extend for a very short distance below them; for when they were cut-off with a sharp pair of scissors without being themselves touched, the tentacles often became inflected. These headless tentacles frequently re-expanded; and when afterwards drops of the two most powerful known stimulants were placed on the cut-off ends, no effect was produced. Nevertheless these headless tentacles are capable of subsequent inflection if excited by an impulse sent from the disc. We secceded on several occasion in crushing glands between fine pincers, but this did not excite any movement; not did raw meat and salts of ammonia, when placed on such crushed glands. It is probable that they were killed so instantly that they were not able to transmit any motor impulse; for in six observed cases (in two of which, however, the gland was quite pinched off) the protoplasm within the cells of the tentacles did not become aggregated; whereas in some adjoining tentacles, which were inflected from having been roughly touched by the pincers, it was well aggregated. In like manner the protoplasm does not become aggregated when a leaf is instantly killed by being dipped into boiling water. On the other hand, in several cases in which tentacle became inflected after their glands had been cut off with sharp scissors a distinct though moderate degree of aggregation supervened.

The pedicels of the tentacles were roughly and repeatedly rubbed; raw meat or other exciting substances were placed on them, both on the upper surface near the base and else where, but no distinct movement ensued. Some bits of meat, after being left for a considerable time on the pedicels, were pushed upwards, so as just to touch the glands, and in a minute the tentacles began to bend. We believed that the blade of the leaf is not sensitive to any stimulant. We drove the point of a lancet through the blades of several leaves, and a needle three or four times through nineteen leaves: in the former case no movement ensued; but about a dozen of the leaves which were repeatedly pricked had a few tentacles irregularly inflected. As, however, their backs had to be supported during the operation, some of the outer glands, as well as those on the disc may have been touched; and this perhaps sufficed to cause the slight degree of movement observed. Nitschke says that cutting and pricking the leaf does not excite movement. The petiole of the leaf is quite insensible.

The backs of the leaves bear numerous minute papillae, which do not secrete, but have the power of absorption. These papillae are rudiments of formerly existing tentacles, together with their glands. Many experiments were made to ascertain whether the backs of the leaves could be irritated in any way, thirty-seven leaves being thus tried. Some were rubbed for a long time with a blunt needle, and drops of milk and other exciting fluids, raw meat, crushed flies, and various substances, placed on other. These substances were apt soon to become dry, showing that no secretion had been excited. Hence we moistened them with saliva, solutions of ammonia, weak hydrochloric acid, and frequently with the secretion from the glands of other leaves. We also kept some leaves, on the backs of which exciting objects had been placed, under a damp bell-glass; but with all care we never saw any true movement. We were led to make so many trials because, contrary to our previous experience, Nitschke states that, after affixing objects to the backs of leaves by the aid of the viscid secretion, he repeatedly saw the tentacles (and in one instance the blade) become reflexed. This movement, if a true one, would be most anomalous; for it implies that the tentacles receive a motor impulse from an unnatural source, and have the power of bending in a direction exaction exactly the reverse of that which is habitual to them; this power not being of the least use to the plant, as insects cannot adhere to the smooth backs of the leaves.

We have said that no effect was produced in the above cases; but this is not strictly true, for in three instances a little syrup was added to the bits of raw meat on the backs of leaves, in order to keep them damp for a time; and after 36 hrs. there was trace of reflection in the tentacles of one leaf, and certainly in the blade of another. After twelve additional hours the glands began to dry, and all three leaves seemed much injured. Four leaves were then placed under a bell-glass, with their fool-stalks in water, with drops of syrup on their backs, but without any meat. Two of these leaves, after a day, had a few tentacles reflexed. The drops had now increased considerably in size, from having imbibed moisture, so as to trickle down in backs of the tentacles and footstalks. On the second day, one leaf had its blade much reflexed; on the third day the tentacles of two were much reflexed, as well as the blades of all four to a greater or less degree. The upper side of one leaf, instead of being, as at first, slightly concave, now presented a strong convexity upwards. Even on the fifth day the leaves did not appear dead. Now, as sugar does not in the last excite Drosera, we may safely attribute the reflexion of the blades and tentacles of the above leaves to exosmose from the cells which were in contact with the syrup, and their consequent contraction. When drops of syrup are placed on the leaves of plant with their roots still in damp earth, no inflection ensues, for the roots, no doubt, pump up water as quickly as it is lost by exosmose. But if cut-off leaves are immersed in syrup, or in any dense fluid, the tentacles are greatly, though irregularly, inflected, some of them assuming the shape of corkscrews; and the leaves soon become flaccid. If they are now immersed in a fluid of low specific gravity, the tentacles re-expand. From these facts we may conclude that drops of syrup place on the backs of leaves do not act by exciting a motor impulse which is transmitted to the tentacles; but that they cause reflection by inducing exosmose. Dr. Nitschke used the secretion for sticking insects to the backs of the leaves; and we suppose that he used a large quantity, which from being dense probably caused exosmose. Perhaps he experimented on cut-off leaves, or on plants with their roots not supplied with enough water.

As far, therefore, as our present knowledge, we may conclude that the glands, together with the immediately underlying cells of the tentacles, are the exclusive seats of that irritability or sensitiveness with which the leaves are endowed. The degree to which a gland is excited can be measured only by the number of the surrounding tentacles which ae inflected, and by the amount and rate of their movement. Equally vigorous leaves, exposed to the same temperature (and this is an important condition), are excited in various degrees under the following circumstances. A minute quantity of weak solution produces no effect; add more, or give a rather stronger solution, and the tentacles bend. Touch a gland once or twice, and the tentacles becomes inflected. But the nature of the substance which is given a very important element: if equal-sized particles of glass (which acts only mechanically), of gelatine, and raw meat are placed on the discs of several leaves, the meat causes far more rapid, energetic, and widely extended movement than the two former substances. The number of glands which are excited also makes a great difference in the result: place a bit of meat on one or two of the discal glands, and only a few of the immediately surrounding short tentacles are inflected; place it on several glands, and many more are acted on; place it on thirty or forty, and all the tentacles, including the extreme marginal ones, become closely inflected. We thus see that the impulses proceeding from a number of glands strengthen one another, spread farther, and act on a larger number of tentacles, than the impulse from any single gland.

Transmission of the Motor Impulse

In every case the impulse from a gland has to travel for at least a short distance to the basal part of the tentacle, the upper part and the gland itself being merely carried by the inflection of the lower part. The impulse is thus always transmitted down nearly the whole length of the pedicel. When the central glands are stimulated, and the extreme marginal tentacles become inflected, the impulse is transmitted across nearly the whole width of the disc. A gland transmits its motor impulse far more easily and quickly down its own tentacle to the bending place than across the disc to neighbouring tentacle. Thus a minute dose of a very weak solution of ammonia, if given to one of the glands of the exterior tentacles, causes it to bend and reach the centre; whereas a large drop of the same solution, given to a score of glands on the disc, will not cause through their combined influence the least inflection of the exterior tentacles. Again, when a bit of meat is placed on the gland of an exterior tentacle, we have seen movement in ten seconds, and repeatedly within a minute; but much larger bit placed on several glands on the disc does not cause the exterior tentacles to bend until half an hour or even several hours have elapsed.

The motor impulse spreads gradually on all sides from one or more excited glands, so that the tentacle which stand nearest are always first affected. Hence, when the glands in the centre of the disc are excited, the extreme marginal tentacles are the last inflected. But the glands on different parts of the leaf transmit their motor power in a somewhat different manner. If a bit of meat be placed on the long-headed gland of a marginal tentacle, it quickly transmits an impulse to its own bending portion; but never to the adjoining tentacles; for these are not affected until the meat has been carried to the central glands, which then radiate forth their conjoint impulse on all the glands from the centre, so that these could not be excited by the bits of meat brought to them by the inflection of the marginal tentacles; and now these marginal tentacles re-expanded after a time without any other tentacle being affected. Other leaves were similarly prepared, and bits of meat were placed on the glands of two tentacles in the third row from the outside, and on the glands of two tentacles in the fifth row. In these four cases the impulse was sent in the first place laterally, that is, in the same concentric row of tentacles, and then towards the centre; but not centrifugally, or towards the exterior tentacles. In one of these cases only a single tentacle on each side of the one with meat was affected. In the three other cases, from half a dozen to a dozen tentacles, both laterally and towards the centre, were well inflected, or sub-inflected. Lastly, in ten other experiments, minute bits of meat were placed on a single gland or on two glands in the centre of the disc. In order that no other glands should touch the meat, through the inflection of the closely adjoining short tentacles, about half a dozen glands had been previously removed round the selected ones. On eight of these leaves from sixteen to twenty-five of the short surrounding tentacles were inflected in the course of one or two days; so that the motor impulse radiating from one or two of the discal glands is able to produce this much effect. The tentacles which had been removed are included in the above numbers; for from standing so close they would certainly have been affected. On the two remaining leaves, almost all the short tentacles of the disc were inflected. With a more powerful stimulus than meat, namely a little phosphate of lime moistened with saliva, we have seen the inflection spread still farther from a single gland thus treated; but even in this case the three or four outer rows of tentacles were not affected. From these experiments it appears that the impulse from a single gland on the disc acts on a greater number of tentacles than that from a glands of one of the exterior elongated tentacles; and this probably follows, at least in part, from the impulse having to travel a very short distance down the pedicels of the central tentacles, so that it is able to spread to a considerable distance all round.

Whilst examining these leaves, we were struck with the fact that in six, perhaps seven, of them the tentacles were much more inflected at the distal and proximal ends of the leaf (*i.e.,* towards the apex and base) than on either side; and yet the tentacles on the sides stood as near to the glands where the bit of meat lay as did those at the two ends. It thus appeared as if the motor impulse was transmitted from the centre across the disc more readily in a longitudinal than in a transverse direction; and as this appeared a new and interesting fact in the physiology of plants, thirty-five fresh experiments were made to test its truth. Minute bits of met were placed on a single gland or on a few glands, on the right or left side of the discs of eighteen leaves; other bits of the same size being placed on the distal or proximal ends of seven another leaves. Now if the motor impulse were transmitted with equal force or at an equal rate through the blade in all directions, bit of meat placed at one side or at one end of the disc ought to affect equally all the tentacles situated at an equal distance from it; but this certainly is not the case. Before giving the general results, it may be well to describe three or four rather unusual cases.

1. A minute fragment of a fly was placed on one side of the disc, and after 32 m. seven of the other tentacles near the fragment were inflected: after 10 hrs. several more become so, and after 23 hrs. a still greater

number; and now the blade of the leaf on this side was bent inwards so as to stand up at right angles to the other side. Neither the blade of the leaf nor a single tentacle or the opposite side was affected; the line of separation between the two halves extending from the footstalk to the apex. The leaf remained in this state for three days, and on the fourth day began to re-expand; not a single tentacles having been inflected on the opposite side.

2. We will here give a case not included in the above thirty-five experiments. A small fly was found adhering by its feet to the left side of the disc. The tentacles on this side soon closed in and killed the fly: and owing probably to its struggle whilst alive, the leaf was so much excited that in about 24 hrs. all the tentacles on the opposite side became inflected; but as they found no prey, for their glands did not reach the fly, they re-expanded in the course of 15 hrs.; the tentacles on the left side remaining clasped for several days.

3. A bit of meat, rather larger than those commonly used, was placed in a medial line at the balsa end of the disc, near the footstalk; after 2 hrs. 30 m. some neighbouring tentacles were inflected; after 6 hrs. the tentacles on both sides of the footstalk, and some way up both sides, were moderately inflected; after 8 hrs. the tentacles at the further or distal end were more inflected than those on either side; after 23 hrs. the meat was well clasped by all the tentacles, excepting by the exterior ones on the two sides.

4. Another bit of meat was placed at the opposite or distal end of another leaf, with exactly the same relative results.

5. A minute bit of meat was placed one side of the disc; next day the neighbouring short tentacles were inflected, as well as in a slight degree three or four on the opposite side of the disc were inflected. As soon as these began to re-expand, I added another bit of meat, and next day all the tentacle on the opposite side of the disc were inflected towards the meat; whereas we have seen that those on the same side were affected by the first bit of meat which was given.

Now for the general results. Of the eighteen leaves on which bits of meat were placed on the right or left sides of the disc, eight had a vast number of tentacles inflected on the same side, and in four of them the blade itself on this side was likewise inflected; whereas not a single tentacle nor the blade was affected on the opposite side. These leaves presented a very curious appearance, as if only the inflected side was active, and the other paralyzed. In the remaining ten cases, a few tentacles became inflected beyond the medial line, on the side opposite to that where the meat lay; but, in some of these cases, only at the proximal or distal ends of the leaves. The inflection on the opposite side always occurred considerably after that on the same side, and in one instance not until the forth day. We have also seen with No. 5 that bits of meat had to be added thrice before all the short tentacles on the opposite side of the disc were inflected.

The result was widely different when bits of meat were placed in a medial line at the distal or proximal ends of the disc. In three of the seventeen experiments thus made, owing either to the state of the leaf or to the smallness of the bits of meat, only the immediately adjoining tentacles were affected; but in the other fourteen cases the tentacles at the opposite end of the leaf were inflected, though these were as distant from where the meat lay as were those on one side of the disc from the meat on the opposite side. In some of the present cases the tentacles on the sides were not at all affected, or in a less degree, or after a longer interval of time, then those at the opposite end. One set of experiment is worth giving in fuller detail. Cubes of meat, not quite so small as those usually employed, were placed on one side of the disc of four leaves, and cubes of the same size at the proximal or distal end of four other leaves. Now, when these two sets of leaves were compared after an interval of 24 hrs., they presented a striking difference. Those having the cubes on one side were very slightly affected on the opposite side; whereas those with the cubes at either end has almost every tentacles at the opposite end, even the marginal ones, closely inflected. After 48 hrs. the contrast in the state of the two sets was still great; yet those with the meat on one side now had their discal and submarginal tentacles on the opposite side somewhat inflected, this being due to the large size of the cubes. Finally we may conclude from these thirty-fives experiments, not to mention the six or seven previous ones, that the motor impulse is transmitted from any single gland or small group of glands through the blade to the other tentacles more readily and effectually in a longitudinal than in a transverse direction.

As long as the glands remain excited, and this may last for many days, even for eleven, as when in contact with phosphate of lime, they continue to transmit a motor impulse to the basal and bending parts of their own pedicels, for otherwise they would re-expand. The great difference in the length of time during which tentacles remain inflected over inorganic objects, and over objects of the same size cotaining soluble nitrogenous matter, proves the same fact. But the intensity of the impulse transmitted from an excited glands, which has begun to pour forth its acid secretion and is at the same time absorbing, seems to be very small compared with that which it transmits when first excited. Thus, when moderately large bits of meat were placed on one side of the disc, and the discal and submarginal tentacles on the opposite side became inflected, so that their glands at last touched the meat and observed matter from it. They did not transmit any motor influence to the exterior rows of tentacles on the same side, for these never became inflected. If, however, meat had been placed on the glands of these same tentacles before they had begun to secrete copiously and to absorb, they undoubtedly would have affected the exterior row. Nevertheless, when we gave some phosphate of lime, which is a most powerful stimulant, to several submarginal tentacles already considerably inflected, but not yet in contact with some phosphate previously placed on two glands in the centre of the disc, the exterior tentacles on the same side were acted on.

When a gland is first excited, the motor impulse is discharged within a few seconds, as we know for the bending of the tentacle; and it appears to be discharged at first with much greater force than afterwards. Thus in the case above given of small fly naturally caught by a few glands on one side of a leaf, an impulse was slowly transmitted from them across the whole breadth of the leaf, causing the opposite tentacles to be temporarily inflected, but the glands which remained in contact with the insect, though they continued for several days to send an impulse down their own pedicels to bending place, did not prevent the tentacles on the opposite side from quickly re-expanding; so that the motor discharge must at first have been more powerful than afterwards.

When an object of any kind is placed on the disc, and the surrounding tentacles are inflected, their glands secrete more copiously and the secretion becomes acid, so that some influence is sent to them from the discal glands.

This change in the nature and amount of the secretion cannot depend on the bending of the tentacles, as the glands of the short central tentacles secrete acid when an object is placed on them, though they do not themselves bend. Therefore we inferred that the glands of the disc sent some influence up the surrounding tentacles to their glands, and that these reflected back a motor impulse to their basal parts; but this view was soon proved erroneous. It was found by many trials that tentacles with their glands closely cut off by sharp scissors often become inflected and again re-expand, still appearing healthy. One which was observed continued healthy for ten days after the operation. We therefore cut the glands off twenty-five tentacles, at different times and on different leaves, and seventeen of these soon became inflected, and afterwards re-expanded. The re-expansion commenced about 8 hrs. or 9 hrs. and was completed in from 22 hrs. to 30 hrs. from the time of inflection. After an interval of a day or two, raw meat with saliva was placed on the disc of these seventeen leaves, and when observed next day, seven of the headless tentacles were inflected over the meat as closely as the uninjured ones on the same leaves; and an eight headless tentacle became inflected after three additional days. The meat was removed from one of these leafs, and the surface washed with a little stream of water, and after three days the headless tentacle re-expanded for the second time. These tentacles without glands were, however, in a different state from those provided with glands and which had absorbed matter from the meat, for the protoplasm within the cells of the former had undergone far less aggregation. From these experiments with headless tentacles it is certain that the glands do not, so far as the motor impulse is concerned, act in a reflex manner like the nerve-ganglia of animals.

But there is another action, namely that of aggregation, which in certain cases may be called reflex, and it is the only known instance in the vegetable kingdom. We should bear in mind that the process does not depend on the previous bending of the tentacles, as we clearly see when leaves are immersed in certain strong solution. Nor does it depend on increased secretion from the glands, and this is shown by several facts, more especially by the papillae, which do not secrete, yet undergoing aggregation, if given carbonate of ammonia or an infusion of raw meat. When a gland is directly stimulated in any way, as by the pressure of a minute particle of glass, the protoplasm within the cells of the glands first becomes aggregated, then that in the cells immediately beneath the glands, and so lower and lower down the tentacles to their bases—that is, if the stimulus had been sufficient and not injurious. Now, when the glands of the disc are excited, the exterior tentacles are affected in exactly the same manner: the aggregation always commences in these glands, though these have not been directly excited, but have only received some influence from the disc, as shown by their increased acid secretion. The protoplasm within the cells immediately beneath the glands are next affected, and so downwards from cell to cell to the bases of the tentacles. This process apparently deserves to be called a reflex action, in the same manner as when a sensory nerve is irritated, and carried an impression to a ganglion which sends back some influence to a muscle or gland, causing movement or increased secretion; but the action in the two cases is probably of a widely different nature. After the protoplasm in a tentacle has been aggregated, its redissolution always begins in the lower part, and slowly travels up the pedicel to the glands, so that the protoplasm last aggregated is first redissolved. This probably depends merely on the protoplasm being less and less aggregated, lower and lower down in the tentacles, as can be seen plainly when the excitement has been slight. As soon, therefore, as the aggregating action altogether ceases, redissolution naturally commences in the less strongly aggregated matter in the lowest part of the tentacle, and is there first completed.

Direction of the Inflected Tentacles

When a particle of any kind is placed on the gland of one of the outer tentacles, this invariably moves towards the centre of the leaf; and so it is with all the tentacles of a leaf immersed in any exciting fluid. The glands of the exterior tentacles then form a ring round the middle part of the disc, as shown in a previous figure (fig. 4). The short tentacles within this ring still retain their vertical position, as they likewise do when a large object is placed on their glands, or when an insect is caught by them. In this latter case we can see that the inflection of the short central tentacles would be useless, as their glands are already in contact with prey.

The result is very different when a single gland on one side of the disc is excited, or a few in a group. These send an impulse to the surrounding tentacles, which do not now bend towards the centre of the leaf, but to the point of excitement. If a minute bit of meat be placed by the aid of a needle on a single gland, or on three or four together, halfway between the centre and the circumference of the disc, the directed movement of the surrounding tentacles is well exhibited. An accurate drawing of a leaf with meat in this position is here reproduced (fig. 10), and we see the tentacles, including some of the exterior ones, accurately directed to the point where the meat lay. But a much better plan is to place a particle of the phosphate of lime moistened with saliva on a single gland on one side of the disc of a large leaf, and another particle on a single gland on the opposite side. In four such trials the excitement was not sufficient to affect the outer tentacles, but all those near the two points were directed to them, so that two wheels were formed on the disc of the same leaf; the pedicels of the tentacles forming the spokes and the glands united in a mass over the phosphate representing the axles. The precision with which each tentacle pointed to the particle was wonderful; so that in some cases we could detect no deviation from perfect accuracy. Thus, although the short tentacles in the middle of the disc do not bend when their glands are excited in a direct manner, yet if they receive a motor impulse from a point on one side, they direct themselves to the point equally well with the tentacles on the borders of the disc.

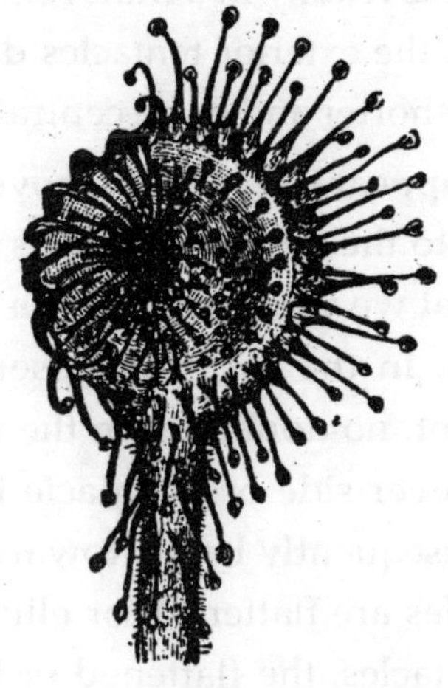

Fig. 10
(*Drosera rotundifolia.*)
Leaf (enlarged) with the tentacles inflected over bit of meat placed on one side of the disc.

In these experiments, some of the short tentacles on the disc, which would have been directed to the centre, had the leaf been immersed in an exciting fluid, were now inflected in an exactly opposite direction, *viz.*, towards the circumference. These tentacles, therefore, had deviated as much as 180° from the direction which they would have assumed if their own glands had been stimulated, and which may be considered as the normal one. Between this, the greatest possible and no deviation from the normal direction, every degree could be observed in the tentacles on these several leaves. Notwithstanding the precision with which the tentacles generally were directed, those near the circumference of one leaf were not accurately directed towards some phosphate of lime at a rather distant point on the opposite side of the disc. It appeared as if the motor impulse in passing transversely across nearly the whole width of the disc had departed somewhat from a true course. This accords with what we have already seen of the impulse travelling less readily in a transverse than in a longitudinal direction. In some other cases, the exterior tentacles did not seem capable of such accurate movement as the shorter and more central ones.

Nothing could be more striking than the appearance of the above four leaves, each with their tentacles pointing truly to the two little masses of the phosphate on their discs. We might imagine that we were looking at a lowly organised animal seizing prey with its arms. In the case of Drosera the explanation of this accurate power of movement, no doubt lies in the motor impulse radiating in all directions, and whichever side of a tentacle it first strikes, that side contracts, and the tentacle consequently bends towards the point of excitement. The pedicels of the tentacles are flattened, or elliptic in section. Near the bases of the short central tentacles, the flattened or broad face is formed of about five longitudinal rows of cells; in the outer tentacles of the disc, it consists of about six or seven rows; and in the extreme marginal tentacles of above a dozen rows. As the flattened bases are thus formed of only a few rows of cells, the precision of the movement of the tentacles is the more remarkable; for when the motor impulse strikes the base of a tentacle in a very oblique direction relatively to its broad face, scarcely more than one or two cells towards one end can be affected at first, and the contraction of these cells must draw the whole tentacle into the proper direction. It is, perhaps, owing to the exterior pedicels being much flattened that they do not bend quite so accurately to the point of excitement as the more central ones. The properly directed movement of the tentacles is not an unique case in the vegetable kingdom, for the tendrils of many plants curve towards the side which is touched; but the case of Drosera is far more interesting, as here the tentacles are not directly excited, but receive an impulse from a distant point; nevertheless, they bend accurately towards this point.

Nature of the Tissues through which the Motor Impulse is Transmitted

It will be necessary first to describe briefly the course of the main fibro-vascular bundles. These are shown in the accompanying sketch (fig. 11) of a small leaf. Little vessels from the neighbouring bundles enter all the many

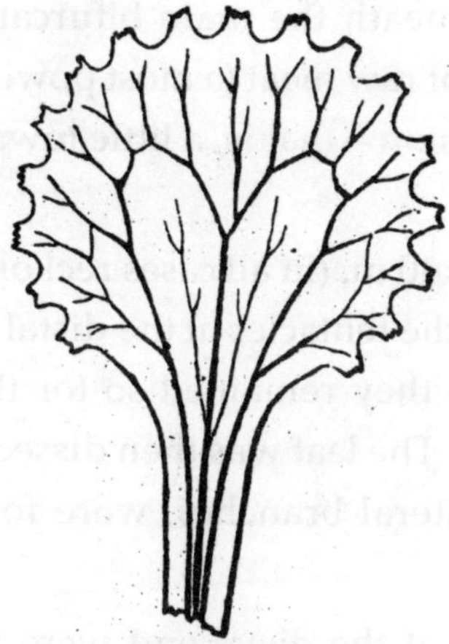

Fig. 11
(*Drosera rotundifolia.*)
Diagram showing the distribution of the vascular tissue in a small leaf.

tentacles with which the surface is studded; but these are not here represented. The central trunk, which runs up the footstalk, bifurcates near the centre of the leaf, each branch bifurcating again and again according to the size of the leaf. This central trunk sends off, low down on each disc, a delicate branch, which may be called the sublateral branch. There is also, on each side, a main lateral branch or bundle, which bifurcates in the same manner as the others. Bifurcation does not imply that any single vessel divides, but that a bundle divides into two. By looking to either side of the leaf, it will be seen that a branch from the great central bifurcation inosculates with a branch from the lateral bundle, and that there is a smaller inosculation between the two chief of the lateral bundle. The course of the vessels is very complex at the larger inosculation; and here vessels, retaining the same diameter, are often formed by the union of the bluntly pointed ends of two vessels. By means of the two inosculations all the vessels on the same side of the leaf are brought into some sort of connection. Near the circumference of the larger leaves the bifurcating branches also come into close union, and then separate again, forming a continuos zigzag line of vessels round the whole circumference. But the union of the vessels in the zigzag line seems to be much less intimate than at the main inosculation. It should be added that the course of the vessels differs somewhat in different leaves, and even on opposite sides of the same leaf, but the main inosculation is always present.

Now in my first experiments with bits of meat placed on one side of the disc, it so happened that not a single tentacle was inflected on the opposite side; and when we saw that the vessels on the same side were all connected together by the two inosculation, whilst not a vessel passed over to the opposite side, it seemed probable that the motor impulse was conducted exclusively along them.

In order to test this view, we divided transversely with the point of a lancet the central trunks of four leaves, just beneath the main bifurcation; and two days afterwards placed rather large bits of raw meat (a most powerful stimulant) near the centre of the discs above incision—that is, a little towards the apex—with the following results:

1. This leaf proved rather torpid: after 4 hrs. 40 m. (in all cases reckoning from the time when the meat was given) the tentacles at the distal end were a little inflected, but nowhere else; they remained so for three days, and re-expanded on the fourth day. The leaf was then dissected, and the trunk, as well as the two sublateral branches, were found divided.
2. After 4 hrs. 30 m. many of the tentacles at the distal end were well inflected. Next day the blade and all the tentacles at this end were strongly inflected, and were separated by a distinct transverse line from the basal half of the leaf, which was not in the least affected. On the third day, however, some of the short tentacles on the disc near the base were very slightly inflected. The incision was found on dissection to extend across the leaf as in the last case.
3. After 4 hrs. 30 m. Strong inflection of the tentacles at the distal end, which during the next two days never extended in the east to the basal end. The incision as before.
4. This leaf was not observed until 15 hrs. had elapsed, and then all the tentacles, except the extreme marginal ones, were found equally well inflected all round the leaf. On careful examination the spiral vessels of the central trunk were certainly divided; but the incision on one side had not passed through the fibrous tissue surrounding these vessels, through it had passed though the tissue on the other side.

The appearance presented by the leaves (2) and (3) was very curious, and might be aptly compared with that of a man with his backbone and lower extremities paralyzed. Excepting that the line between the two halves was here transverse instead of longitudinal, these leaves were in the same state as some of those in the former experiments, with bits of meat placed on one side of the disc. The case of leaf (4) proves that the spiral vessels of the central trunk may be divided, and yet the motor impulse be transmitted from the distal to the basal end; and this led me at first to suppose that the motor force was sent through the closely surrounding fibrous tissue; and that if one half of this tissue was left undivided, it sufficed for complete transmission. But opposed to this conclusion is the fact that no vessels pass directly from one side of the leaf to the other, and yet, as we have seen, if a rather large bit of meat is placed on one side, the motor impulse is sent, though slowly and imperfectly, in a transverse direction across the whole breadth of the leaf. Nor can this latter fact be accounted for by supposing that the transmission is affected through the two inosculation, or through the circumferential zigzag line of union. For had this been the case, the exterior tentacles on the opposite side of this would have been affected before the more central ones, which never occurred. We have also seen that the extreme marginal tentacles appear to have no power to transmit an impulse to the adjoining tentacles; yet the little bundle of vessels which enters each marginal tentacles sends off a minute branch to those on both sides, and this we have not observed in any other tentacles; so that the marginal ones are more closely connected together by spiral vessels than are the others, and yet have much less power of communicating a motor impulse to one another.

But besides these several facts and arguments we have conclusive evidence that the motor impulse is not send, at least exclusively, through the spiral vessels, or through the tissue immediately surrounding them. We know that if a bit of meat is placed on a gland (the immediately adjoining ones having been removed) on any part of the disc, all the short surrounding tentacles bend almost simultaneously with great precision towards it. Now there are tentacles on the disc, for instance near the extremities of the sublateral bundles (fig. 11), which are supplied with vessels that do not come into contact with the branches that enter the surrounding tentacles, except by a very long and extremely circuitous course. Nevertheless, if a bit of meat is placed on the glands of a tentacle of this kind, all the surrounding ones are inflected towards it with great precision. It is, of course, possible that an impulse might be sent through a long and circuitous course, but it is obviously impossible that the direction of the movement could be thus communicated, so that all the surrounding tentacles should bend precisely to the point of excitement. The impulse no doubt is transmitted in straight radiating lines from the excited gland to the surrounding tentacles; it cannot, therefore, be sent along the fibro-vascular bundles. The effect of cutting the central vessels, in the above cases, in preventing the transmission of the motor impulse from the distal to the basal end of a leaf, maybe attributed to a considerable space of the cellular tissue having been divided. We shall hereafter see, when we treat of Dionaea, that this same conclusion, namely that the motor impulse is not transmitted by the fibro-vascular bundle, is plainly confirmed; and Professor Cohn has come to the same conclusion with respect to Aldrovanda—both members of the Droseraceae.

As the motor impulse is not transmitted along the vessels, there remains for its passage only the cellular tissue; and the structure of this tissue explains to a certain extent how it travels so quickly down the long exterior tentacles, and much more slowly across the blade of the leaf. We shall also see why it crosses the blade more quickly in a longitudinal than in a transverse direction; though with time it can pass in any direction. We know that the same stimulus causes movement of the tentacles and aggregation of the protoplasm, and that both influences originate in and proceed from the glands within the same brief space of time. It seems therefore probable that the motor impulse consists of the first commencement of a molecular change in the protoplasm, which, when, well developed, is plainly visible, and has been designated aggregation; but to this subject I shall return. We further know that in the transmission of the aggregating process the chief delay is caused by the passage of the transverse cell-walls; for as the aggregation travels down the tentacles, the contents of each successive cell seem almost to flash into a cloudy mass. We may therefore infer that motor impulse is in like manner delayed chiefly by passing through the cell-walls.

The greater celerity with which the impulse is transmitted down the long exterior tentacles than across the disc may be largely attributed to its being closely confined within the narrow pedicel, instead of radiating forth on all sides as on the disc. But besides this confinement, the exterior cells of the tentacles are fully twice as long as those of the disc; so that only half the number of transverse partitions have to be traversed in a given length of a tentacle, compared with an equal space on the disc; and there would be in the same proportion less retardation of the impulse. Moreover, in section of the exterior tentacles given by Dr. Warming, the parenchymatous cells are shown to be still more clongated; and these would from the most direct line of communication from the glands to the bending place of the tentacles. If the impulse travels down the exterior cells, it would have to cross from between twenty to thirty transverse partitions: but rather fewer if down the inner parenchymatous tissue. In either case it is remarkable that the impulse is able to pass through so many partitions down nearly the whole length of the pedicel, and to act on the bending place, in ten seconds. Why the impulse, after having passes so quickly down one of the extreme marginal tentacles (about $\frac{1}{20}$ of an inch in length), should never, as far as we have seen, affect the adjoining tentacles? It may be in part accounted for by much energy being expended in the rapidity of the transmission.

Most of the cells of the disc, both the superficial ones and the larger cells which form the five or six underlying layers, are about four times as long as broad. They are arranged almost longitudinally, radiating from the footstalk. The motor impulse, therefore, when transmitted across the disc, has to cross nearly four times as many cell-walls as when transmitted in a longitudinal direction, and would consequently be much delayed in the former case. The cells of the disc converge towards the bases of the tentacles, and are thus fitted to convey the motor impulse to them from all sides. On the whole, the arrangement and shape of the cells, both those of the disc and tentacles, throw much light on the rate and manner of diffusion of the motor impulse. But why the impulse proceeding from the glands of the exterior rows of tentacles tends to travel and towards the centre of the leaf, but not centrifugally, is by no means clear.

Mechanism of the Movements and Nature of the Motor Impulse

What ever may be the means of movement, the exterior tentacles, considering their delicacy, are inflected with much force. A bristle, held so that a length of 1 inch projected from a handle, yielded when we tried to lift with it an inflected tentacle, which was somewhat thinner than the bristle. The amount or extent, also, of the movement is great. Fully expanded tentacles in becoming inflected sweep through an angle of 180°; and if they are beforehand reflexed, as often occurs, the angle is considerably greater. It is probably the superficial cells at the bending place which chiefly or exclusively contract; for the interior cells have very delicate walls, and are so few in number that they could hardly cause a tenable to bend with precision to a definite point. Though we carefully looked, we could never detect any wrinkling of the surface at the bending place, even in the case of tentacles abnormally curved into a complete circle, under circumstances hereafter to be mentioned.

All the cells are not acted on, though the motor impulse passes through them. When the gland of the long exterior tentacles is excited, the upper cells are not in the least affected; about half-way down there is a slight bending, but the chief movement is confined to a short space near the base; and no part of the inner tentacle bends except the basal portion. With respect to the blade of the leaf, the motor impulse may be transmitted through many cells, from centre to the circumference, without their being in the least greatly inflected. In the latter case the movement seems to depend partly on the strength of the stimulus, and partly on its nature, as when leaves are immersed in certain fluids.

The power of movement which various plants possess, when irritated, has been attributed by high authorities to the rapid passage of fluid out of certain cells, which, from their previous state of tension immediately contract. Whether or not this is the primary cause of such movement, fluid must pass out of close cells when they contract or are pressed together in one direction, unless they, at the same time, expand in some other direction. For instance, fluid can be seen if slowly bent into a semi-circle. In the case of Drosera there is certainly much movement of the fluid through out the tentacles whilst they are undergoing inflection. Many leaves can be found in which the purple fluid within the cells is of an equally dark tint on the upper and lower sides of the tentacles, extending also downwards on both sides to equally near their bases. If the tentacles of such a leaf are excited into movement, it will generally be found after some hours that the cells on the conclave side are much paler than they were before, or are quite colourless, those on the convex side having become much darker. In two instances, after particles of hair had been placed on glands, and when in the course of 1 hr. 10 m. the tentacles were incurved halfway towards the centre of the leaf, this change of colour in the two sides was conspicuously plain. In another case, after a bit of meat had been placed on a glands, the purple colour was observed at intervals to be slowly travelling from the upper to the lower part, down the convex side of the bending tentacle. But it does not follow from these observations that the cells on the convex side become filled with more fluid during the act of inflection than they contained before; for fluid may all the time be passing into the disc or into the glands which then secrete freely.

The bending of the tentacles, when leaves are immersed in a dense fluid, and their subsequent re-expansion in a less dense fluid, show that the passage of fluid from or into the cells can cause movements like the natural ones. But the inflection thus caused is often irregular; the exterior tentacles being sometimes spirally curved. Other unnatural movements are likewise by the application of dense fluids, as in the case of drops of syrup placed on the backs of leaves and tentacles. Such movements may be compared with the contortions which many vegetable tissue under go when subjected to exosmose. It is therefore doubtful whether they throw any light on the natural movements.

If we admit that the outward passage of fluid is the cause of the bending of the tentacle, we must suppose that the cells, before the act of inflection, are in a high state of tension, and that they are elastic to an extraordinary degree; for otherwise their contraction could not cause the tentacles often to sweep through an angle of above 180°. Professor Cohn, in his interesting paper on the movements of the stamens of certain Çompositae, states that these organs, when dead, are as elastic as they were when alive. He believes that the living protoplasm within their cells is ordinarily in a state of expansion, but is paralysed by irritation, or may be said to suffer temporary death; the elasticity of the cell-walls then coming into play, and causing the contraction of the stamens. Now the cells on the upper or concave side of the bending part of the tentacles of Drosera do not appear to be in a state of tension, nor to be highly elastic; for when a leaf is suddenly killed, or dies slowly, it is not the upper but the lower sides of the tentacles which contract from elasticity. We may therefore conclude that their movements cannot be accounted for by the inherent elasticity of certain cells, opposed as long as they are alive and not irritated by the expanded state of their contents.

A somewhat different view has been advanced by other physiologists, namely, that the protoplasm, when irritated, contracts like the soft sarcode of the muscles of animals. In Drosera the fluid within the cells of the tentacles at the bending place appears under the microscope thin and homogeneous, and after aggregation consists of small, soft masses in almost colourless fluid. These masses are completely redissolved when the tentacles re-expand. Now it seems scarcely possible that such matter should have any direct mechanical power; but if through some molecular change it were to occupy less space than it did before, no doubt the cell walls would close up and contract. But in this case it might be expected that the walls would exhibit wrinkles, and none could ever be seen. Moreover, the contents of all the cells seem to be of exactly the same nature, both before and after aggregation; and yet only a few of the basal cells contract, the rest of the tentacle remaining straight.

A third view maintained by some physiologists, though rejected by most others, is that the whole cell, including the walls, actively contracts. If the walls are composed solely of non-nitrogenous cellulose, this view is highly improbable; but it can hardly be doubted that they must be permeated by proteid matter, at least whilst they are growing. Nor does there seem any inherent improbability in the cell-walls of Drosera contracting, considering their high state of organisation; as shown in the case of the glands by their power of absorption and secretion, and by being exquisitely sensitive so as to be affected by the pressure of the most minute particles. The cell-walls of the pedicels also allow various impulses to pass through them, inducing movement, increased secretion and aggregation. On the whole the belief that the walls of certain cells contract, some of their contained fluid being at the same time forced outwards, perhaps accords best with the observed facts. If this view is rejected, the next most probable one is that the fluid contents of the cells shrink, owing to a change in their molecular state, with the consequent closing in of the walls. Anyhow, the movement can hardly be attributed to the elasticity of the walls, together with a previous state of tension.

With respect to the nature of the motor impulse which is transmitted from the glands down the pedicels and across the disc, it seems not improbable that it is closely allied to that influence which causes the protoplasm within the cells of the glands and tentacles to aggregate. We have seen that both forces originate in and proceed from the glands within a few seconds of the same time, and are excited by the same causes. The aggregation of the protoplasm lasts almost as long as the tentacles remain inflected, even though this be for more than a week; but the protoplasm is redissolved at the bending place shortly before the tentacles re-expand, showing that the exciting cause of the aggregating process has then quite ceased. Exposure to carbonic acid causes both the latter process and the motor impulse to travel very slowly down the tentacles. We know that the aggregating process is delayed in passing through the cell-walls, and we have good reason to believe that this holds good with the motor impulse; for we can thus understand the different rates of its transmission in a longitudinal and transverse line across the disc. Under a high power the first sign of aggregation is the appearance of a cloud, and soon afterwards of extremely fine granules, in the homogeneous purple fluid within the cells; and this apparently is due to the union of molecules of protoplasm. Now it does not seem an improbable view that the same tendency—namely, for the molecules to approach each other—should be communicated to the inner surface of the cell-walls which are in contact with the protoplasm; and if so, their molecules would approach each other, and the cell-wall would contract.

To this view it may with truth be objected that when leaves are immersed in various strong solution, or are subjected to a heat of above 130° F (54°.4 C), aggregation ensues, but there is no movement. Again, various acids and some other fluids cause rapid movement, but no aggregation, or only of an abnormal nature, or only after a long interval of time; but as most of these fluids are more or less injurious, they may check or prevent the aggregating process by injuring or killing the protoplasm. There is another and more important difference in the two processes; when the glands on the disc are excited, they transmit some influence up the surrounding tentacles, which acts on the cells at the bending place, but does not induce aggregation until it has reached the glands; these then send back some other influence, causing the protoplasm to aggregate first in the upper and then in the lower cells.

The Re-expansion of the Tentacles

This movement is always slow and gradual. When the centre of the leaf is excited, or a leaf is immersed in a proper solution, all the tentacles bend directly towards the centre, and afterwards directly back from it. But when the point of excitement is on one side of the disc, the surrounding tentacles bend towards it, and therefore obliquely with respect to their normal direction; when they afterwards re-expand, they bend obliquely back, so as to recover their original position. The tentacles farthest from an excited point, wherever that may be, are the last and the least affected, and probably in consequence of this they are the first to re-expand. The bent portion of a closely inflected tentacles is in a state of active contraction, as shown by the following experiment. Meat was placed on a leaf, and after the tentacles were closely inflected and has quite ceased to move, narrow strips of the disc, with a few of the outer tentacles attached to it, were cut off and laid on one side under the microscope. After several failures, we succeeded in cutting off the convex surface of the bent portion of a tentacle. Movement immediately re-commenced, and the already greatly bent portion went on bending until it formed a perfect circle; the straight distal portion of the tentacle passing on one side of the strip. The convex surface must therefore have previously been in a state of tension, sufficient to counterbalance that of the concave surface, which, when free, curled into a complete ring.

The tentacles of an expanded and unexcited leaf are moderately rigid and elastic; if bent by a needle, the upper end yields more easily than the basal and thicker part, which alone is capable of becoming inflected. The rigidity of this basal part seems due to the tension of the outer surface balancing a state of active and persistent contraction of the cells of the inner surface. We believe that this is the case, because, when a leaf is dipped into boiling water, the tentacles suddenly become reflexed, and this apparently indicates that the tension of the outer surface is mechanical, whilst that of the inner surface is vital, and is instantly destroyed by the boiling water. We can thus also understand why the tentacles as they grow old and feeble slowly become much reflexed. If a leaf with its tentacles closely inflected is dipped into boiling water, these rise up a little, but by no means fully re-expand. This may be owing to the heat quickly destroying the tension and elasticity of the cells of the convex surface; but we can hardly believe that their tension, at any one time, would suffice to carry back the tentacles to their original position, often through an angle of above 180°. It is more probable that fluid, which we know travels along the tentacles during the act of inflection, is slowly re-attracted into the cells of the convex surface, their tension being thus gradually and continually increased.

A recapitulation of the chief facts and discussions in this chapter will be given at the close of the next chapter.

11

Recapitulation of the Chief Observations on *Drosera Rotundifolia*

As summaries have been given to most of the chapters, it will be sufficient here to recapitulate the chief points. In the first chapter a preliminary sketch was given of the structure of the leaves, and of the manner in which they capture insects. This is effected by drops of extremely viscid fluid surrounding the glands and by the inward movement of the tentacles. As the plants gain most of their nutriments by this mans, their roots are very poorly developed; and they often grow in places where hardly any other plant except mosses can exist. The glands have the power of absorption, besides that of secretion. They are extremely sensitive to various stimulants, nearly repeated touches, the pressure of minute particles, the absorption of animal matter and of various fluids, heat, and galvanic action. A tentacle with a bit of raw meat on the gland has been seen to begin bending in 10 s., to be strongly incurved in 5 m., and to reach the centre of the leaf in half an hour. The blade of the leaf often becomes so much inflected that it forms a cup, enclosing any object placed on it.

A gland, when excited, not only sends some influence down its own tentacle, causing it to bend, but likewise to the surrounding tentacles, which become incurved; so that the bending place can be acted on by an impulse received from opposite directions, namely, from the gland on the summit of the same tentacle, and from one or more glands of the neighbouring tentacles. Tentacles, when inflected, re-expand after a time, and during this process the glands secrete less copiously or become dry. As soon as they begin to secrete again, the tentacles are ready to react; and this may be repeated at least three, probably many more times.

It was shown in the second chapter that animal substances placed on the discs cause much more prompt and energetic inflection than do inorganic bodies of the same size, or mere mechanical irritation; but there is a still more marked difference in the greater length of time during which the tentacles remain inflected over bodies yielding soluble and nutritious matter, than over those which do not yield such matter. Extremely minute particles of glass, cinders, hair, thread, precipitated chalk, & c., when placed on the glands of the outer tentacles, cause them to bend. A particle, unless it sinks through the secretion and actually touches the surface of the glands with some one point, does not produce any effect. A little bit of thin human hair $\frac{8}{1000}$ of an inch (.203 mm.) in length, and weighing only $\frac{1}{78740}$ of a grain (.000822 mg.), though largely supported by the dense secretion, suffices to induce movement. It is not probable that the pressure in this case could have amounted to that from the millionth of a grain. Even smaller particles cause a slight movement, as could be seen through a lens. Larger particles than those of which the measurements have been given cause no sensation when placed on the tongue, one of the most sensitive parts of the human body.

Movement ensues if a gland is momentarily touched three or four times; but if touched only once or twice, though with considerable force and with a hard object, the tentacle does not bend. The plant is thus saved from much useless movement, as during a high wind the glands can hardly escape being occasionally brushed by the leaves of surrounding plants. Though insensible to a single touch, they are exquisitely sensitive, as just stated, to the slightest pressure if prolonged for a few seconds; and this capacity is manifestly of service to the plant in capturing small insects. Even gnats, if they rest on the glands with their delicate feet, are quickly and securely embraced. The glands are insensible to the weight and repeated blows of drops of heavy rain, and the plants are thus likewise saved from much useless movement.

The description of the movements of the tentacles was interrupted in the third chapter for the sake of describing the process of aggregation. This process always commences in the cells of the glands, the contents of which first become cloudy; and this has been observed within 10 s. after a gland has been excited. Granules just resolvable under a very high power soon appear, sometimes within a minute, in the cells beneath the glands; and these then aggregate into minute spheres. The process afterwards travels down the tentacles, being arrested for a short time at each transverse partition. The small spheres coalesce into larger spheres, or into oval, club-headed, thread or necklace–like, or otherwise shaped masses of protoplasm, which, suspended in almost colourless fluid, exhibit incessant spontaneous changes of form. These frequently coalesce and again separate. If a gland has been powerfully excited all the cells down to the base of the tentacle are affected. In cells, especially if filled with dark red fluid, the first step in process often is the formation of a dark red, bag–like mass of protoplasm which afterwards divides and undergoes the unusual repeated changes of form. Before any aggregation has been excited, a sheet of colourless protoplasm, including granules (the primordial utricle of Mohl), flows round the walls of the cells; and this becomes more distinct after the contents have been partially aggregated into spheres or bag–like masses. But after a time the granules are drawn towards the central masses and unite with them; and then the circulating sheet can no longer be distinguished, but there is still a current of transparent fluid within the cells.

Aggregation is excited by almost all the stimulants which induce movement; such as the glands being touched two or three times, the pressure of minute inorganic particles, the absorption of various fluids, even long immersion in distilled water, exosmose, and heat of the many stimulants tried, carbonate of ammonia is the most energetic and acts the quickest; a dose of $\frac{1}{134400}$ of a grain (.00048 mg.) given to a single gland suffices to cause in one hour well-marked aggregation in the upper cells of the tentacle. The process goes on only as long as the protoplasm is in a living, vigorous, and oxygenated condition.

The result is in all respects exactly the same, whether a gland has been excited directly, or has received an influence from other and distant glands. But there is one important difference; when the central glands are irritated, they transmit centrifugally an influence up the pedicels of the exterior tentacles to their glands; but the actual process of aggregation travels centripetally, from the glands of the exterior tentacles down their pedicels. The exciting influence, therefore, which is transmitted from one part of the leaf to another must be different from that which actually induces aggregation. The process does not depend on the glands secreting more copiously than they did before; and is independent of the inflection, and as soon as these are fully re-expanded, the little masses of protoplasm are all redissolved; the cells becoming filled with homogeneous purple fluid, as they were before the leaf was excited.

As the process of aggregation can be excited by a few touches, or by the pressure of insoluble particles, it is evidently independent of the absorption of any matter, and must be of a molecular nature. Even when caused by the absorption of the carbonate or other salt of ammonia, or an infusion of meat, the process seems to be of exactly the same nature. The protoplasmic fluid must, therefore, be in a singularly unstable condition, to be acted on by such slight and varied causes. Physiologists believe that when a nerve is touched, and it transmits an influence to other parts of the nervous system, molecular change is induced in it, though not visible to us. Therefore it is a very interesting spectacle to watch the effects on the cells of gland, of the pressure of a bit of hair, weighing only $\frac{1}{78700}$ of a grain and largely supported by the dense secretion, for its excessively slight pressure soon causes a visible change in the protoplasm, which change is transmitted down the whole length of the tentacle, giving it at last a mottled appearance, distinguishable even by the naked eye.

In the fourth chapter it was shown that leaves placed for a short time in water at a temperature of 110° F (43.3° C) become somewhat inflected; they are thus also rendered more sensitive to the action of meat than they were before. If exposed to a temperature of between 115° and 125° F (46.1°-51.6° C), they are quickly inflected, and their protoplasm undergoes aggregation; when afterwards placed in cold water, they re-expand. Expose to 130° F (54.4° C), no inflection immediately occurs, but the leaves are only temporarily paralysed, for on being left in cold water, they often become inflected and afterwards re-expand. In one leaf thus treated, we distinctly saw the protoplasm in movement. In other leaves treated in the same manner, and then immersed in a solution of carbonate of ammonia, strong aggregation ensued. Leaves placed in cold water, after an exposure to so high a temperature as 145° F (62.7° C), sometimes become slightly, though slowly inflected; and afterwards have the contents of their cells strongly aggregated by carbonate of ammonia. But the duration of the immersion is an importation element, for if left in water at 145° F (62.7° C) or only at 140° ° (60° C), until it becomes cool, they are killed, and the contents of the glands are rendered white and opaque. This latter result seems to be due to the coagulation of the albumen, and was almost always caused by even short exposure to 150° F (65.5° C); but different leaves, and even the separate cells in the same tentacle, differ considerably in their power of resisting heat. Unless the heat has been sufficient to coagulate the albumen, carbonate of ammonia subsequently induces aggregation.

In the fifth chapter, the results of placing drops of various nitrogenous and non-nitrogenous organic fluids on the discs of leaves were given, and it was shown that they detect with almost unerring certainty the presence of nitrogen. A decoction of green peas or of fresh cabbage-leaves acts almost as powerfully as an infusion of raw meat, whereas an infusion of cabbage-leaves made by keeping them for a long time in merely warm water is far less efficient. A decoction of grass-leaves is less powerful than one of green peas or cabbage-leaves.

These results led me to inquire whether Drosera possessed the power of dissolving solid animal matter. The experiments proving that the leaves are capable of true digestion, and that the glands absorb the digested matter, are given in detail in the sixth chapter. These are, perhaps the most interesting of all my observations on Drosera, as no such power was before distinctly known to exist in the vegetable kingdom. It is likewise interesting fact that the glands of the disc, when irritated, should transmit some influence to the glands of the exterior tentacles, causing them to secrete more copiously and the secretion to become acid, as if they had been directly excited by an object placed on them. The gastric juice of animals contains, as is well known, an acid and a ferment, both of which are indispensable for digestion, and so it is with the secretion of Drosera. When the stomach of an animal is mechanically irritated, it secretes an acid, and when particles of glass or other such objects were placed on the glands of Drosera, the secretion, and that of the surrounding and untouched glands, was increased in quantity and became acid. But according to Schiff, the stomach of an animal does not secrete its proper ferment, pepsin, until certain substances, which he calls peptogenes, are absorbed; and it appears from my experiments that some matter must be absorbed by the glands of Drosera before they secrete their proper ferment. That the secretion does contain a ferment which acts only in the presence on an acid on solid animal matter, was clearly proved by adding minute doses of an alkali, which entirely arrested the process of digestion, this immediately recommencing as soon as the alkali was neutralized by a little weak hydrochloric acid. From trials made with a large number of substances, it was found that those which the secretion of Drosera dissolves completely, or partially, or not at all, are acted on in exactly the same manner by gastric juice. We may therefore conclude that the ferment of Drosera is closely analogous to, or identical with, the pepsin of animals.

The substances which are digested by Drosera act on the leaves very differently. Some cause much more energetic and rapid inflection of the tentacles, and keep them inflected for a much longer time, than do others. We are thus led to believe that the former are more nutritious than the latter, as is known to be case with some of these same substances when given to animals; for instance, meat in comparison with gelatine. As cartilage is so tough a substance and is so little acted on by water, its prompt dissolution by the secretion of Drosera, and subsequent adsorption, is, perhaps, one of the most striking cases. But it is not really more remarkable than the digestion of meat, which is dissolved by this secretion in the same manner and by the same stages as by gastric juice. The secretion dissolves bone, and even the enamel of teeth, but this is simply due to the large quantity of acid secreted, owing, apparently, to the desire of the plant for phosphorus. In the case of bone, the ferment does not come into play until all the phosphate of lime has been decomposed and free acid is present, and then the fibrous basis is quickly dissolved. Lastly, the secretion attacks and dissolves matter out of living seeds, which it sometimes kills, or injures, as shown by the diseased state of the seedlings. It also absorbs matter from pollen, and from fragments of leaves.

The seventh chapter was devoted to the action of the salts of ammonia. These all cause the tentacles, and often the blade of the leaf, to be inflected, and the protoplasm to be aggregated. They act with very different power; the citrate being the least powerful, and the phosphate, owing, no doubt, to the presence of phosphorus and nitrogen, by far the most powerful. But the relative efficiency of only three salts of ammonia was carefully determined, namely the carbonate, nitrate, and phosphate. The experiments were made by placing half-minims (.0296 c.c.) of solutions of different strengths on the discs of the leaves, and by the immersion of whole leaves in a measured quantity. In relation to these experiments it was necessary first to ascertain the effects of distilled water, and it was found, as described in detail, that the more sensitive leaves are affected by it, but only in a slight degree.

A solution of the carbonate is absorbed by the roots and induces aggregation in their cells, but does not affect the leaves. The vapour is absorbed by the gland, and causes inflection as well as aggregation. A drop of a solution containing $\frac{1}{960}$ of a grain (.0675 mg.) is the least quantity which, when placed on the glands of the disc, excited the exterior tentacles to bend inwards. But a minute drop, containing $\frac{1}{14400}$ of a grain (.00445 mg.) if applied for a few seconds to the secretion surrounding a gland, causes the inflection of the same tentacle. When a highly sensitive leaf is immersed in a solution, and there is ample time for absorption, the $\frac{1}{268800}$ of a grain (.00024 mg.) is sufficient to excite a single tentacle into movement.

The nitrate of ammonia induces aggregation of the protoplasm much less quickly than the carbonate, but is more potent in causing inflection. A drop containing $\frac{1}{2400}$ of a grain (.027 mg.) placed on the disc acts powerfully on all the exterior tentacles, which have not themselves received any of the solution; whereas a drop with $\frac{1}{2800}$ of a grain caused only a few of these tentacle to bend, but affected rather more plainly the blade. A minute drop applied as before, and containing $\frac{1}{28800}$ of a grain (.0025 mg.), caused the tentacle bearing this gland to bend. By the immersion of whole leaves, it was proved that the adsorption by a single gland of $\frac{1}{691200}$ of a grain (.0000937 mg.) was sufficient to set to same tentacle into movement.

The phosphate of ammonia is much more powerful than the nitrate. A drop containing $\frac{1}{3840}$ of a grain (.0169mg.) placed on the disc of a sensitive leaf causes most of the exterior tentacles to be inflected, as well as the blade of the leaf. A minute drop containing $\frac{1}{153600}$ of a grain (.000423 mg.), applied for a few seconds to a gland, acts, as shown by the movement of the tentacle. When a leaf is immersed in thirty minims (1.7748 c.c.) of a solution of one part by weight of the salt to 21, 875,000 of water, the absorption by a gland of only the $\frac{1}{19760000}$ of a grain (.00000328 mg.), that is, a little more than the one-twenty-millionth of a grain, is sufficient to cause the tentacle bearing this gland to bend to the centre of the leaf. In this experiment, owing to the presence of the water of crystallisation, less than the one-thirty-millionth of a grain of the efficient elements could have been absorbed. There is nothing remarkable in such minute quantities being absorbed by the glands, for all physiologists admit that the salts of ammonia, which must be brought in still smaller quantity by a single shower of rain to the roots, are absorbed by them. Nor is it surprising that Drosera should be enabled to profit by the absorption of these salts, for yeast and other low fungoid forms flourish in solution of ammonia, if the outer necessary elements are present. But it is an astonishing fact that so inconceivably minute a quantity as the one-twenty-million that a grain of phosphate of ammonia should induce some change in a gland of Drosera, sufficient to cause a motor impulse to be sent down the whole length of the tentacle; this impulse exciting movement often through an angle of above 180°. We know not whether to be most astonished at this fact, or that the pressure of a minute bit of hair, supported by the dense secretion, would quickly cause conspicuous movement. Moreover, this extreme sensitiveness, exceeding that of the most delicate part of the human body, as well as the power of transmitting various impulses from one part of the leaf to another, have been acquired without the intervention of any nervous system.

As few plants are at present known to possess glands especially adapted for absorption, it seemed worthwhile to try the effects on Drosera of various other salts, besides those of ammonia, and of various acids. Their action, as described in the eighth chapter, does not correspond at all strictly with their chemical affinities, as inferred from the classification commonly followed. The nature of the base is far more influential than that of the acid; and this is known to hold good with animal. For instance, nine salts of sodium all caused well-marked inflection, and none of them were poisonous in small doses; whereas seven of the nine corresponding salts of potassium produced no effect, two causing slight inflections. Small doses, moreover, of some of the latter salts were poisonous. The salts of sodium and potassium, when injected into the veins of animals, likewise differ widely in their action. The so-called earthy salts produce hardly any effect on Drosera. On the other hand, most of the metallic salts cause rapid and strong inflection, and are highly poisonous; but there are some odd exception to this rule; thus chloride of lead and zinc, as well as two salts of barium, did not cause inflection, and were not poisonous.

Most of the acids which were tried, though much diluted (one part to 437 of water), and given in small doses, acted powerfully on Drosera; nineteen, out of the twenty-four, causing the tentacles to be more or less inflected. Most of them, even the organic acids, are poisonous, often highly so; and this is remarkable, as the juices of so many plants contain acids. Benzoic acid, which is innocuous to animals, seems to be as poisonous to Drosera as hydrocyanic. On the other hand, hydrochloric acid is not poisonous either to animals or to Drosera, and induces only a moderate amount of inflection. Many acids excite the glands to secrete an extraordinary quantity of mucus; and the protoplasm within their cells seems to be often killed, as may be inferred from the surrounding fluid soon becoming pink. It is strange that allied acids act very differently: formic acid induces very slight inflection, and is not poisonous; whereas acetic acid of the same strength acts most powerfully and is poisonous. Lactic acid is also poisonous, but causes inflection only after a considerable lapse of time. Malic acid acts slightly, whereas citric and tartaric acids produce no effect.

In the ninth chapter the effects of the absorption of various alkaloids and certain other substances were described. Although some of these are poisonous, yet as several, which act powerfully on the nervous system of animals, produce no effect on Drosera, we may infer that the extreme sensibility of the glands, and their power of transmitting an influence to other parts of the leaf, causing movement, or modified secretion, or aggregation, does not depend on the presence of a diffused element, allied to nerve tissue. One of the most remarkable facts is that long immersion in the poison of the cobra-snake does not in the least check, but rather stimulates, the spontaneous movement of the protoplasm in the cells of the tentacles. Solution of various salts and acids behave very differently in delaying or in quite arresting the subsequent action of a solution of phosphate of ammonia. Camphor dissolved in water acts as a stimulant, as do small doses of certain essential oils, for they cause rapid and strong inflection. Alcohol is not a stimulant. The vapours of camphor, alcohol, chloroform, sulphuric and nitric ether, are poisonous in moderately large doses, but in small doses serve as narcotics or anaesthetics, greatly delaying the subsequent action of meat. But some of these vapours also act as stimulants, exciting rapid, almost spasmodic movement in the tentacles. Carbonic acid is likewise a narcotic, and retards the aggregation of the protoplasm when carbonate of ammonia is subsequently given. The first access of air to plants which have been immersed in this gas sometimes acts as a stimulant and induces movement. But, as before remarked, a special pharmacopoeia would be necessary to describe the diversified effects of various substances on the leaves of Drosera.

In the tenth chapter it was shown that the sensitiveness of the leaves appears to be wholly confined to the glands and to the immediately underlying cells. It was further shown that the motor impulse and other forces or influences, proceeding from the glands when excited, pass through the cellular tissue, and not along the fibro-vascular bundles. A gland sends its motor impulse with great rapidity down the pedicel of the same tentacles to the basal part which alone bends. The impulse, then passing onwards, spreads on all sides to the surrounding tentacles, first affecting those which stand nearest and then those farther off. But by being thus spread out, and from the cells of the disc not being so much elongated as those of the tentacles, it loses force, and here travels much more slowly than down the pedicels. Owing also to the direction and form of the cells, it passes with greater ease and celerity in a longitudinal than in a transverse line across the disc. The impulse proceeding from the glands of the extreme marginal tentacles does not seem to have force enough to affect the adjoining tentacles; and this may be in part due to their length. The impulse from the glands of the next few inner rows spreads chiefly to the tentacles on each side and towards the centre of the leaf; but that proceeding from the glands of the shorter tentacles on the disc radiates almost equally on all sides.

When a gland is strongly excited by the quantity or quality of the substance placed on it, the motor impulse travels farther then from one slightly excited; and if several glands are simultaneously excited, the impulses from all unite and spread still farther. As soon as a gland is excited, it discharges an impulse which extends to a considerable distance; but afterwards, whilst the gland is secreting and absorbing, the impulse suffices only to keep the same tentacle inflected; though the inflection may last for many days.

If the bending place of a tentacle receives an impulse from its own gland, the movement is always towards the centre of the leaf; and so it is with all the tentacles, when their glands are excited by immersion in a proper fluid. The short ones in the middle part of the disc must be excepted, as these do not bend at all when thus excited. On the other hand, when the motor impulse comes from one side of the disc, the surrounding tentacles, including the short ones in the middle of the disc, all bends with precision towards the point of excitement, wherever this may be seated. This is in every way a remarkable phenomenon; for the leaf falsely appears as if endowed with the senses of an animal. It is all the more remarkable, as when the motor impulse strikes the base of a tentacle obliquely with respect to its flattened surface, the contraction of the cells must be confined to one, two, or a very few rows at one end. And different sides of the surrouding tentacles must be acted on, in order that all should bend with precision to the point of excitement.

The motor impulse, as it spreads from one or more glands across the disc, enters the bases of the surrounding tentacles, and immediately acts on the bending place. It does not in the first place proceed up the tentacles to the glands, exciting them to reflect back an impulse to their bases. Nevertheless, some influence is sent up to the glands, as their secretion is soon increased and rendered acid; and then the glands, being thus excited, send back some other influence (not dependent on increased secretion, nor on the inflection of the tentacles), causing the protoplasm to aggregate in cell beneath cell. This may be called a reflex action, though probably very different from that proceeding from the nerve-ganglion of an animal; and it is the only known case of reflex action in the vegetable kingdom.

About the mechanism of the movements and the nature of the motor impulse we know very little. During the act of inflection fluid certainly travels from one part to another of the tentacles. But the hypothesis which agrees best with the observed facts is that the motor impulse is allied in nature to the aggregating process; and that this causes the molecules of the cell-wells to approach each other, in the same manner as do the molecules of the protoplasm within the cells; so that the cell-walls contract, but some strong objections may be urged against this view. The re-expansion of the tentacles is largely due to the elasticity of their outer cells which comes into play as soon as those on the inner side cease contracting with prepotent force; but we have reason to suspect that fluid is continually and slowly attracted into the outer cells during the act of re-expansion, thus increasing their tension.

We have now given a brief recapitulation of the chief points observed by us, with respect to the structure, movements constitution, and habits of *Drosera rotundifolia;* and we see how little has been made out in comparison with what remains unexplained and unknown.

❑❑❑

12

On the Structure and Movements of Some Other Species of Drosera

Drosera anglica – Drosera intermedia – Drosera capensis – Drosera spathulata – Drosera filiformis – Drosera binata-Concluding remarks.

We examined six other species of Drosera, some of them inhabitants of distant countries, chiefly for the sake of ascertaining whether they caught insects. This seemed the more necessary as the leaves of some of the species differ to an extraordinary degree in shape from the rounded ones of *Drosera rotundifolia*. In functional powers, however, they differ very little.

Drosera anglica (Hudson)

The leaves of this species are much elongated, and gradually widen from the footstalk to the bluntly pointed apex. They stand almost erect, and their blades sometimes exceed 1 inch in length, whilst their breadth is only the 1/5 of an inch. The glands of all the tentacles have the same structure, so that the extreme marginal ones do not differ from the other, as in the case of *Drosera rotundifolia*. When they are irritated by being roughly touched, or by the pressure of minute inorganic particles, or by contact with animal matter, or by the absorption of carbonate of ammonia, the tentacles become inflected; the basal portion being the chief seat of movement. Cutting or pricking the blade of the leaf did not excite any movement. They frequently capture insects, and the glands of the inflected tentacles pour forth much acid secretion. Bits of roast meat were placed on some glands, and the tentacles began to move in 1 m. or 1 m. 30 s.; and in 1 hr. 10 m. reached the centre. Two bits of boiled cork, one of boiled thread, and two of coal-cinders taken from the fire, were placed, by the aid of an instrument which had been immersed in boiling water, on five glands; these superfluous precautions having been taken on account of M. Ziegler's statements. One of the particles of cinder caused some inflection in 8 hrs. 45 m., as did after 23 hrs. the other particle of cinder, the bit of thread, and both bits of cork. Three glands were touched half a dozen times with a needle; one of the tentacles became well inflected in 17 m., and re-expanded after 24 hrs.; the two others never moved. The homogeneous fluid within the cells of the tentacles undergoes aggregation after these have become inflected; especially if given a solution of carbonate of ammonia; and we observed the usual movements in the masses of protoplasm. In one case, aggregation ensued in 1 hr. 10m. after a tentacle had carried a bit of meat to the centre. From these facts it is clear that the tentacles of Drosera anglica behave like those of *Drosera rotundifolia.*

If an insect is placed on the central glands, or has been naturally caught there, the apex of the leaf curls inwards. For instance, dead flies were placed on three leaves near their bases, and after 24 hrs. the previously straight apices were curled completely over, so as to embrace and conceal the flies; they had therefore moved through an angle of 180°. After three days the apex of one leaf, together with the tentacles, began to re-expand.

Drosera intermedia (Hayne)

This species is quite as common in some parts of England as *Drosera rotundifolia.* It differs from *Drosera anglica,* as far as the leaves are concerned, only in their smaller size, and in their tips being generally a little reflexed. They capture a large number of insects. The tentacles are excited into movement by all the causes above specified; and aggregation ensues, with movement of the protoplasmic masses. We have seen, through a lens, a tentacle beginning to bend in less than a minute after a particle raw meat had been placed on the gland. The apex of the leaf curls over an exciting object as in the case of *Drosera anglica.* Acid secretion is copiously poured over captured insects. A leaf which has embraced a fly with all its tentacles re-expanded after nearly three days.

Drosera capensis

This species is a native of the Cape of Good Hope. The leaves are elongated, slightly concave along the middle and taper towards the apex, which is bluntly pointed and reflexed. They rise from an almost woody axis, and their greatest peculiarity consists in their fallacious green footstalks, which are almost as broad and even longer than the gland-bearing blade. This species, therefore, probably draws more nourishment from the air, and less from captured insects, than the other species of the genus. Nevertheless, the tentacles are crowded together on the disc, and are extremely numerous; those on the margins being much longer than the central ones. All the glands have the same form; their secretion is extremely viscid and acid.

The specimen which we examined has only just recovered from a weak state of health. This may account for the tentacles moving very slowly when particles of meat were placed on the glands, and perhaps for my never succeeding in causing any movement by repeatedly touching them with a needle. But with all the species of the genus this latter stimulus is the least effective of any particles of glass, cork, and coal-cinders, were placed on the glands of six tentacles; and one alone moved after an interval of 2 hrs. 30 m. Nevertheless, two glands were extremely sensitive to very small doses of the nitrate of ammonia, namely to about 1/20 of a minim of a solution (one part to 5250 of water), containing only 1/115200 of a grain (.000562 mg.) of the salt. Fragments of flies were placed on two leaves near their tip, which became incurved in 15 hrs. A fly was also placed in the middle of the leaf; in a few hours the tentacles on each side embraced it, and in 8 hrs. the whole leaf directly beneath fly was a little bent transversely. By the next morning, after 23 hrs. the leaf was curled so completely over that the apex rested on the upper end of the footstalk. In no case did the sides of the leaves become inflected. A crushed fly was placed on the foliaceous footstalk, but produced no effect.

Drosera spathulata

We made only a few observations on this Australian species, which has long, narrow leaves, gradually widening towards their tips. The glands of the extreme marginal tentacles are elongated and differ from the other, as in the case of *Drosera rotundifolia*. A fly was placed on a leaf, and in 18 hrs. it was embraced by the adjoining tentacles. Gum-water dropped on several leaves produced no effect. A fragment of a leaf was immersed in a few drop of a solution of one part of carbonate of ammonia to 146 of water; all the glands were instantly blackened; the process of aggregation could be seen travelling rapidly down the cells of the tentacles; and the granules of protoplasm soon united into spheres and variously shaped masses, which displayed the usual movements. Half a minim of a solution of one part of nitrate of ammonia to 146 of water was next placed on the centre of a leaf; after 6 hrs. some marginal tentacles on both sides were inflected, and after 9 hrs. they met in the centre. The lateral edges of the leaf also became incurved, so that it formed a half-cylinder; but the apex of the leaf in none of my few trials was inflected. The above dose of the nitrate (*viz.*, $\frac{1}{320}$ of a grain or .202 mg.) was too powerful, for in the course of 23 hrs. the leaf died.

Drosera filiformis

This North American species grows in such abundance in part of New Jersey as almost to cover the ground. It catches, according to Mrs. Treat, an extraordinary number of small and large insects—even great flies of the genus Asilus, moths, and butter-flies. The specimen which we examined had thread-like leaves, from 6 to 12 inches in length, with the upper surface convex and the lower flat and slightly channelled. The whole convex surface, down to the roots—for there is no distinct footstalk—is covered with short gland-bearing tentacles, those on the margins being the longest and reflexed. Bits of meat placed on the glands of some tentacles caused them to be slightly inflected in 20 m.; but the plant was not in a vigorous state. After 6 hrs. they moved through an angle of 90°, and in 24 hrs. reached the centre. The surrounding tentacles by this time began to curve inwards. Ultimately a large drop of extremely viscid, slightly acid secretion was poured over the meat from the united glands. Several other glands were touched with a little saliva, and the tentacles became incurved in under 1 hr., and re-expand after 18 hrs. Particles of glass, cork, cinders, thread, and gold-leaf, were placed on numerous glands on two leaves; in about 1 hr. four tentacles became curved, and four others after an additional interval of 2 hrs. 30 m. We never once succeeded in causing any movement by repeatedly touching the glands with a needle; and Mrs. Treat made similar trials for us with no success. Small flies were placed on several leaves near their tips, but the thread-like blade became only on occasion very slightly bent, directly beneath the insect. Perhaps this indicates that the blades of vigorous plants would bend over captured insects, and Dr. Candy informed that this is the case; but the movement cannot be strongly pronounced, as it was not observed by Mrs. Treat.

Drosera binata (or *dichotoma*)

This almost gigantic Australian species differs in some interesting points from those previously described. In this specimen the rush-like footstalks of the leaves were 20 inches in length. The blade bifurcates at its junction with the footstalk, and twice or thrice afterwards, curling about in an irregular manner. It is narrow, being only $\frac{3}{20}$ of an inch in breadth. One blade was 7½ inches long, so that the entire leaf, including the footstalk, was above 27 inches in length. Both surfaces are slightly hollowed out. The upper surface is covered with tentacles arranged in alternate rows; those in the middle being short and crowded together, those towards the margins longer, even twice or thrice as long as the blade is broad. The glands of the exterior tentacles are of a much darker red then those of the central ones. The pedicels of all are green. The apex of the blade is attenuated, and bears very long tentacles. Mr. Copland informed that the leaves of a plant which he kept for some years were generally covered with captured insects before they withered.

The leaves do not differ in essential points of structure or of function from those of the previously described species. Bits of meat or a little saliva placed on the glands of the exterior tentacles caused well-marked movement in 3 m., and particles of glass acted in 4 m. The tentacles with the latter particles re-expanded after 22 hrs. A piece of leaf immersed in a few drops of a solution of one part of carbonate of ammonia to 437 of water had all the glands blackened and all the tentacles inflected in 5 m. A bit of raw meat, placed on several glands in the medial furrow, was well clasped in 2 hrs. 10 m. by the marginal tentacles on both sides. Bits of roast meat and small flies did not act quite so quickly; and albumen and fibrin still less quickly. One of the bits of meat excited so much secretion (which is always acid) that it flowed some way down the medial furrow, causing the inflection of the tentacles on both sides as far as it extended. Particles of glass placed on the glands in the medial furrow did not stimulate them sufficiently for any motor impulse to be sent to the outer tentacles. In no case was the blade of the leaf, even the attenuated apex, at all inflected.

On both the upper and lower surface of the blade there are numerous minute, almost sessile glands, consisting of four, eight, or twelve cells. On the lower surface they are pale purple, on the upper, greenish. Nearly similar organs occur on the foot-stalks, but they are smaller and often in a shrivelled condition. The minute glands on the blade can absorb rapidly: thus, a piece of leaf was immersed in a solution of one part of carbonate of ammonia to 218 of water (2 gr. to 1 oz.) and in 5 m. they were all so much darkened as to be almost black, with their contents aggregated. They do not, as far as we could observe, secrete spontaneously; but in between 2 and 3 hrs. after a leaf had been rubbed with a bit of raw meat moistened with saliva, they seemed to be secreting freely; and this conclusion was afterwards supported by other appearances. They are, therefore, homologous with the sessile glands hereafter to be described on the leaves of Dionaea and Drosophyllum. In this latter genus they are associated, as in the present case, with glands which secrete spontaneously, that is, without being excited.

Drosera binata presents another and more remarkable peculiarity, namely, the presence of a few tentacles on the backs of the leaves, near their margins. These are perfect in structure; spiral vessels run up their pedicels; their glands are surrounded by drops of viscid secretion, and they have the power of absorbing. This latter fact was shown by the glands immediately becoming black, and the protoplasm aggregated, when a leaf was placed in a little solution of one part of carbonate of ammonia to 437 of water. These dorsal tentacles are short, not being nearly so long as the marginal ones on the upper surface; some of them are so short as almost to graduate into the minute sessile glands. Their presence, number and size, vary on different leaves, and they are arranged rather irregularly. On the back of one leaf we counted as many as twenty-one along one side.

These dorsal tentacles differ in one important respect from these on the upper surface, namely, in not possessing any power of movement, in whatever manner they may be stimulated. Thus, portions of four leaves were placed at different times in solution of carbonate of ammonia(one part to 437 or 218 of water), and all the tentacles on the upper surface soon became closely inflected; but the dorsal ones did not move, though the leaves were left in the solution for many hours, and though their glands from their blackened colour had obviously absorbed some of the salt. Rather young leaves should be selected for such trials, for the dorsal tentacles, as they grow old and begin to wither often spontaneously incline towards the middle of the leaf. If these tentacles had possessed the power of movement, they would not have been thus rendered more serviceable to the plant; for they are not long enough to bend round the margin of the leaf so as to reach an insect caught on the upper surface. Nor would it have been of any use if these tentacles could have moved towards the middle of the lower surface, for there are no viscid glands there by which insects can be caught. Although they have no power of movement, they are probably of some use by absorbing animal matter from any minute insect which may be caught by them, and by absorbing ammonia from the rain-water. But their varying presence and size, and their irregular position, indicate that they are not of much service, and that they are tending towards absorption. In a future chapter we shall see that Drosophyllum, with its elongated leaves, probably represents the condition of an early progenitor of the genus Drosera; and none of the tentacles of Drosophyllum, neither those on the upper nor lower surface of the leaves, are capable of movement when excited, though they capture numerous insects, which serve as nutriment. Therefore it seems that *Drosera binata* has retained remnants of certain ancestral characters – namely, a few motionless tentacles on the backs of the leaves, and fairly well developed sessile glands – which have been lost by most or all the other species of the genus.

Concluding Remarks

From what we have now seen, there can be little doubt that most or probably all the species of Drosera are abapted for catching insects by nearly the same means. Besides the two Australian species above described, it is said that two other species from this country, namely, *Drosera pallida* and *Drosera sulplurea,* "close their leaves upon insects with great rapidity: and the same phenomenon is manifested by an Indian species, *D. lunata,* and by several of those of the cape of Good Hope, especially by *D. trinervis.*" Another Australian species, *Drosera heterophylla* (made by Lindley into a distinct genus, Sondera) is remarkable from its peculiarly shaped leaves. The leaves form minute flattened cups, with the footstalk attached not to one margin, but to the bottom. The inner surface and the edges of the cups are studded with tentacles, which include fibro-vascular bundles, rather different from those seen by me in any other species: for some of the vessels are barred and punctured, instead of being spiral. The glands secrete copiously, judging from the quantity of dried secretion adhering to them.

❑❑❑

13

Dionaea Muscipula

Structure of the leaves – Sensitiveness of the filaments – Rapid movement of the lobes caused by irritation of the filaments – Glands, their power of secretion – Slow movement caused by the absorption of animal matter – Digestive power of the secretion – Action of chloroform, ether, and hydrocyanic acid – The manner in which insects are captured – Use of the marginal spikes – Kinds of insects captured – The transmission of the motor impulse and mechanism of the movements – Re-expansion of the lobes.

Structure of the Leaves

This plant commonly called Venus' fly-trap, from the rapidity and force of its movements, is one of the most wonderful in the world. It is a member of the small family of the Droseraceae, and is found only in the eastern part of North Carolina, growing in damp situations. The roots are small: those of a moderately fine plant which we examined consisted of two branches about 1 inch in length, springing from a bulbous enlargement. They probably serve, as in the case of Drosera, solely for the absorption of water; for a gardener, who has been very successful in the cultivation of this plant; grows it, like an epiphytic orchid, in well-drained damp moss without any soil. The form of the bilobed leaf, with its foliaccous footstalk, is shown in the accompanying drawing (fig. 12). The two lobes stand at rather less than a right angle to each

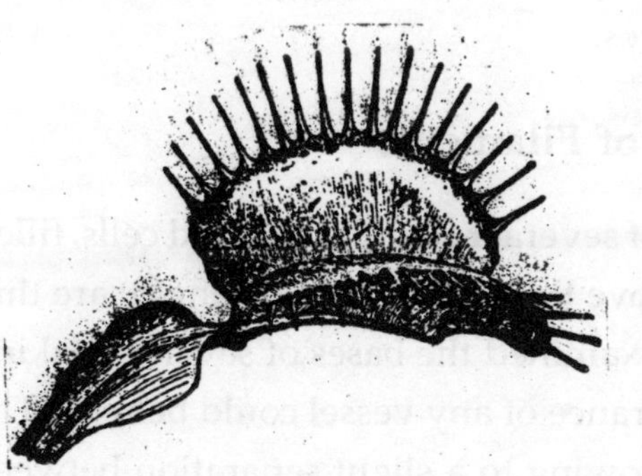

Fig. 12

(Dionaea muscipula)

Leaf viewed laterally in its expanded state.

other. Three minute pointed processes of filaments, placed triangularly, project from the upper surfaces of both; but we have seen two leaves with four filaments on each side, and another with only two. These filaments are remarkable from their extreme sensitiveness to a touch, as shown not by their own movement, but their own movement, but by that of the lobes. The margins of the leaf are prolonged into sharp rigid projections which we will call spikes, into each of which a bundle of spiral vessels enters. The spikes stand in such a position that, when the lobes close, they interlock like the teeth of a rat-trap. The midrib of the leaf, on the lower side, is strongly developed and prominent.

The upper surface of the leaf is thickly covered, excepting towards the margins, with minute glands of a reddish or purplish colour, the rest of the leaf being green. There are no glands on the spikes, or on the fallacious footstalk. The glands are formed of from twenty polygonal cells, filled with purple fluid. Their upper surface is convex. They stand on very short pedicels, into which spiral vessels do not enter, in which respect they differ from the tentacles of Drosera. They secrete, but only when excited by the absorption of certain matter; and they have the power of absorption. Minute projections, formed of eight divergent arms of a reddish-brown or orange colour, and appearing under the microscope like elegant little flowers, are scattered in considerable numbers over the footstalk, the backs of the leaves, and the spikes, with a few on the upper surface of the lobes. These octofid projections are no doubt homologous with the papillae on the leaves of *Drosera rotundifolia*. There are also a few very minute, simple, pointed hairs, about $\frac{7}{12000}$ of an inch (.0148 mm.) in length on the backs of the leaves.

Sensitiveness of Filaments

The sensitive filaments are formed of several rows of elongated cells, filled with purplish fluid. They are a little above the $\frac{1}{20}$ of an inch in length; are thin and delicate, and taper to a point. We examined the bases of several, making sections of them, but no trace of the entrance of any vessel could be seen. The apex is sometimes bifid or even trifid, owing to a slight separation between the terminal pointed cells. Towards the base there is constriction, formed of broader cells, beneath which there is an articulation, supported on an enlarge base, consisting of differently shaped polygonal cells. As the filaments project at right angles to the surface of the leaf, they would have been liable to be broken whenever the lobes closed together, had it not been for the articulation which allows them to bend flat down.

These filaments from their tips to their bases are exquisitely sensitive to a momentary touch. It is scarcely possible to touch them ever so lightly or quickly with any hard object without causing the lobes to close. A piece of very delicate human hair, 2½ inches in length, held dangling over a filament, and swayed to and fro so as to touch it, did not excite any movement. But when a rather thick cotton thread of the same length was similarly swayed, the lobes closed. Pinches of fine wheaten flour, dropped from a height, produced no effect. The above-mentioned hair was then fixed into a handle, and cut off so that 1 inch projected; this length being sufficiently rigid to support itself in a nearly horizontal line. The extremity was then brought by a slow movement laterally into contact with the tip of a filament, and the leaf instantly closed. On another occasion two or three touches of the same kind were necessary before any movement ensued. When we consider how flexible a fine hair is, we may form some idea how slight must be the touch given by the extremity of a piece, 1 inch in length, moved slowly.

Although these filaments are so sensitive to a momentary and delicate touch, they are far less sensitive than the glands of Drosera to prolonged pressure. Several times we succeeded in placing on the tip of a filament, by the aid of a needle moved with extreme slowness, bits of rather thick human hair, and these did not excite movement, although they were more than ten times as long as those which caused the tentacles of Drosera to bend; and although in this latter case they were largely supported by the dense secretion. On the other hand, the glands of Drosera may be struck with a needle or any hard object, once, twice, or even thrice, with considerable force, and no movement ensues. This singular difference in the nature of the sensitiveness of the filaments of Dionaea and of the glands of Drosera evidently stands in relation to the habits of the two plants. If a minute insect alights with its delicate feet on the glands of Drosera, it is caught by the viscid secretion, and the slight, though prolonged pressure, gives notice of the presence of prey, which is secured by the slow bending of the tentacles. On the other hand, the sensitive filaments of Dionaea are not viscid, and the capture of insects can be assured only be their sensitiveness to a momentary touch, followed by the rapid closure of the lobes.

As just stated, the filaments are not glandular, and do not secrete. Nor have they the power of absorption, as may be inferred from drops of a solution of carbonate of ammonia (one part to 146 of water), placed on two filaments, not producing any effect on the contents of their cells, nor causing the lobes to close. When, however, a small portion of a leaf with an attached filament was cut off and immersed in the same solution, the fluid within the basal cells became almost instantly aggregated into purplish or colourless, irregularly shaped masses of matter. The process of aggregation gradually travelled up the filaments from cell to cell in their extremities, that is in a reverse course to what occurs in the tentacles of Drosera when their glands have been excited. Several other filaments were cut off close to their bases, and left for 1 hr. 30 m. in a weaker solution of one part of the carbonate to 218 of water, as this caused aggregation in all the cells, commencing as before at the bases of the filaments.

Long immersion of the filaments in distilled water likewise causes aggregation. Nor is it rare to find the contents of a few of the terminal cells in a spontaneously aggregated condition. The aggregated masses undergo incessant slow changes of form, uniting and again separating; and some of them apparently revolve round their own axes. A current of colourless granular protoplasm could also be seen travelling round the walls of the cells. This current ceases to be visible as soon as the contents are well aggregated; but it probably still continues, though no longer visible, owing to all the granules in the flowing layer having become united with the central masses. In all these respects the filaments of Dionaea behave exactly like the tentacles of Drosera.

Notwithstanding this similarity there is one remarkable difference. The tentacles of Drosera, after their glands have been repeatedly touched, or a particle of any kind has been placed on them, become inflected and strongly aggregated. No such effect is produced by touching the filaments of Dionaea; we compared, after an hour or two, some which had been touched and some which had not, and others after twenty-five hours, and there was no difference in the contents of the cells. The leaves were kept open all the time by clips; so that the filaments were not pressed against the opposite lobe.

Drops of water, or a thin broken stream, falling from a height on the filaments, did not cause the blades to close; though these filaments were afterwards proved to be highly sensitive. No doubt, as in the case of Drosera, the plant is indifferent to the heaviest shower of rain. Drops of a solution of half an ounce of sugar to a fluid ounce of water were repeatedly allowed to fall from a height on the filaments, but produced no effect, unless they adhered to them. Again, we blew many times through a fine pointed tube with my utmost force against the filaments without any effect; such blowing being received with as much indifference as no doubt is a heavy gale of wind. We thus see that the sensitiveness of the filaments is of a specialised nature, being related to a momentary touch rather than to prolonged pressure; and the touch must not be from fluids, such as air or water, but from some solid object.

Although drops of water and of a moderately strong solution of sugar, falling on the filaments, does not excite them, yet the immersion of a leaf in pure water sometime caused the lobes to close. One leaf was left immersed for 1 hr. 10 m. and three other leaves for some minutes, in water at temperatures varying between 59° and 65° F (15° to 18.3° C) without any effect. One, however, of these four leaves, on being gently withdrawn from the water, closed rather quickly. The three closed leaves were proved to be in good condition, as they closed when their filaments were touched. Nevertheless two fresh leaves on being dipped into water at 75° and 62½° F (23.8° and 16.9° C) instantly closed. These were then place with their footstalk in water, and after 23 hrs. partially re-expanded; on touching their filaments one of them closed. This latter leaf after an additional 24 hrs. again re-expanded, and now, on the filaments of both leaves being touched, both closed. We thus see that a short immersion in water does not at all injure the leaves, but sometimes excites the lobes to close. The movement in the above cases was evidently not caused by the temperature of the water, It has been shown that long immersion causes the purple fluid within the cells of the sensitive filaments to become aggregated; and the tentacles of Drosera are acted on in the same manner by long immersion, often being somewhat inflected. In both cases the result is probably due to a slight degree of exosmose.

We are confirmed in this belief by the effects of immersing a leaf of Dionaea in a moderately strong solution of sugar; the leaf having been previously left for 1 hr. 10 m. in water without any effect; for now the lobes closed rather quickly. The tips of the marginal spikes crossing in 2 m. 30 s., and the leaf being completely shut in 3 m. Three leaves were then immersed in a solution of half an ounce of sugar to a fluid ounce of water, and all three leaves closed quickly. As we were doubtful whether this was due to the cells on the upper surface of the lobes, or to the sensitive filaments, being acted on by exosmose, one leaf was first tried by pouring a little of the same solution in the furrow between the lobes over the midrib, which is the chief seat of movement. it was left there for some time, but no movement ensued. The whole upper surface of leaf was then painted (except close round the bases of the sensitive filaments, which we could not do without risk of touching them) with the same solution, but no effect was produced. So that cells on the upper surface are not thus affected. But when, after many trials, we succeeded in getting a drop of the solution into cling to one of the filaments, the leaf quickly closed. Hence we may conclude that the solution causes fluid to pass out of the delicate cell of the filaments by exosmose; and that this sets up some molecular change in their contents, analogous to that which must be produced by a touch.

The immersion of leaves in a solution of sugar affects them for a much longer time then does than does an immersion in water, or a touch on the filaments; for in these latter cases the lobes begin to re-expand in less then a day. On the other hand, of the three leaves which were immersed for a short time in the solution, and were then washed by means of a syringe inserted between the lobes, one re-expanded after two days; a second after seven days; and the third after nine days; the leaf which closed, owing to a drop of the solution having adhered to one of the filaments, opened after two days.

We were surprised to find on two occasions that the heat from the rays of the sun, concentrated by a lens on the bases of several filaments, so that they were scorched and discoloured, did not cause any movement; though the leaves were active, as they closed, though rather slowly, when a filament on the opposite side was touched. On a third trial, a fresh leaf closed after a time, though very slowly; the rate not being increased by one of the filaments, which had not been injured, being touched. After a day these three leaves opened, and were fairly sensitive when the uninjured filaments were touched. The sudden immersion of a leaf into boiling water does not cause it to close. Judging from the analogy of Drosera, the heat in these several cases was too great and too suddenly applied. The surface of the blade is very slightly sensitive; it may be freely and roughly handled, without any movement being caused. A leaf was scratched rather hard with a needle, but did not close; but when the triangular space between the three filaments on another leaf was similarly scratched, the lobes closed. They always closed when the blade or midrib was deeply pricked or cut. Inorganic bodies, even of large size, such as bits of stone glass, &c.- or organic bodies not containing soluble nitrogenous matter, such as bits of wood, cork, moss, or bodies containing soluble nitrogenous matter, if perfectly dry, such as bits of meat, albumen, gelatine, &c., may be long left (and many were tried) on the lobes, and no movement is excited. The result, however, is widely different, as we shall presently see, if nitrogenous organic bodies which are at all damp, are left on the lobes; for these then close by a slow and gradual movement, very different from that caused by touching one of the sensitive filaments. The footstalk is not in the least sensitive; a pin may be driven through it, or it may be cut off, and no movement follows.

The Rapid Movement of Lobes

The upper surface of the lobes, as already, is thickly covered with small purplish, almost sessile glands. These have the power both of secretion and absorption; but unlike those of Drosera, they do not secrete until excited by the absorption of nitrogenous matter. No other excitement, as far as we have seen, produces this effect. Objects, such as bits of wood, cork, moss, paper, stone, or glass, may be left for a length of time on the surface of a leaf, and it remains quite dry. Nor does it make any difference if the lobes close over such objects. For instance, some little balls of blotting-paper were placed on a leaf, and a filament was touched; and when after 24 hrs. the lobes began to re-open, the balls were removed by the aid of thin pincers, and were found perfectly dry. On the other hand, if a bit of damp meat or a crushed fly is placed on the surface of an expanded leaf, the glands after a time secrete freely. In one such case there was a little secretion directly beneath the meat in 4 hrs; and after an additional 3 hrs. there was a considerable quantity both under and close round it. In another case, after 3 hrs. 40 m., the bit of meat was quite wet. But none of the glands secreted, excepting those which actually touched the meat or the secretion containing dissolved animal matter.

If, however, the lobes are made to close over a bit of meat or an insect, the result is different, for the glands over the whole surface of the leaf now secrete copiously. As in this case the glands on both sides are pressed against the meat or insect, the secretion from the first is twice as great as when a bit of meat is laid on the surface of one lobe; and as the two lobes come into almost close contact, the secretion, containing dissolved animal matter, spreads by capillary attraction, causing fresh glands on both sides to begin secreting in a continually widening circle. The secretion is almost colourless, slightly mucilaginous, and, judging by the manner in which it coloured litmus paper, more strongly acid than that of Drosera. It is so copious that on one occasion, when a leaf was cut open, on which a small cube of albumen had been placed 45 hrs. before, drops rolled off the leaf. On another occasion, in which a leaf with an enclosed bit of roast meat spontaneously opened after eight days, there was so much secretion in the furrow over the midrib that it trickled down. A large crushed fly (Tipula) was placed on a leaf from which a small portion at the base of one lobe had previously been cut away, so that an opening was left; and through this, the secretion continued to run down the footstalk during nine days,—that is, for as long a time as it was observed. By forcing up one of the lobes, we were able to see some distance between them, and all the glands within sight were secreting freely.

We have seen that inorganic and non-nitrogenous objects placed on the leaves do not excite any movement; but nitrogenous bodies, if in the least degree damp, cause after several hours the lobes to close slowly. Thus bits of quite dry meat and gelatine were placed at opposite ends of the same leaf, and in the course of 24 hrs. excited neither secretion nor movement. They were then dipped in water, their surfaces dried on blotting-paper, and replaced on the same leaf, the plant being now covered with a bell-glass. After 24 hrs. the damp meat had excite some acid secretion, and the lobes at this end of the leaf were almost shut. At the other end, where the damp gelatine lay, the leaf was still quite open, nor had any secretion been excite; so that, as with Drosera, gelatine is not nearly so exciting a substance as meat. The secretion beneath the meat was tested by pushing a strip of litmus paper under it (the filaments not being touched), and this slight stimulus caused the leaf to shut. On the eleventh day it reopened; but the end where the gelatine lay expanded several hours before the opposite end with the meat.

A second bit of roast meat, which appeared dry, though it had not been purposely dried, was left for 24 hrs, on a leaf, caused neither movement nor secretion. The plant in its pot was now covered with a bell-glass, and the meat absorbed some moisture from the air; this sufficed to excite acid secretion, and by the next morning the leaf was closely shut. A third bit of meat, dried so as to be quite brittle, was placed on a leaf under a bell-glass, and this also became in 24 hrs. Slightly damp, and excite some acid secretion, but no movement.

A rather large piece of perfectly dry albumen was left at one end of leaf for 24 hrs. without any effect. It was then soaked for a few minutes in water, rolled about on blotting-paper, and replaced on the leaf; in 9 hrs. some slightly acid secretion was excited, and in 24 hrs. this end of the leaf was partially closed. The bit of albumen, which was now surrounded by much secretion, was gently removed, and although no filament was touched, the lobes closed. In this and the previous case, it appears that the absorption of animal matter by the glands renders the surface of the leaf much more sensitive to a touch than it is in its ordinary state; and this is a curious fact. Two days afterwards the end of the leaf where nothing had been placed began to open, and on the third day was much more open than the opposite end where the albumen had lain.

Lastly, large drops of a solution of one part of carbonate of ammonia is 146 of water were placed on some leaves, but no immediate movement ensued, we did not then know of the slow movement caused by animal matter, otherwise we should have observed the leaves for a longer time, and they would probably have been found closed, though the solution (judging from Drosera) was, perhaps, too strong.

From the foregoing cases it is certain that bits of meat and albumen, if at all damp, excite not only the glands to secrete, but the lobes to close. This movement is widely different from the rapid closure caused by one of the filaments being touched. We shall see its importance when we treat of the manner in which insects are captured. There is a great contrast between Drosera and Dionaea in the effects produced by mechanical irritation on the one hand, and the absorption of animal matter on the other. Particles of glass placed on the glands of the exterior tentacles of Drosera excite movement within nearly the same time, as do particles of meat, the latter being rather the most efficient; but when the glands of the disc have bits of meat given them, they transmit a motor impulse to the exterior tentacles much more quickly than do these glands when bearing inorganic particles, or when irritated by repeated touches. On the other hand, with Dionaea, touching the filaments excited incomparably quicker movement than the absorption of animal matter by the glands. Nevertheless, in certain cases, this latter stimulus is the more powerful of the two. On three occasions leaves were found which from some cause were torpid. So that their lobes closed only slightly, however much their filaments were irritated; but on inserting crushed insects between the lobes, they became in a day closely shut.

The facts just given plainly show that the glands have the power of absorption, for otherwise it is impossible that the leaves should be so differently affected by non-nitrogenous and nitrogenous bodies, and between these latter in a dry and damp condition. It is surprising how slightly damp a bit of meat of albumen need be in order to excite secretion and afterwards slow movement, and equally surprising how minute a quantity of animal matter, when absorbed, suffices to produce these two effects. It seems hardly credible, and yet it is certainly a fact, that a bit of hard-boiled white of egg, first thoroughly dried, then soaked for some minutes in water and rolled on blotting-paper, should yield in a few hours enough animal matter to the glands to cause them to secrete, and afterwards the lobes to close. That the glands have the power of absorption is likewise shown by the very different lengths of time (as we shall presently see) during which the lobes remain closed over insects and other bodies yielding soluble nitrogenous matter, and over such as do not yield any. But there is direct evidence of absorption in the condition of the glands which have remained for some time in contact with animal matter. Thus bits of meat and crushed insects were several times placed on glands, and these were compared after some hours with other glands from distant parts of the same leaf. The latter showed not a trace of aggregation, whereas those which had been in contact with the animal matter were well aggregated. Aggregation may be seen to occur very quickly if a piece of leaf is immersed in a weak solution of carbonate of ammonia. Again, small cubes of albumen and gelatine were left for eight days on a leaf, which was then cut open. The whole surface was bathed with acid secretion, and every cell in the many glands which examined had its contents aggregated in a beautiful manner into dark or pale purple or colourless globular masses of protoplasm. These underwent incessant slow changes of forms; sometimes separating from one another and then reuniting, exactly as in the cells of Drosera. Boiling water makes the contents of the gland-cells white and opaque, but not so purely white and porcelain-like as in the case of Drosera. How living insects, when naturally caught, excite the glands to secrete so quickly as they do, we know not; but we suppose that the great pressure to which they are subjected forces a little excretion from either extremity of their bodies, and we have seen that an extremely small amount of nitrogenous matter is sufficient to excite the glands.

Before passing on the subject of digestion, we may state that we endeavoured to discover, with no success, the functions of the minute octofid processes with which the leaves are studded. From facts hereafter to be given in the chapters on Aldrovanda and Utricularia, it seemed probable that they served to absorb decayed matter left by the capture insects; but their position on the backs of the leaves and on the footstalks rendered this almost impossible. Nevertheless, leaves were immersed in a solution of one part of urea to 437 of water, and after 24 hrs. the orange layer of protoplasm within the arms of these processes did not appear more aggregated then in other specimens kept in water. We then tried suspending a leaf in a bottle over an excessively putrid infusion of raw meat, to see whether they absorbed the vapour, but their contents were not affected.

Digestive Power of the Secretion

When a leaf closes over any object, it may said to form itself into a temporary stomach; and if the object yields ever so little animal matter, this serves, to use Schiff's expression, as a peptogene, and the glands on the surface pour forth their acid secretion, which acts like the gastric juice of animals. As so many experiments were tried on the digestive power of Drosera, only a few were made with Dionaea, but they were amply sufficient to prove that it digests. This plant, moreover, is not so well fitted as Drosera for observation, as the process goes on within the closed lobes. Insects, even beetles, after being subjected to the secretion for several days, are surprisingly softened, though their chitinous coats are not corroded.

Experiment 1: A cube of albumen of $\frac{1}{16}$ of an inch (2.540 mm.) was placed at one end of a leaf, and at the other end an oblong piece of gelatine, $\frac{1}{5}$ of an inch (5.08 mm.) long, and $\frac{1}{10}$ broad; the leaf was then made to close. It was cut open after 45 hrs. The albumen was hard and compressed, with its angles only little rounded; the gelatine was corroded into an oval form; and both were bathed in so much acid secretion that it dropped off the leaf. The digestive process apparently is rather slower than in Drosera, and this agrees with the length of time during which the leaves remain closed over digestible objects.

Experiment 2: A bit of albumen $\frac{1}{10}$ of an inch square, but only $\frac{1}{20}$ in thickness, and a piece of gelatine of the same size as before, were placed on leaf, which eight days afterwards was cut open. The surface was bathed with slightly adhesive, very acid secretion, and the glands were all in an aggregated condition. Not a vestige of the albumen of gelatine was left. Similarly sized pieces were placed at the same time on wet moss on the same pot, so that they were subjected to nearly similar conditions; after eight days these were brown, decayed, and matted with fibres of mould, but had not disappeared.

Experiment 3: A piece of albumen $\frac{3}{20}$ of an inch (3.81 mm.) long, and $\frac{1}{20}$ broad and thick, and a piece of gelatine of the same size as before, were placed on another leaf, which was cut open after seven days; not a vestige of either substance was left, and only a moderate amount of secretion on the surface.

Experiment 4: Pieces of albumen and gelatine, of the same size as in the last experiment, were placed on a leaf, which spontaneously opened after twelve days, and here again not a vestige of either was left, and only a little secretion at one end of the midrib.

Experiment 5: pieces of albumen and gelatine of the same size were placed on another leaf, which after twelve days was still firmly closed, but had begun to wither; it was cut open, and contained nothing except a vestige of brown matter where the albumen had lain.

Experiment 6: A cube of albumen of $\frac{1}{10}$ of an inch and a piece of gelatine of the same size as before were placed on a leaf, which opened spontaneously after thirteen days. The albumen, which was twice as thick as in the latter experiments, was too large; for the glands in contact with it were injured and were dropping off; a film also of albumen of a brown colour, matter with mould, was left. All the gelatine was absorbed, and there was only a little acid secretion left on the midrib.

Expreiment 7: A bit of half roasted meat (not measured) and a bit of gelatine were placed on the two ends of a leaf, which opened spontaneously after eleven days; a vestige of the meat was left, and the surface of the leaf was here blackened; the gelatine had all disappeared.

Experiment 8: A bit of half roasted meat(not measured) was placed on a leaf which was forcibly kept open by a clip, so that it was moistened with the secretion (very acid) only on its lower surface. Nevertheless, after only 22½ hrs. it was surprisingly softened, when compared with another bit of the same meat which had been kept damp.

Experiment 9: A cube of $\frac{1}{10}$ of an inch very compact roasted beef was placed on a leaf, which opened spontaneously after twelve days; so much feebly acid secretion was left on the leaf that it trickled off. The meat was completely disintegrated, but not all dissolved; there was no mould. The little mass was placed under the microscope; some of the fibrillae in the middle still exhibited transverse striae; others showed not a vestige of Striae; and every gradation could be traced between these two states. Globule apparently of fat and some undigested fibro-elastic tissue remained. The meat was thus in the same state as that formerly described, which was half digested by Drosera. Here, again, as in the case of albumen, the digestive process seems slower than in Drosera. At the opposite end of the same leaf, a firmly compressed pellet of bread had been placed; this was completely disintegrated owing to the digestion of the gluten, but seemed very little reduced in bulk.

Experiment 10: A cube of $\frac{1}{20}$ of an inch of cheese and another of albumen were placed at opposite ends of the same leaf. After nine days the lobes opened spontaneously a little at the end enclosing the cheese, but hardly any or none was dissolved, though it was softened and surrounded by secretion. Two days subsequently the end with the albumen also opened spontaneously (*i.e.*, eleven days after it was put on), a mere trace in the blackened and dry condition being left.

Experiment 11: The same experiment with cheese and albumen repeated no another and rather torpid leaf. The lobes at the end with the cheese, after an interval of six days, opened spontaneously a little; the cube of cheese was much softened, but not dissolved, and but little, if at all reduced in size. Twelve hours afterwards the end with the albumen opened, which now consisted of a large drop of transparent, not acid, viscid fluid.

Experiment 12: Some experiment as the two last, and here again the leaf at the end enclosing the cheese opened before the opposite end with the albumen; but no further observations were made.

Experiment 13: A globule of chemically prepared casein, about $\frac{1}{10}$ of an inch in diameter, was placed on a leaf, which spontaneously opened after eight days. The casein now consisted of a soft sticky mass, very little, if at all, reduced in size, but bathed in acid secretion.

These experiments are sufficient to show that the secretion from the glands of Dionaea dissolves albumen, gelatine, and meat if too large are not given. Globules of fat and fibro-elastic tissue are not digested. The secretion, with its dissolved matter, if not in excess, is subsequently absorbed. On the other hand, although chemically prepared casein and cheese (as in the case of Drosera) excite much acid secretion, owing to the absorption of some included albuminous matter, these substances are not digested, and are not appreciably, if at all, reduced in bulk.

Effects of the Vapours of Chloroform, Sulphuric Ether, and Hydrocyanic Acid

A plant bearing one leaf was introduced into a large bottle with a drachm (3.549 c.c.) of chloroform, the month being imperfectly closed with cotton-wool. The vapour caused in 1 m. the lobes to begin moving at an imperceptibly slow rate; but in 3 m. the spikes crossed, and the leaf was soon completely shut. The dose, however, was much too large, for in between 2 and 3 hrs. the leaf appeared as if burnt, and soon died.

Two leaves were exposed for 30 m. in a 2-oz. vessel to the vapour of 30 minims (1.774 c.c.) of sulphuric ether. One leaf closed after a time, as did the other whilst being removed from the vessel without being touched. Both leaves were greatly injured. Another leaf, exposed for 20 m. to 15 minims of ether, closed its lobes to a certain extent, and the sensitive filaments were now quite insensible. After 24 hrs. this leaf recovered its sensibility, but was still rather torpid. A leaf exposed in a large bottle for only 3 m. to ten drops was rendered insensible. After 52 m. it recovered its sensibility, and when one of the filaments was touched, the lobes closed. It began to reopen after 20 hrs. Lastly, another leaf was exposed for 4 m. to only four drops of the ether; it was rendered insensible, and did not close when its filaments were repeatedly touched, but closed when the end of the open leaf was cut off. This shows either that the internal parts had not been rendered insensible, or that an incision is a more powerful stimulus then repeated touches on the filaments. Whether the larger doses of chloroform and ether, which caused the leaves to close slowly, acted on the sensitive filaments or on the leaf itself, we do not know.

Cyanide of potassium, when left in a bottle, generates prussic or hydrocyanic acid. A leaf was exposed for 1 hr. 35 m. to the vapour thus formed;

and the glands became within this time so colourless and shrunken as to be scarcely visible. The leaf was not rendered insensible; for as soon as one of the filaments was touched it closed. It had, however, suffered, for it did not reopen until nearly two days had passed, and was not even then in the least sensitive. After an additional day it recovered its powers, and closed on being touched and subsequently re-opened. Another leaf behaved in nearly the same manner after a shorter exposuree to this vapour.

Manner in which Insects are Caught

We will now consider the action of the leaves when insects happen to touch one of the sensitive filaments. They are caught in large numbers by the plant in its native country. As soon as a filament is touched, both close with astonishing quickness; and as they stand at less than a right angle to each other, they have a good chance of catching any intruder. The angle between the blade and footstalk does not change when the lobes close. The chief seat of movement is near the midrib, but is not confined to this part; of, as the lobes come together, each curves inwards across its whole breadth; the marginal spikes, however, not becoming curved. This movement of the whole lobe was well seen in a leaf to which a large fly had been given, and from which a large portion had been cut off the end of one lobe; so that the opposite lobe, meeting with no resistance in this part, went on curing inwards much beyond the medial line. The whole of the lobe, from which portion had been cut, was afterwards removed, and the opposite lobe now curled completely over, passing through an angle of from 120° to 130°, so as to occupy a position almost at right angles to that which it would have held had the opposite lobe been present.

From the curving inwards of the two lobes, as they move towards each other, the straight marginal spikes intercross by their tips at first, and ultimately by their bases. The leaf is then completely shut and encloses a shallow cavity. If it has been made to shut merely by one of the sensitive filaments having been touched, or if it includes an object not yielding soluble nitrogenous matter, the two lobes retain their inwardly concave form until they re-expand. The re-expansion under these circumstances – that is when no organic matter is enclosed – was observed in ten cases. In all of these, the leaves re-expanded to about two-thirds of the full extent in 24 hrs. From the time of closure. Even the leaf from which a portion of one lobe had been cut off opened to a slight degree within this same time. In one case a leaf re-expanded to about two-thirds of the full extent in 7 hrs., and completely in 32 hrs; but one of its filaments had been touched merely with a hair just enough to cause the leaf to close. Of these ten leaves only a few re-expanded completely in less than two days, and two or three required even a little longer time. Before, however, they fully re-expand, they are ready to close to close instantly if their sensitive filaments are touched. How many times a leaf is capable of shutting and opening if no animal matter is left enclosed, we do not know; but one leaf was made to close four times, reopening afterwards, within six days. On the last occasion it caught a fly, and then remained closed for many days.

This power of re-opening quickly after the filaments have been accidentally touched by blades of grass, or by objects blown on the leaf by the wind, as occasionally happens in its native place, must be of some importance to the plant;for as long as a leaf remains closed, it cannot of course capture an insect.

When the filaments are irritated and a leaf is made to shut over an insect, a bit of meat, albumen, gelatine, casein, and, no doubt, any substance containing soluble nitrogenous matter, the lobes, instead of remaining concave, thus including a concavity, slowly press closely together throughout their whole breadth. As this takes place, the margins gradually become a little everted, so that the spikes, which at first intercrossed, at last project in two parallel rows. The lobes press against each other with such force that we have seen a cube of albumen much flattened, with distinct impressions of the little prominent glands; but this latter circumstance may have been partly caused by the corroding action of the secretion. So firmly do they become pressed together that, if any large in sector other object has been caught, a corresponding projection on the outside of the leaf is distinctly visible. When the two lobes are thus completely shut, they resist being opened, as by a thin wedge being driven between them, with astonishing force, and are generally ruptured rather than yield. If not ruptured, they close again, as Dr. Candy informed in a letter, "with quite a loud flap". But if the end of a leaf is held firmly between the thumb and finger, or by a clip, so that the lobes cannot begin to close, they exert, whilst in this position, very little force.

We thought at first that the gradual pressing together of the lobes was caused exclusively by captured insects crawling over the repeatedly irritating the sensitive filaments; and this view seemed the more probable when we learnt from Dr. Burdon Sanderson that whenever the filaments of a closed leaf are irritated, the normal electric-current is disturbed. Nevertheless, such irritation is by no means necessary, for a dead insect, or a bit of meat, or of albumen, all act equally well; proving that in these cases it is the absorption of animal matter which excited the lobes slowly to press close together. We have seen that the absorption of an extremely small quantity of such matter also causes a fully expanded leaf to close slowly; and this movement is clearly analogous to the slow pressing together of the concave lobes. This latter action is of high functional importance to the plant, for the glands on both sides are thus brought into contact with a captured insect, and consequently secrete. The secretion with animal matter in solution is then drawn by capillary attraction over the whole surface of the leaf, causing all the glands to secrete and allowing them to absorb the diffused animal matter. The movement, excited by the absorption of such matter, though slow, suffices for its final purpose, whilst the movement excited by one of the sensitive filaments being touched is rapid, and this is indispensable for the capturing of insects. These two movements, excited by two such widely different means, are thus both well adapted, like all the other functions of the plant, for the purposes which they subserve.

There is another wide difference in the action of leaves which enclose objects, such as bits of wood, cork, balls of paper, or which have had their filaments merely touched, and those which enclose organic bodies yielding soluble nitrogenous matter. In the former case the leaves, as we have seen, open in under 24 hrs. and are then ready, even before being fully expanded, to shut again. But if they have closed over nitrogen yielding bodies, they remain closely shut for many days; and after re-expanding are torpid, and never act again, or only after a considerable interval of time. In four instances, leaves after catching insects never re-opened, but began to wither, remaining closed—in one case for fifteen days over a fly; in a second, for twenty-days, though the fly was small; in a third for twenty-four days over a wood louse; and in a fourth, for thirty-five days over a large Tipula. In two other cases leaves remain closed for at least nine days over flies, and for how many more we do not know. It should, however, be added that in two instances in which very small insects had been naturally caught the leaf opened as quickly as if nothing has been caught; and this was due to such small insects not having been crushed or not having excreted any animal matter, so that the glands were not excited. Small angular bits of albumen and gelatine were placed at both ends of three leaves, two of which remained closed for thirteen and the other for twelve days. Two other leaves remained closed over bits of meat for eleven days, a third leaf for eight days, and a fourth (but this had been cracked and injured) for only six days. Bits of cheese, or casein, were placed at one end and albumen at the other end of three leaves; and the ends with the former opened after six, eight, and nine days, whilst the opposite ends opened a little later. None of the above bits of meat, albumen, & c., exceeded a cube of $\frac{1}{10}$ of an inch (2.54 mm.) in size, and were sometimes smaller; yet these small portions sufficed to keep the leaves closed for many days. Dr. Canby informed that leaves remain shut for a longer time over insects than over meat; and from what we have seen, we can well believe that this is the case, especially if the insects are large.

In all the above cases, and in many others in which leaves remained closed for a long but unknown period over insects naturally caught, they were more or less torpid when they re-opened. Generally they were so torpid during many succeeding days that no excitement of the filaments caused the least movement. In one instances, however, on the day after a leaf opened which had clasped a fly, it closed with extreme slowness when one of its filaments was touched; and although no object was left enclosed, it was so torpid that it did not re-open for the second time until 44 hrs. had elapsed. In a second case a leaf which had expanded after remaining closed for at least nine days over a fly, when greatly irritated, moved one alone of its two lobes, and retained this unusual position for the next two days. A third case offers the strongest exception which I have observed; leaf, after remaining clasped for an unknown time over a fly, opened, and when one of its filaments was touched, closed, though rather slowly. Dr. Canby, who observed in the United States a large number of plants which, although not in their native site, were probably more vigorous than our plants, informed that he has "several times known vigorous leaves to devour their prey several times; but ordinarily twice, or quite often, once was enough to render them unserviceable." Mrs. Treat, who cultivated many plants in New Jersey, also informed that "several leaves caught successively three insects each, but most of them were not able to digest the third fly, but dies in the attempt. Five leaves, however, digested each three flies, and closed over the fourth, but dies soon after the fourth capture. Many leaves did not digest even one large insect." It thus appears that the power of digestion in somewhat limited, and it is certain that leaves always remain clasped for many days over an insect, and do not recover their power of closing again for many subsequent days. In this respect Dionaea differs from Drosera, which catches and digest many insects after shorter intervals of time.

Use of Marginal Spikes

We are now prepared to understand the use of the marginal spikes, which form so conspicuous a feature in the appearance of the plant (fig. 12), and which at first seemed to me in my ignorance useless appendages. From the inward curvature of the lobes as they approach each other, the tips of the marginal spikes first intercross, and ultimately their bases. Until the edges of the lobes come into contact, elongated spaced between the spikes, varying from the 1/15 to the $\frac{1}{10}$ of an inch (1.693 to 2.540 mm.) in breadth, according to the size of the leaf, are left open. Thus an insect, if its body is not thicker than these measurements, can easily escape between the crosse spikes, when disturbed by the closing lobes and increasing darkness; and one of my sons actually saw a small insect thus escaping. A moderately large insect, on the other hand, if it tries to escape between the bars will surely be pushed back again into its horrid prison with closing walls, for the spikes continue to cross more and more until the edges of the lobes come into contact. A very strong insect, however, would be able to free itself, and Mrs. Treat saw this effected by a rose chafer (*Macrodactylus subspinosus*) in the United states. Now it would manifestly be a great disadvantage to the plant to waste many days in remaining clasped over a minute insect, and several additional days or weeks in afterwards recovering its sensibility; in as much as a minute insect would afford but little nutriment. It would be far better for the plant to wait for a time until a moderately large insect was capture, and to allow all the little ones to escape; and its advantage is secured by the slowly inter-crossing marginal spikes, which act like the large meshes of a fishing net, allowing the small and useless fry to escape.

To know whether this view is correct, we experimented on fourteen leaves, containing naturally captured insects. Four of these had caught rather small insects, *viz.*, three of them ants, and the fourth a rather small fly, but the other ten had all caught large insects, namely, five elaters, two chrysomelas, a curculio, a thick and broad spider, and a scolopendra. Out of these ten insects, no less than eight were beetles, and out of the whole fourteen there was only one, *viz.*, a dipterous insect, which could readily take flight. Drosera, on the other hand, lives chiefly on insects which are good flyers, especially Diptera, caught by the aid of its viscid secretion. But what most concerns us is the size of the ten larger insects. Their average length from head to tail was .256 of an inch, the lobes of the leaves being on an average .53 of an inch in length, so that the insects were very nearly half as long as the leaves within which they were enclosed. Only a few of these leaves, therefore, had wasted their powers by capturing small prey, though it is probable that many small insects had crawled over them and been caught, but then escaped through the bars.

The Transmission of the Motor Impulse, and Means of Movement

It is sufficient to touch any one of the six filaments to cause both lobes to close, these becoming at the same time incurved throughout their whole breadth. The stimulus must therefore radiate in all directions from any one filament. It must also be transmitted with must rapidity across the leaf, for in all ordinary cases both lobes close simultaneously, as far as the eye can judge. Most physiologists believe that in irritable plants the excitements transmitted a long, or in close connection with, the fibro-vascular bundles. In Dionaea, the course of these vessels (composed of spiral and ordinary vascular tissue) seems at first sight to favour this belief; for they run up the midrib in a great bundle, sending off small bundles almost at right angles on each side. These bifurcate occasionally as they extend towards the margin, and close to the margins all branches from adjoining vessels unite and enter the marginal spikes. At some of these points of union the vessels from curious loops, like those described under Drosera. A continuous zigzag line of vessels thus runs round the whole circumference of the leaf, and in the midrib all the vessels are in close contact; so that all parts of the leaf seem to be brought into some degree of communication. Nevertheless, the presence of vessels is not necessary for the transmission of the motor impulse, for it is transmitted from the tips of the sensitive filaments (these being about the $\frac{1}{20}$ of an inch in length), into which no vessels enter; and these could not have been overlooked, as I made thin vertical sections of the leaf at the bases of the filaments.

On several occasions, slits about the $\frac{1}{10}$ of an inch in length were made with a lancet, close to the bases of the filaments, parallel to the midrib, and, therefore, directly across the course of the vessels. These were made sometimes on the inner and sometimes on the outer side of the filaments; and after several days, when the leaves had reopened, these filaments were touched roughly (for they were always rendered in some degree torpid by the operation), and the lobes then closed in the ordinary manner, though slowly, and sometimes not until after a considerable interval of time. These cases show that motor impulse is not transmitted along the vessels, and they further show that there is no necessity for a direct line of communication from the filament which is touched towards the midrib and opposite lobe, or towards the outer parts of the same lobe.

Two slits near each other, both parallel to the midrib, were next made in the same manner as before, one on each side of the base of a filament, on five distinct leaves, so that a little slip bearing a filament was connected with the rest of the leaf only at its two ends. These slips were nearly of the same size; one was carefully measured: it was 12 of an inch (3.048 mm.) in length, and .08 of an inch (2.032 mm.) in breadth; and in the middle stood the filament. Only one of these slips withered and perished. After the leaf had recovered from the operation, though the slits were still open, the filaments thus circumstanced were roughly touched, and both lobes, or one alone, slowly closed. In two instances touching the filament produced no effect; but when the point of a needle was driven into the slip at the base of the filament, the lobes slowly closed. Now in these cases the impulse must have proceeded along the slip in a line parallel to the midrib, and then have radiated fourth, either from both ends of from one end alone of the slip, over the whole surface of the two lobes.

Again, two parallel slits, like the former ones, were made, one on each of the base of a filament, at right angles to the midrib. After the leaves (two in number) had recovered, the filaments were roughly touched, and the lobes slowly closed; and here the impulse must have travelled for a short distance in a line at right angles to the midrib, and then have radiated fourth on all sides over both lobes. These several cases prove that the motor impulse travels, in all directions through the cellular tissue, independently of the course of the vessels.

With Drosera we have seen that the motor impulse is transmitted in like manner in all directions through the cellular tissue; but that its rate is largely governed by the length of the cells and the direction of their longer axes. Thin sections of a leaf of Dionaea were made by my son, and the cells, both these of the central and of the more superficial layers, were found much elongated, with their longer axes directed towards the midrib; and it is in this direction that the motor impulse must be sent with great rapidity from one lobe to the other, as both close simultaneously. The central parenchymatous cells are larger, more loosely attached together, and have more delicate walls than the more superficial cells. A thick mass of cellular tissue forms the upper surface of the midrib over the great central bundle of vessels.

When the filaments were roughly touched, at the bases of which slits had been made, either on both sides or on one side, parallel to the midrib or at right angles to it, the two lobes, or only one, moved. In one of these cases, the lobe on the side which bore the filament that was touched moved, but in three other cases the opposite lobe alone moved: so that an injury which was sufficient to prevent a lobe moving did not prevent the transmission from it of a stimulus which excited the opposite lobe to move. We thus also learn that, although normally both lobes move together, each has the power of independent movement. A case, indeed, has already been given of a torpid leaf that had lately re-opened after catching an insect, of which one lobe alone moved when irritated. Moreover, one end of the same lobe can close and re-expand, independently of the other end, as was seen in some of the forgoing experiments.

When the lobes, which are rather thick, close, no trace of wrinkling can be seen on any part of their upper surfaces. It appears therefore that the cells must contract. The chief seat of the movement is evidently in the thickly mass of cells which overlies the central bundle of vessels in the midrib. To ascertain whether this part contracts, a leaf was fastened on the stage of the microscope in such a manner that the two lobes could not become quite shut, and having made two minute black dots on the midrib, in a transverse line and a little towards one side, they were found by the micrometer to be $\frac{17}{1000}$ of an inch apart, so that a small portion of the upper surface of the midrib had contracted in a transverse line $\frac{2}{1000}$ of an inch (.0508 mm.).

We know that the lobes, whilst closing, become slightly incurved throughout their whole breadth. This movement appears to be due to the contraction of the superficial layers of cells over the whole upper surface. In order to observe their contraction, a narrow strip was cut out of one lobe at right angles to the midrib, so that the surface of the opposite lobe could be seen in this part when the leaf was shut. After the leaf had recovered from the operation and had re-expanded, three minute black dots were made on the surface opposite to the slit or window, in a line at right angles to the midrib. The distance between the dots was found to be $\frac{40}{1000}$ of an inch, so that the two extreme dots were $\frac{80}{1000}$ of an inch apart. One of the filaments was now touched and the leaf closed. On again measuring the distances between the dots, the two next to the midrib were nearer together by $\frac{1 \text{ to } 2}{1000}$ of an inch, and the two further dots by $\frac{3 \text{ to } 4}{1000}$ of an inch, than they were before; so that the two extreme dots now stood about $\frac{5}{1000}$ of an inch (.127 mm.) nearer together than before. If we suppose the whole upper surface of the lobe, which was $\frac{400}{1000}$ of an inch in breadth, to have contracted in the same proportion, the total contraction will have amounted to about $\frac{25}{1000}$ or $\frac{1}{40}$ of an inch (.635 mm.); but whether this is sufficient to account for the slight inward curvature of the whole lobe, we are unable to say.

Finally, with respect to the movement of the leaves the wonderful discovery made by Dr. Burdon Sanderson is now universally known; namely, that there exists a normal electrical current in the blade and footstalk; and that when the leaves are irritated, the current is disturbed in the same manner as takes place during the contraction of the muscle of an animal.

The Re-expansion of the Leaves

This is effected at an in-sensibly slow rate, whether or not any object is enclosed. One lobe can re-expand by itself, as occurred with the torpid leaf of which one lobe alone had closed. We have also seen in the experiments with cheese and albumen that the two ends of the same lobe can re-expand to a certain extent independently of each other. But in all ordinary cases both lobes open at the same time. The re-expansion is not determined by the sensitive filaments; all three filaments on one lobe were cut off close to their bases; and the three leaves thus treated re-expanded—one to a partial extent in 24 hrs.—a second to the same extent in 48 hrs.—and the third, which had been previously injured, not until the sixth day. These leaves after their re-expansion closed quickly when the filaments on the other lobe were irritated. These were then cut off one of the leaves, so that none were left. This mutilated leaf, notwithstanding the loss of all its filaments, re-expanded in two days in the usual manner. When the filaments have been excited by immersion in a solution of sugar, the lobes do not expand so soon as when the filaments have been merely touched; and this is due to their having been strongly affected through exosmose, so that they continue for some time to transmit a motor impulse to the upper surface of the leaf.

The following facts make us believe that the several layers of cells forming the lower surface of the leaf are always in a state of tension; and that it is owing to this mechanical state, aided probably by fresh fluid being attracted into the cells, that the lobes begin to separate or expand as soon as the contraction of the upper surface diminishes. A leaf was cut off and suddenly plunged perpendicularly into boiling water; we expected that the lobes would have closed, but instead of doing so, they diverged a little. We then took another fine leaf with the lobes standing at an angle of nearly 80° to each other; and on immersing it as before, the angle suddenly increased to 90°. A third leaf was torpid from having recently re-expanded after having caught a fly, so that repeated touches of the filaments caused not the least movement; nevertheless when similarly immersed, the lobes separated a little. As these leaves were inserted perpendicularly into the boiling water, both surfaces and the filaments must have been equally affected; and we can understand the divergence of the lobes only by supposing that the cells on the lower side, owing to their state of tension, acted mechanically and thus suddenly drew the lobes a little apart, as soon as the cells on the upper surface were killed and lost their contractile power. We have seen that boiling water in like manner causes the tentacles of Drosera to curve backwards; and this is an analogous movement to the divergence of the lobes of Dionaea.

In some concluding remarks in the fifteenth chapter on the Droseraceae, the different kinds of irritability possessed by the several genera, and the different manner in which they capture insects, will be compared.

14
Aldrovanda Vesiculosa

Captures crustaceans—Structure of the leaves in comparison with those of Dionaea—Absorption by glands, by the quadrifid processes, and points on the infolded margins—*Aldrovanda vesiculosa,* var. *australis*—Captures prey—Absorption of animal matter—*Aldrovanda vesiculosa,* var. *verticillata*—Concluding remarks.

Structure of the Leaves

This plant may be called a miniature aquatic Dionaea. Stein discovered in 1873 that the bilobed leaves, which are generally found closed in Europe, open under a sufficiently high temperature, and when touched, suddenly close. They re-expand in from 24 to 36 hrs., but only, as it appears, when inorganic objects are enclosed. The leaves sometimes contain bubbles of air, and were formally supposed to be bladders; hence the specific name of *vesiculosa.* Stein observed that water-insects were sometimes caught, and Prof. Cohn found within the leaves of naturally growing plants many kinds of crustaceans and larvae. Plants which have been kept in filtered water were placed by him in a vessel containing numerous crustaceans of the genus Cyprus, and next morning many were found imprisoned and alive, still swimming about within the closed leaves, but doomed to certain death.

Aldrovanda is destitute of roots and floats freely in the water. The leaves are arranges in whorls round the stem. Their broad petioles terminate in from four to six rigid projections, each tipped with a stiff, short bristle. The boiled leaf, with the midrib likewise tipped with a bristle, stands in the midst of these projections, and in evidently defended by them. The lobes are formed of very delicate tissue, so as to be translucent; they open, according to cohn, about as much as the two valves of living mussel-shell, therefore even less than the lobes of Dionaea; and this must make the capture of aquatic animals more easy. The outside of the leaves and the petioles are covered with minute two-armed papillae. Evidently answering to the eight-rayed papillae of Dionaea.

Each lobe rather exceeds a semi-circle in convexity, and consists of two very different concentric portions; the inner and lesser portion, or that next to the midrib, is slightly concave, and is formed, according to Cohn, of three layers of cells. Its upper surface is studded with colorless glands like, but more simple then, those of Dionaea; they are supported on distinct footstalks, consisting of two rows of cells. The outer and broader portion of the lobe is flat and very thin, being formed of only two layers of cells. Its upper surface does not bear any glands, but, in their place, small quadrifid processes, each

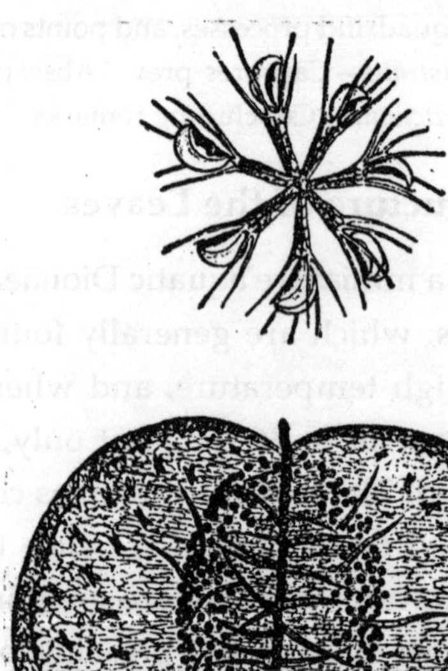

Fig. 13
(Aldrovanda vesiculosa)
Upper figure, whorl of leaves (from Prof. Cohn.)
Lower figure, leaf pressed flat open and greatly enlarged.

consisting of four tapering projections, which rise from a common prominence. These processes are formed of very delicate membrane lined with a layer of protoplasm; and they sometimes contain aggregated globules of hyaline matter. Two of the slightly diverging arms are directed towards the circumference, and two towards the midrib, forming together a sort of Greek cross. Occasionally two of the arms are replaced by one, and then the projection is trifled. We shall see in a future chapter that these projections curiously resemble those found within the bladders of Utricularia, more especially of *Utricularia montana*, although this genus is not related to Aldrovanda.

A narrow rim of the broad flat exterior part of each lobe is turned inwards, so that, when the lobes are closed, the exterior surfaces of the infolded portions come into contact. The edge itself bears a row of conical, flattened, transparent points with broad bases, like the prickles on the stem of a bramble or Rubus. As the rim is infolded, these points are directed towards the midrib and they appear at first as if they were adapted to revert to escape of prey; but this can hardly be their chief function, for they are composed of very delicate and highly flexible membrane, which can be easily bend or quite doubled back without being cracked. Nevertheless, the infolded rims, together with the points, must somewhat interfere with the retrograde movement of any small creature, as soon as the lobes begin to close. The circumferential part of the leaf of aldrovanda thus differs greatly from that of Dionaea; nor can the points on the rim be considered as homologous with the spikes round the leaves of Dionaea, as these latter are prolongations of the blade, and not mere epidemic production. They appear also to serve for a widely different purpose.

On the concave gland-bearing portion of the lobes, and especially on the midrib, there are numerous long, finely pointed hairs, which, as Prof. Cohn remarks, there can be little doubt are sensitive to touch, and, when touched, cause the leaf to close. They are formed of two rows of cells, or, according to Cohn, sometimes of four, and do not include any vascular tissue. They differ also from the six sensitive filaments of Dionaea in being colourless, and in having a medial as well as a basal articulation. No doubt it is owing to these two articulations that, not withstanding their length, they escape being broken when the lobes close.

Absorption by Glands

A leaf was cut open along the midrib, and the glands examined under a high power. It was then placed in a few drops of an infusion or raw meat. After 3 hrs. 20 m. there was no change, but when next examined after 23 hrs. 20 m., the outer cells of the glands contained, instead of limpid fluid, spherical masses of a granular substance, showing that matter had been absorbed from the infusion. That these glands secrete a fluid which dissolves or digests animal matter out of the bodies of the creatures which the leaves capture, is also highly probable from the analogy of Dionaea. If we may trust to the same analogy, the concave and inner portions of the two lobes probably close together by a slow movement, as soon as the glands have absorbed a slight amount of already soluble animal matter. The included water would thus be pressed out, and the secretion consequently not be too much diluted to act. With respect to the quadrifid processes on the outer parts of the lobes we were not able to decide whether they had been acted on by the infusion; for the lining of protoplasm was somewhat shrunk before they were immersed. Many of the points on the infolded rims also had their lining of protoplasm similarly shrunk and contained spherical granules of hyaline matter.

A solution of urea was next employed. This substance was chosen partly because it is absorbed by the quadrifid processes and more especially by the glands of Utricularia-plant which, as we shall hereafter see, feeds on decayed animal matter. As urea is one of the last products of the chemical changes going on in the living body, it seems fitted to represent the early stages of the decay of the dead body. We were also led to try urea from a curious little fact mentioned by Prof. Cohn, namely, that when rather large crustaceans are caught between the closing lobes, they are pressed so hard whilst making their escape that they often void their sausage-shaped masses of excrement, which were found within most of the leaves. These masses, no doubt, contain urea. They would be left either on the broad outer surfaces of the lobes where the quadrifids are situated, or within the closed concavity. In the latter case, water charges with excrementitious and decaying matter would be slowly forced outwards, and would bathe the quadrifids, if we were right in believing that the concave lobes contract after a time like those of Dionaea. Foul water would also be apt to ooze out at all times, especially when bubbles of air were generated within the concavity.

A leaf was cut open and examined, and the other cells of the glands were found to contain only limpid fluid. Some of the quadrifids included a few spherical granules, but several were transparent and empty, and their positions were marked. This leaf was now immersed in a little solution of one part of urea to 146 of water, or three grains to the ounce. After 3 hrs. 40 m., there was no change either in the glands or quadrifids; nor was there any certain change in the glands after 24 hrs.; so that, as far as one trial goes, urea does not act on them in the same manner as an infusion of raw meat. It was different with the quadrifids; for the lining of protoplasm instead of presenting a uniform texture, was now slightly shrunk, and exhibited in many placed minute, thickened, irregular, yellowish specks and ridges, exactly like those which appear within the quadrifids of Utricularia when treated with this same solution. Moreover, several of the quadrifids, which were before empty, now contained moderately sized or very small, more or less aggregated, globules of yellowish matter, as likewise occurs under the same circumstances with Utricularia. Some of the points on the infolded margins of the lobes were similarly affected; for their lining of protoplasm was little shrunk and included yellowish specks; and those which were before empty now contained small spheres and irregular masses of hyaline matter more or less aggregated; so that both the points on the margins and the quadrifids had absorbed matter from the solution in the course of 24 hrs. In another rather old leaf, to which nothing had been given, but which had been kept in foul water, some of the quadrifids contained aggregated translucent globules. These are not acted on by a solution of one part of carbonate of ammonia to 218 of water: and this negative result agrees with what we have observed under similar circumstances with Utricularia.

Aldrovanda vesiculosa, var. *Australis*

Whether it ought to be considered as a distinct species or a variety, cannot be told until the flowers are examined by a botanist. The projections at the upper end of the petiole (from four to six in number) are considerably longer relatively to the blade, and much more attenuated than those of the European form. They are thickly covered for a considerable space near their extremities with the upcurved pricks, which are quite absent in the latter form; and they generally beat on their tips two or three straight prickles instead of one. The bilobed leaf appears also to be rather larger and somewhat broader, with the pedicel by which it is attached to the upper end of the petiole a little longer. The points on the infolded margins likewise differ; they have narrower bases, and are more pointed; long and short points also alternate with much more regularity than in the European form. The glands and sensitive hairs are similar in the two forms. No quadrifid processes could be seen on several of the leaves, but we do not doubt that they were present, though indistinguishable from their delicacy and from having shrivelled; for they were quite distinct on one leaf under circumstances presently to be mentioned.

Some of the closed leaves contained no prey, but in one there was rather a large beetle, which from its flattened tibiae was an aquatic species, but was not allied to Colymbetes. All the softer tissues of this beetle were completely dissolved, and its chitinous integument's were as clean as if they had been boiled in caustic potash; so that it must have been enclosed for a considerable time. The glands were browner and more opaque then those on other leaves which had caught nothing; and the quadrifid processes, from being partly filled with brown granular matter, could be plainly distinguished, which was not the case, as already stated, on the other leaves. Some of the points on the infolded margins likewise contained brownish granular matter. We thus gain additional evidence that the glands, the quadrifid processes, and the marginal points, all have the power of absorbing matter, though probably of a different nature.

Within another leaf disintegrated remnants of a rather small animal, not a crustacean, which had simple, strong, opaque mandibles, and a large unarticulated chitinous coat, were present. Lumps of black organic matter, possibly of a vegetable nature, were enclosed in two other leaves; but in one of these there was also a small worm much decayed. But the nature of partially digested and decayed bodies, which have been pressed flat, long dried, and then soaked in water, cannot be recognised easily. All the leaves contained unicellular and other Algae, still of a greenish colour, which had evidently lived as intruders, in the same manner as occurs, according to Cohn, within the leaves of this plant in Germany.

Aldrovanda vesiculosa, var. *verticillata*

This form was considered by Wallich as a distinct species, under the name of *Verticillata.* It resembles the Australian form much more nearly then the European; namely in the projection at the upper end of the petiole being much attenuated and covered with upcurved prickles; they terminate also in two straight little prickles. The bilobed leaves are larger and certainly broader even than those of the Australian form; so that the greater convexity of their margins was conspicuous. The length of an open leaf being taken at 100, the breadth of the Bengal form is nearly 173, of the Australian form 147, and of the German 134. The points on the infolded margins are like those in the Australian form. Of the few leaves which were examined, three constrained entomostracan curstaceans.

Concluding Remarks

The leaves of the three foregoing closely allied species or varieties are manifestly adapted for catching living creature. With respect to the functions of the several parts, there can be little doubt that the long jointed hairs are sensitive, like those of Dionaea, and that, when touched, they cause the lobes to close. That the glands secrete a true digestive fluid and afterwards absorb the digested matter, is highly probable from the analogy of Dionaea—from the limpid fluid within their cells being aggregated into spherical masses, after they had absorbed an infusion of raw meat—from their opaque and granular condition in the leaf, which had enclosed a beetle for a long time—and from the clean condition of the integument's of this insect, as well as of crustaceans (as described by Cohn), which have been long captured. Again, from the effect produced on the quadrifid processes by an immersion for 24 hrs. in a solution of urea—from the presence of brown granular matter within the quadrifids of the leaf in which the beetle had been caught—and from the analogy of Utricularia—it is probable that these processes absorb excementitious and decaying animal matter. It is a more curious fact that the points on the infolded margins apparently serve to absorb decayed animal matter in the same manner as the quadrifids. We can thus understand the meaning of the infolded margins of the lobes furnished with delicate points directed inwards, and of the broad, flat, outer portions, bearing quadrifid processes; for these surfaces must be liable to be irrigated by foul water flowing from the concavity of the leaf when it contains dead animals. This would follow from various causes—from fluid in excess being secreted—and from the generation of bubbles of air. More observations are requisite on this head; but if this view is correct, we have the remarkable case of different parts of the same leaf serving for very different purposes—one part for true digestion, and another for the absorption of decayed animal matter. We can thus also understand how, by the gradual loss of either power, a plant might be gradually adapted for the one function to the exclusion of the other: and it will hereafter be shown that two genera, namely, Pinguicula and Utricularia, belonging to the same family, have been adapted for these two different functions.

15

Drosophyllum—Roridula—Byblis—Glandular Hairs of Other Plants—Concluding Remarks on the Droseraceae

Drosophyllum—Structure of leaves—Nature of the secretion—Manner of catching insects—Power of absorption—Digestion of animal substances—Summary on Drosophyllum—Byblis—Glandular hairs of other plants, their power of absorption—Saxifraga—Primula—Pelargonium—Erica—Mirabilis—Nicotiana—Summary on glandular hairs—Concluding remarks on the Droseraceae.

Drosophyllum lusitanicum

This rare plant has been found only in Portugal, and in Morocco. Vast numbers of flies adhere to the leaves. This latter fact is well known to the villagers, who call the plant the "fly-catcher," and hang it up in their cottages for this purpose. A plant in our hot-house caught so many insects during the early part of April, although the weather was cold and insects scare, that it must have been in some manner strongly attractive to them. On four leaves of young and small plant, 8, 10, 14 and 16 minute insects, chiefly Diptera, were found in the autumn adhering to them. We neglected to examine the roots, but we heard from Dr. Hooker that they are very small, as in the case of the previously mentioned members of the same family of the Droseraceae.

The leaves arise from an almost woody axis; they are linear, much attenuated towards their tips, and several inches in length. The upper surface is concave, the lower convex, with a narrow channel down the middle. Both surfaces, with the exception of the channel, are covered with glands, supported on pedicels and arranged in irregular longitudinal rows. These organs we shall call tentacles, from their close resemblance to those of Drosera, though they have no power of movement. Those on the same leaf differ much in length. The glands also differ in size, and are of a bright pink or of a purple colour; their upper surfaces are convex, and the lower flat or even concave, so that they resemble miniature mushrooms in appearance. They are formed of two layers of delicate angular cells, enclosing eight or ten larger cells with thicker zigzag walls. Within these larger cells there are others marked by spiral lines, and apparently connected with the spiral vessels which run up the green multicellular pedicels. The glands secrete large drops of viscid secretion. Other glands, having the same general appearance, are found on the flower – peduncles and calyx.

Besides the glands which are borne on longer or shorter pedicels, there are numerous ones, both on the upper and lower surfaces of the leaves, so small as to be scarcely visible to the naked eye. They are colourless and almost sessile, either circular or oval in outline; the latter occurring chiefly on the backs of the leaves (fig. 14). Internally they have exactly the same structure as the larger glands which are supported on pedicels; and indeed the two sets almost graduate into one another. But the sessile glands differ in one important respect, for they never secrete spontaneously, as far as we have seen, though we have examined them under a high power on a hot day, whilst the glands on pedicels were secreting copiously. Nevertheless, if little bits of damp albumen or fibrin are placed on these sessile glands, they begin after a time to secrete, in the same manner as do the glands of Dionaea when similarly treated. When they were merely rubbed with a bit of raw meat they likewise secreted. Both the sessile glands and the taller ones on pedicels have the power of rapidly absorbing nitrogenous matter.

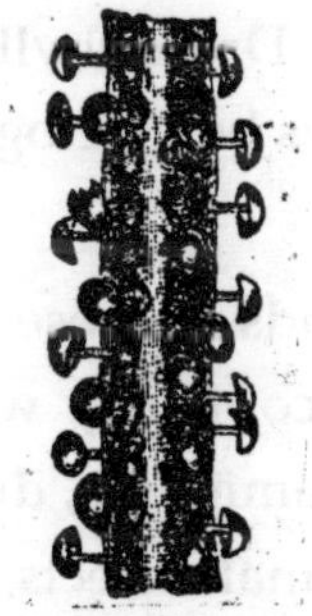

Fig. 14
(Drosophyllum lusitanicum)
Part of leaf, enlarged seven times, showing lower surface.

The secretion from the taller glands differs in a remarkable manner from that of Drosera, in being acid before the glands have been in any way excited; and judging from the changed colour of litmus paper, more strongly acid than that of Drosera. This fact was observed repeatedly; on one occasion we chose a young leaf, which was not secreting freely, and has never caught an insect, yet the section on all the glands coloured litmus paper of a bright red.

From the quickness with which the glands are able to obtain animal matter from such substances as well-washed fibrin and cartilage, we suspect that a small quantity of the proper ferment must be present in the secretion before the glands are excited, so that a little animal matter is quickly dissolved.

Owing to the nature of the secretion or to the shape of the glands, the drops are removed from them with singular facility. It is even somewhat difficult, by the aid of a finely pointed polished needle, slightly damped with water, to place a minute particle of any kind on one of the drops; for on a withdrawing the needle, the drop is generally withdrawn; whereas with Drosera there is no such difficulty, though the Drops are occasionally withdrawn. From this peculiarity, when a small insect alights on a leaf of Drosophyllum, the drops adhere to its wings, feet, or body, and are drawn from the gland; the insect then crawls onward and other drops adhere to it; so that at last, bathed by the viscid secretion, it sinks down and dies, resting on the small sessile glands with which the surface of the leaf is thickly covered. In the case of Drosera, an insect sticking to one or more of the exterior glands is carried by their movement to the centre of the leaf; with Drosophyllum, this is effected by the crawling of the insect, as from its wings being clogged by the secretion it cannot fly away.

There is another difference in function between the glands of these two plants: we know that the glands of Drosera secrete more copiously when properly excited. But when minute particles of carbonate of ammonia, drops of a solution of this salt or of the nitrate of ammonia, saliva, small insects, bits of raw or roast meat, albumen, fibrin or cartilage, as well as inorganic particles, were placed on the glands of Drosophyllum, the amount of secretion never appeared to be in the least increased. As insects do not commonly adhere to the taller glands, but withdraw the secretion, we can see that there would be little use in their having acquired the habit of secreting copiously when stimulated; whereas with Drosera this is of use, and the habit has been acquired. Nevertheless, the glands of Drosophyllum, without being stimulated, continually secrete, so as to replace the loss by evaporation. Thus when a plant was placed under a small bell-glass with its inner surface and support thoroughly wetted, there was no loss by evaporation, and so much secretion was accumulated in the course of a day that it ran down the tentacles and covered large spaces of the leaves.

The glands to which the above named nitrogenous substances and liquids were given did not, as just stated, secrete more copiously; on the contrary, they absorbed their own drops of secretion with surprising quickness. Bits of damp fibrin were place on five glands, and when they were looked at after an interval of 1 hr. 12 m., the fibrin was almost dry, the secretion having been all absorbed. So it was with three cubes of albumen after 1 hr. 19 m., and with four other cubes, though these latter were not looked at until 2 hrs. 15 m. had elapsed. The same result followed in between 1 hr. 15 m. and 1 hr. 30 m. when particles both of cartilage and meat were placed on several glands. Lastly, a minute drop (about 1/20 of a minim) of a solution of one part of nitrate of ammonia to 146 of water was distributed between the secretion surrounding three glands, so that the amount of fluid surrounding each was slightly increased; yet when looked at after 2 hrs. All three were dry. On the other and, seven particles of glass and three of coal-cinders, of nearly the same size as those of the above-named organic substances, were placed on ten glands; some of them being observed for 18 hrs., and others for two or three days; but there was not the least sign of the secretion being absorbed. Hence, in the former cases, the absorption of the secretion must have been due to the presence of some nitrogenous matter, which was either already soluble or was rendered so by the secretion. As the fibrin was pure, and had been well washed in distilled water after being kept in glycerine, and as the cartilage had been soaked in water, we suspect that these substances must have been slightly acted on and rendered soluble within the above stated short periods.

The glands have not only the power of rapid absorption, but likewise of secreting again quickly; and this latter habit has perhaps been gained, in as much as insects, if they touch the glands, generally withdraw the drops of secretion, which have to be restored. The exact period of re-secretion was recorded in only a few cases. The glands on which bits of meat were placed, and which were nearly dry after about 1 hr. 30 m., when looked at after 22 additional hours, were found secreting; so it was after 24 hrs. with one gland on which a bit of albumen had been placed. The three glands to which a minute drop of a solution of nitrate of ammonia was distributed, and which became dry after 2 hrs., were beginning to re-secrete after only 12 additional hours.

Tentacles Incapable of Movement – Many of the tall tentacles, with insects adhering to them, were carefully observed; and fragments of insects, bits of raw meat, albumen, &c., drops of a solution of two salts of ammonia and of saliva, were placed on the glands of many tentacles; but not a trace of movement could ever be detected. I also repeatedly irritated the glands with a needle, and scratched and pricked the blades, but neither the blade nor the tentacles became at all inflected. We may therefore conclude that they are incapable of movement.

On the Power of Absorption Possessed by the Glands – It has already been indirectly shown that the glands on pedicels absorb animal matter; and this is further shown by their changed colour, and by the aggregation of their contents, after they have been left in contact with nitrogenous substances or liquids. The following observations apply both to the glands supported on pedicels and to the minute sessile ones. Before a gland has been in anyway stimulated, the exterior cells commonly contain only limpid purple fluid; the more central ones including mulberry – like masses of purple granular matter. A leaf was placed in a little solution of one part of carbonate of ammonia to 146 of water (3 grs. to 1 oz.), and the glands were instantly darkened and very soon became black; this change being due to the strongly marked aggregation of their contents, more especially of the inner cells. Another leaf was placed in a solution of the same strength of nitrate of ammonia, and the glands were slightly darkened in 25 m., more so in 50 m., and after 1 hr. 30 m. were of so dark a red as to appear almost black. Other leaves were placed in a weak infusion of raw meat and in human saliva, and the glands were much darkened in 25 m., and after 40 m. were so dark as almost to deserve to be called black. Even immersion for a whole day in distilled water occasionally induces some aggregation within the glands, so that they become of a darker tint. In all these cases the glands are affected in exactly the same manner as those of Drosera. Milk, however, which acts so energetically on Drosera, seems rather less effective on Drosophyllum, for the glands were only slightly darkened by an immersion of 1 hr. 20 m., but became decidedly darker after 3 hrs. Leaves which had been left for 7 hrs. in an infusion of raw meat or in saliva were placed in the solution of carbonate of ammonia, and the glands now became greenish; whereas, if they had been first placed in the carbonate, they would have become black. In this latter case, the ammonia probably combines with the acid of the secretion, and therefore does not act on the colouring matter; but when the glands are first subjected to an organic fluid, either the acid is consumed in the work of digestion or the cell-walls are rendered more permeable, so that the undecom-posed carbonate enters and acts on the colouring matter. If a particle of the dry carbonate is placed on a gland, the purple colour is quickly discharged, owing probably to an excess of the salt. The gland, moreover, is killed.

Turning now to the action of organic substances, the glands on which bits of raw meat were placed became dark-coloured; and in 18 hrs. their contents were conspicuously aggregated. Several glands with bits of albumen and fibrin were darkened in between 2 hrs. and 3 hrs.; but in one case the purple colour was completely discharged. Some glands which had caught flies were compared with others close by; and though they did not differ much in colour, there was a marked difference in their state of aggregation. In some few instances, however, there was no such difference, and this appeared to be due to the insects having been caught long ago, so that glands had recovered their pristine state. In one case, group of the sessile colourless glands, to which a small fly adhered presented a peculiar appearance; for they had become purple, owing to purple granular matter coating the cell-walls. We may here mention as a caution that, soon after some of my plants arrived in the spring from Portugal, the glands were not plainly acted on by bits of meat, or insects, or a solution of ammonia.

Digestion of Solid Animal Matter—Whilst we were trying to place on two of the taller glands little cubes of albumen, these slipped down, and, besmeared with secretion, were left resting on some of the small sessile glands. After 24 hrs. one of these cubes was found completely liquefied, but with a few white streaks still visible; the other was much rounded, but not quite dissolved. Two other cubes were left on tall glands for 2 hrs. 45 m., by which time all the secretion was absorbed; but they were not perceptibly acted on, though no doubt some slight amount of animal matter had been absorbed from them. They were then placed on the small sessile glands, which being thus stimulated secreted copiously in the course of 7 hrs. One of these cubes was much liquefied within this short time; and both were completely liquefied after 21 hrs. 15 m.; the little liquid masses, however, still showing some white streaks. These streaks disappeared after an additional period of 6 hrs. 30 m.;

and by next morning (*i.e.,* 48 hrs. from the time when the cubes were first placed on the glands) the liquefied matter was wholly absorbed. A cube of albumen was left on another tall gland, which first absorbed the secretion and after 24 hrs. poured forth a fresh supply. This cube, now surrounded by secretion, was left on the glands for an additional 24 hrs., but was very little,if at all, acted on. We may therefore conclude, either that the secretion from the tall glands has little power of digestion, though strongly acid, or that the amount poured forth from a single gland is insufficient to dissolve a particle of albumen which within the same time would have been dissolved by the secretion from several of the small sessile glands.

Four minute shreds of pure fibrin were placed, each resting on one, two, or three of the taller glands. In the course of 2 hrs. 30 m. the secretion was all absorbed, and the shreds were left almost dry. They were then pushed on to the sessile glands. One shred, after 2 hrs 30 m., seemed quite dissolved, but this may have been a mistake. A second, when examined after 17 hrs. 15 m., was liquefied, but the liquid as seen under the microscope still contained floating granules of fibrin. The other two shreds were completely liquefied after 21 hrs. 30 m.; but in one of the drops a very few granules could still be detected. These, however, were dissolved after an additional interval of 6 hrs. 30 m.; and the surface of the leaf for some distance all round was covered with limpid fluid. It thus appears that Drosophyllum digests albumen and fibrin rather more quickly than Drosera can; and this may perhaps be attributed to the acid, together probably with some small amount of the ferment, being present in the secretion, before the glands have been stimulated; so that digestion begins at once.

Concluding Remarks – The linear leaves of Drosophyllum differ but slightly from those of certain species of Drosera; the chief differences being, firstly, the presence of minute, almost sessile glands, which, like those of Dionaea, do not secrete until they are excited by the absorption of nitrogenous matter. But glands of this kind are present on the leaves of *Drosera binata,* and appear to be represented by the papillae on the leaves of *Drosera rotundifolia.* Secondly, the presence of tentacles on the backs of the leaves; but we have seen that a few tentacles, irregularly placed and tending towards abortion, are retained on the backs of the leaves of *Drosera binata.* There are greater differences in function between the two genera. The most important one is that the tentacles of Drosophyllum have no power of movement; this loss being partially replaced by the drops of viscid secretion being readily withdrawn from the glands; so that, when an insect comes into contact with a drop, it is able to crawl away, but soon touches other drops, and then, smothered by the secretion, sinks down on the sessile glands and dies. Another difference is, that the secretion from the tall glands, before they have been in any way excited, is strongly acid, and perhaps contains a small quantity of the proper ferment. Again these glands do not secrete more copiously from being excited by the absorption of nitrogenous matter; on the contrary, they then absorb their own secretion with extraordinary quickness. In a short time they begin to secrete again. All these circumstances are probably connected with the fact that insects do not commonly adhere to the glands with which they first come into contact, though this does sometimes occur; and that it is chiefly the secretion from the sessile glands which dissolves animal matter out of their bodies.

Roridula

Roridula dentata—This plant, a native of the western parts of the Cape of Good Hope has an almost woody stem and branches, and apparently grows to a height of some feet. The leaves are linear, with their summits much attenuated. Their upper and lower surfaces are concave, with a ridge in the middle, and both are covered with tentacles, which differ greatly in length; some being very long, especially those on the tips of the leaves, and some very short. The glands also differ much in size and are somewhat elongated. They are supported on multicelluair pedicels.

This plant, therefore, agrees in several respects with Drosophyllum, but differs in the following points. We could detect no sessile glands; nor would these have been of any use, as the upper surface of the leaves is thickly clothed with pointed, unicellular hairs directed upwards. The pedicels of the tentacles do not include spiral vessels; nor are there any spiral cells within the glands. The leaves often arise in tufts and are pinnatifid, the divisions projecting at right angles to the main linear blade. These lateral divisions are often very short and bear only a single terminal tentacle, with one or two short ones on the sides. No distinct line of demarcation can be drawn between the pedicels of the long terminal tentacles and the much attenuated summits of the leaves.

We may, indeed, arbitrarily fix on the point to which the spiral vessels proceeding from the blade extend; but there is no other distinction.

It was evident from the many particles of dirt sticking to the glands that they secrete much viscid matter. A large number of insects of many kinds also adhered to the leaves. We could nowhere discover any signs of the tentacles having been inflected over the captured insects; and this probably would have been seen even in the dried specimens, had they possessed the power of movement. Hence, in this negative character, Roridula resembles its northern representative, Drosophyllum.

Byblis

Byblis gigantea **(Western Australia)** – A dried specimen about 18 inches in height, with a strong stem, the leaves are some inches in length, linear, slightly flattened, with a small projecting rib on the lower surface. They are covered on all sides by glands of two kinds – sessile ones arrange in rows, and others supported on moderately long pedicels. Towards the arrow summits of the leaves the pedicels are longer then else where, and here equal the diameter of the leaf. The glands are purplish, much flattened, and formed of a single layer of radiating cells, which in the larger glands are from forty to fifty in number. The pedicels consist of single elongated cells, with colourless, extremely delicate walls, marked with the finest intersecting spiral lines. Whether these lines are the result of contraction from the drying of the walls, but the whole pedicel was often spirally rolled up. These glandular hairs are far more simple in structure than the so-called tentacles of the preceding genera, and they do not differ essentially from those borne by innumerable other plants. The flower-peduncles bear similar glands. The most singular character about the leaves is that apex is enlarged into a little knob, covered with glands, and about a third broader than the adjoining part of the attenuated leaf. In two places dead fluid adhered to the glands. As no instance is known of unicellular structures having any power of movement, Byblis, no doubt, catches insects solely by the aid of its viscid secretion. These probably sink down besmeared with the secretion and rest on the small sessile glands, which, if we may judge by the analogy of Drosophyllum, then pour forth their secretion and afterwards absorb the digested matter.

Supplementary Observations on the Power of Absorption by the Glandular Hairs of Other Plant

A few observations on this subject may be here conveniently introduced. As the glands of many, probably of all, the species of Droseraceae absorb various fluids or at least allow them readily to enter, it seemed desirable to ascertain how far the glands of other plants which are not specially adapted for capturing insects, had the same power. Plants were chosen for trial at hazard, with the exception of two species of Saxifrage, which were selected from belonging to a family allied to the Droseraceae. Most of the experiments were made by immersing the glands either in an infusion of raw meat or more commonly in a solution of carbonate of ammonia, this latter substance acts so powerfully and rapidly on protoplasm. It seemed also particularly desirable to ascertain whether ammonia was absorbed, as a small amount is contained in rain-water. With the Droseraceae the secretion of a viscid fluid by the glands does not prevent their absorbing; so that the glands of other plants might excrete superfluous matter, or secrete an odoriferous fluid a protection against the attacks of insects or for any other purpose, and yet have the power of absorbing.

Saxifraga umbrosa – The flower-peduncles and petioles of the leaves are clothed with short hairs, bearing pink-coloured glands, formed of several polygonal cells, with their pedicels divided by partitions into distinct cells, which are generally colourless, but sometimes pink. The glands secrete a yellowish viscid fluid, by which minute Diptera are sometimes, though not often, caught. The cells of the glands contain bright pink fluid, charged with granules or with globular masses or pinkish pulpy matter. This matter must be protoplasm, for it is seen to undergo slow but incessant changes of form if a glands be placed in a drop of water and examined. Similar movements were observed after glands had been immersed in water for 1, 3, 5, 18 and 27 hrs. Even after this latter period the glands retained their bright pink colour; and the protoplasm within their cells did not appear to have become more aggregated. The continually changing forms of the little masses of protoplasm are not due to the absorption of water, as they were seen in glands kept dry.

A flower-stem, still attached to a plant, was bent so as to remain immersed for 23 hrs. 30 m. in a strong infusion of raw meat. The colour of the contents of the glands was slightly changed, being now of duller and more purple tint than before. The contents also appeared more aggregated, for the spaces between the little masses of protoplasm were wider; but this latter result did not follow in some other and similar experiments. The masses seemed to change their forms more rapidly than did those in water; so that the cells had a different appearance every four or five minutes. Elongated masses became in the course of one or two minutes spherical; and spherical ones drew themselves out and united with others. Minute masses rapidly increased in size, and three distinct ones were seen to unite. The movement was, in short, exactly like those described in the case of Drosera. The cells of the pedicels were not affected by the infusion; nor were they in the following experiment.

Another flower-stem was placed in the same manner and for the same length of time in a solution of one part of nitrate of ammonia to 146 if water (or 3 grs. to 1 oz), and the glands were discoloured in the exactly the same manner as by the infusion of raw meat.

Another flower-stem was immersed, as before, in a solution of one part carbonate of ammonia to 109 of water. The glands, after 1 hr. 30 m., were not discoloured, but after 3 hrs. 45m. most of them had become dull purple, some of them blackish-green, a few being still unaffected. The little masses of protoplasm within the cells were seen in movement. The cells of the pedicels were unaltered. The experiment was repeated, and afresh flower-stem was left for 23 hrs. in the solution, and now a great effect was produced; all the glands were much blackened, and the previously transparent fluid in the cells of the pedicels, even down to their bases, contained spherical masses of granular matter. By comparing many different hairs, it was evident that the glands first absorb the carbonate, and that the effect thus produced travels down the hairs from cell to cell. The first change which could be observed is a cloudy appearance in the fluid, due to the formation of very fine granules, which afterwards aggregate into larger masses. Altogether, in the darkening of the glands, and in the process of aggregation travelling down the cells of the pedicels, there is the closest resemblance to what takes place when a tentacle of Drosera is immersed in a weak solution of the same salt. The glands, however, absorb very much more slowly then those of Drosera. Besides the glandular hairs, there are star-shaped organs which do not appear to secrete, and which were not in the least affected by the above solution.

Although in the case of uninjured flower-stems and leaves the carbonate seems to be absorbed only by the glands, yet it enters a cut surface much more quickly than a gland. Strips of the rind of a flower-stem were torn off, and the cells of the pedicels were seen to contain only colourless transparent fluid; those of the glands including as usual some granular matter. These strips were them immersed in the same solution as before (one part of the carbonate to 109 of water), and in a few minutes granular matter appeared in the lower cells of all the pedicels. The action invariably commenced (for we tried the experiment repeatedly) in the lowest cells, and therefore close to the torn surface, and then gradually travelled up the hairs until it reached the glands, in a reversed direction to what occurs in uninjured specimens. The glands then became discoloured, and the previously contained granular matter was aggregated into larger masses. Two short bits of a flower-stem were also left for 2 hrs. 40m. in a weaker solution of one part of the carbonate of 218 of water; and the both specimens the pedicels of the hairs near the cut ends now contained much granular matter; and the glands were completely discoloured.

Lastly, bits of meat were placed on some glands; these were examined after 23 hrs., as were others, which had apparently not long before caught minute flies; but they did not present any difference from the glands of other hairs. Perhaps there may not have been time enough for absorption. We think so, as some glands, on which dead flies had evidently long lain, were of a pale dirty purple colour of even almost colourless, and the granular matter within them presented an unusual and somewhat peculiar appearance. That these glands had absorbed animal matter from the flies, probably by exosmose into the viscid secretion, we may infer, not only from their changed colour, but because, when placed in solution of carbonate of ammonia, some of the cells in their pedicels become filled with granular matter; whereas the cells of other hairs, which had not caught files, after being treated with the same quantity of granular matter. But more evidence is necessary before we fully admit that the glands of this saxifrage can absorb, even with ample time allowed, animal matter from the minute insects which they occasionally and accidentally capture.

***Saxifraga rotundifolia*—**The hairs on the flower-stems of this species are longer than those just described, and bear brown glands. Many were examined, and the cells of the pedicels were quite transparent. A bent stem was immersed for 30 m. in a solution of one part of carbonate of ammonia in 109 of water, and two or three of the uppermost cells in the pedicels now contained granular or aggregated matter; the glands having become of bright yellowish-green. The glands of this species therefore absorb the carbonate much more quickly than do those of *Saxifraga umbrosa*, and the upper cells of the pedicels are likewise affected much more quickly. Pieces of the stem were cut off and immersed in the same solution; and now the process of aggregation travelled up the hairs in a reversed direction; the cells close to the cut surfaces being first affected.

Primula sinensis—The flower-stems, the upper and lower surfaces of the leaves and their footstalks, are all clothed with a multitude of longer and shorter hairs. The pedicels of the longer hairs are divided by transverse partitions into eight or nine cells. The enlarged terminal cell is globular, forming a gland, which secretes a variable amount of thick, slightly viscid, nor acid, brownish-yellow matter.

A piece of a young flower-stem was first immersed in distilled water for 2 hrs. 30 m., and the glandular hairs were not at all affected. Another piece, bearing twenty-five short and nine long hairs, was carefully examined. The glands of the latter contained no solid or semi-solid matter; and those of only two of the twenty-five short hairs in a solution of one part of carbonate of ammonia to 109 of water, and now the glands of the twenty-five shorter hairs, with two to five smaller spherical masses of semi-solid masses. In a few hairs there were also globules in the cells immediately beneath the glands. Looking to all thirty-four hairs, there could be no doubt that the glands had absorbed some of the carbonate. Another piece was left for only 1 hr. in same solution, and aggregated matter appeared in all the glands. We examined some glands of the longer hairs, which contained little masses of matter, before they were immersed in any solution; and these masses slowly change their forms, so that no doubt they consisted of protoplasm. He then irrigated these hairs for 1 hr. 15 m., whilst under the microscope, with a solution of one part of the carbonate to 218 of water; the glands were not perceptibly affected, nor could this have been expected, as their contents were already aggregated. But the cells of the pedicels numerous, almost colourless, spheres of matter appeared, which changed their forms and slowly coalesced; the appearance of the cells being thus totally changed at successive intervals of time.

The glands on a young flower-stem, after having been left for 2 hrs. 45 m. in a strong solution of one part of the carbonate to 109 of water contained an abundance of aggregated masses, but whether generated by the action of the salt, we do not know. This piece was again placed in the solution, so that it was immersed altogether for 6 hrs. 15 m., and now there was a great change; for almost all the spherical masses within the gland-cells had disappeared, being replaced by granular matter of a darker brown. The experiment was thrice repeated with nearly the same result. On one occasion the piece was left immersed for 8 hrs. 30 m., and though almost all the spherical masses were changed into the brown granular matter, a few still remained. If the spherical masses of aggregated matter had been originally produced merely by some chemical or physical action it seems strange that a somewhat longer immersion in the same solution should so completely alter their character. But as the masses which slowly and spontaneously changed their forms must have consisted of living protoplasm, there is nothing surprising in its being injured or killed, and its appearance wholly changed by long immersion in so strong a solution of the carbonate as that employed. A solution of this strength paralyses all movement in Drosera, but does not kill the protoplasm; a still stronger solution prevents the protoplasm from aggregating into the ordinary full-sized globular masses, and these, though they do not disintegrate, become granular and opaque. In nearly the same manner, too, hot water and certain solution (for instance, of the salts of soda and potash) cause at first an imperfect kind of aggregation in the cells of Drosera; the little masses afterwards breaking up into granular or pulpy brown matter. All the foregoing experiments were made on flower-stems, but a piece of a leaf was immersed for 30 m. in a strong solution of the carbonate (one part of 109 of water), and little globular masses of matter appeared in all the glands, which before contained only limpid fluid.

We made also several experiments on the action of the vapour of the carbonate on the glands; but will give only a few cases. The cut end of the footstalk of young leaf was protected with sealing-wax, and was then placed under a small bell-glass, with a large pinch of the carbonate. After 10 m. the gland showed a considerable degree of aggregation, and the protoplasm lining the cells of the pedicels was little separated from the walls. Another leaf was left for 50 m. with the same result, excepting that the became throughout their whole length of a brownish colour. In a third leaf, which was exposed for 1 hr. 50 m., there was much aggregated matter in the glands; and some of the masses showed signs of breaking up into brown granular matter. This leaf was again placed in the vapour, so that it was exposed altogether for 5 hrs. 30 m.; and now though we examined a large number of glands, aggregated masses were found in only two or there; in all the other, the masses, which before had been globular, were converted into brown, opaque, granular matter. We thus see that exposure to the vapour for a considerable time produces the same effects as long immersion in a strong solution. In both cases there could hardly be doubt that the salt had been absorbed chiefly or exclusively by the glands.

On another occasion bits of damp fibrin, drops of weak infusion of raw meat and water, were left for 24 hrs. on some leaves; the hairs were then examined, but to my surprise differed in no respect from others which had not been touched by these fluids. Most of the cells, however, included hyaline, motionless little spheres, which did not seem to consist of protoplasm, but we suppose, of some balsam or essential oil.

***Pelargonium zonale* (var. edged with white)** – The leaves are clothed with numerous multicelluair hairs; some simply pointed; others bearing glandular heads, and differing much in length. The glands on a piece of leaf were examined and found to contain only a limpid fluid; most of the water was removed from beneath the covering glass and a minute drop of one part of carbonate of ammonia to 146 of water was added; so that an extremely small dose was given. After an interval of only 3 m. there were signs of aggregation within the glands of the shorter hairs; and after 5 m. many small, globules of a pale brown tint appeared in all of them; similar globules, but larger, being found in the large glands of the longer hairs. After the specimen had been left for 1 hr. in the solution, many of the smaller globules had changed their positions; and two or three vacuoles or small spheres of a rather darker tint appeared within some of the larger globules. Latter globules could now be seen in some of the uppermost cells of the walls of the lower cells. After 2 hrs. 30 m. from the time of first immersion, the large globules within the glands of the longer hairs were converted into masses of darker brown granular matter. Hence from what we have seen with *Primula sinensis*, there can be little doubt that these masses originally consisted of living protoplasm.

A drop of weak infusion of raw meat was placed on a leaf, and after 2 hrs. 30 m. many spheres could be seen within the glands. These spheres, when looked at again after 30 m., had slightly changed their positions and forms, and one had separated into two; but the changes were not quite like those which the protoplasm of Drosera undergoes. These hairs, moreover, had not been examined before immersion, and there were similar spheres in some glands, which had not been touched by the infusion.

Erica tretralix – A few long glandular hairs project from the margins of the upper surfaces of the leaves. The pedicels are formed of several rows of cells, and support rather large globular herds, secreting viscid matter, by which minute insects are occasionally though rarely, caught. Some leaves were left for 23 hrs. in a weak infusion of raw meat and in water, and the hairs were then compared, but they differed very little or not at all. In both cases the contents of the cells seemed rather more granular than they were before; but the granule did not exhibit any movement. Other leaves were left for 23 hrs. in a solution of one part of carbonate of ammonia to 218 of water, and here again the granular matter appeared to have increased in amount; but one such mass retained exactly the same form as before after an interval of 5 hrs., so that it could hardly have consisted of living absorption, certainly much less than those of the foregoing plants.

Mirabilis longiflora – The stems and both surfaces of the leaves bare viscid hairs. Young plants, from 12 to 18 inches in height in my greenhouse, caught so many minute Diptera, Coleoptera, and larvae, that they were quite dusted with them. The hairs are short, of unequal lengths, formed of a single row of cells, surmounted by an enlarged cell which secretes viscid matter. These terminal cells or glands contain granules and often globules of granular matter. Within a glands which had caught a small insect, one such mass was observed to undergo incessant changes of form, with the occasional appearance of vacuoles. But we do not believe that this protoplasm had been generated by matter absorbed from the dead insect; for, on comparing several glands which had and had not caught insects, not a shade of difference could be perceived between them, and they all contained fine granular matter. A piece of leaf was immersed for 24 hrs. in a solution of one part of carbonate of ammonia to 218 of water, but the hairs seemed very little affected by it, excepting that perhaps the glands were rendered rather more opaque. In the leaf itself, however, the grains of chlorophyll near the cut surfaces had run together, or become aggregated. Nor were the glands on another leaf, after an immersion for 24 hrs. in an infusion of raw meat, in the least affected; but the protoplasm lining the cells of the pedicels had shrunk greatly from the walls. This latter effect may have been due to exosmose, as the infusion was strong. We may therefore conclude that the glands of this plant either have no power of absorption or that the protoplasm which they contain not acted on by a solution of carbonate of ammonia (and this seems scarcely credible) or by an infusion of meat.

Nicotiana tabacum – This plant is covered with innumerable hairs of unequal lengths, which catch many minute insects. The pedicels of the hairs are divided by transverse partitions, and the secreting glands are formed of many cells, containing greenish matter with little globules of some substance. Leaves were left in an infusion of raw meat and in water for 26 hrs., but presented no difference. Some of these same leaves were then left for above 2 hrs. in a solution of carbonate of ammonia, but no effect was produced.

Summary of the Observations on Glandular Hairs – From the foregoing observations, few as they are, we see that the glands of two species of Saxifraga, of a Primula and Pelargonium, have the power of rapid absorption; whereas the glands of an Erica, Mirabilis and Nicotiana, either have no such power, or the contents of the cells are not affected by the fluid employed, namely, a solution of carbonate of ammonia and an infusion of raw meat. As the glands of the Mirabilis contain protoplasm, which did not become aggregated from exposure to the fluids just named, through the contents of the cells in the blade of the leaf were greatly affected by carbonate of ammonia, we may infer that they cannot absorb. We may further infer that the innumerable insects caught by this plant are of no more service to it than are those which adhere to the deciduous and sticky scales of the leaf-buds of the Horse-chestnut.

The most interesting case for us is that of the two species of Saxifraga, as this genus is distantly allied of Drosera. Their glands absorb matter from an infusion of raw meat, from solution of the nitrate and carbonate of ammonia, and apparently from decayed insects. This was shown by the changed dull purple colour of the protoplasm within the cells of the glands, by its state of aggregation, and apparently by its more rapid spontaneous movement. The aggregating process spreads from the glands down the pedicels of the hairs; and we may assume that any matter which is absorbed ultimately reached the tissues of the plant. On the other hand, the process travels up the hairs whenever a surface is cut and exposed to a solution of the carbonate of ammonia.

The glands on the flower-stalks and leaves of *Primula sinensis* quickly a solution of the carbonate of ammonia, and the protoplasm which they contain becomes aggregated. The process was seen in some cases to travel from the glands into the upper cells of the pedicels. Exposure for 10 m. to the vapour of this salt likewise induced aggregation. When leaves were left from 6 hrs. to 7 hrs. in a strong solution or were long exposed to the vapour, the little masses of protoplasm became disintegrated, brown, and granular, and were apparently killed. An infusion of raw meat produced no effect on the glands.

The limpid contents of the glands of *Pelargonium zonale* became cloudy and granular in from 3 m. to 5 m. when they were immersed in a weak solution of the carbonate of ammonia; and in the course of 1 hr. granules appeared in the upper cells of the pedicels. As the aggregated masses slowly changed their forms, and as they suffered disintegration when left for a considerable time in a strong solution there can be little doubt that they consisted of protoplasm. It is doubtful whether an infusion of raw meat produced any effect.

The glandular hairs of ordinary plants have generally been considered by physiologists to serve only as secreting or excreting organs, but we now know that they have the power, at least in some cases, of absorbing both a solution and the vapour of ammonia. As rain-water contains a small percentage of ammonia, and the atmosphere a minute quantity of the carbonate, this power can hardly fail to be beneficial. Nor can the benefit be quite so insignificant as it might at first be thought, for a moderately fine plant of *Primula sinensis* bears the astonishing number of above two millions and a half of glandular hairs, all of which are able to absorb ammonia brought to them by the rain. It is moreover probable that the glands of some of the above-named plants obtain animal from the insects which are occasionally entangled by the viscid secretion.

Concluding Remarks on the Droseraceae

The six known genera composing this family have now been described in relation to our present subject, as far as my means have permitted. They all capture insects. This is affected by Drosophyllum, Roridula, and Byblis solely by the viscid fluid secreted from their glands; by Drosera, through the same means, together with the movements of the tentacles; by Dionaea and Aldrovanda, through the closing of the blades of the leaf. In these two last genera rapid movement makes up for the loss of viscid secretion. In every case it is some part of the leaf which moves. In aldrovada it appears to be the basal parts alone which contract and carry with them broad, thin margins of the lobes. In Dionaea the whole lobe, with the exception of the marginal prolongation or spikes, curves inwards, though the chief seat of movement is near the midrib. In Drosera the chief seat is in the lower part of the tentacles, which, homologically, may be considered as prolongations of the leaf; but the whole blade often curls inwards, converting the leaf into a temporary stomach.

There can hardly be a doubt that all the plants belonging to these six genera have the power of dissolving animal matter by the aid of their secretion, which contains as acid, together with a ferment almost identical in nature with pepsin; and that they afterwards absorb the matter thus digested. This is certainly the case with Drosera, Drosophyllum, and Dionaea; almost certainly with Aldrovanda; and, from analogy, very probable with Roridula and Byblis. We can thus understand how it is that the three first named genera are provided with such small roots, and that aldrovanda is quite rootless; about the roots of the two other genera nothing is known. It is, no doubt, a surprising fact that a whole group of plants (and, as we shall presently see, some other plants not allied to the Droseraceae) should subsist partly by digesting animal matter, and partly by decomposing carbonic acid, instead of exclusively by this latter means, together with the absorption of matter from the soil by the aid of roots. We have, however, an equally anomalous case in the animal kingdom; the rhizocephalous crustaceans do not feed like other animal by their mouths, for they are destitute of an alimentary canal; but they live by absorbing through root – like processes the juice of the animal on which they are parasitic.

Of the six genera, Drosera has been incomparably the most successful in the battle for life; and a large part of its success may be attributed to its manner of catching insects. It is a dominant form, for it is believed to include about 100 species, which range in the Old World from the Arctic regions to Southern India, to the cape of Good hope, Madagascar, and Australia; and in the New World from Canada to Tierra del Fuego. In this respect it presents a market contrast with the five other genera, which appear to be failing groups. Dionaea includes only a single species, which is confined to one district in Carolina. The three varieties or closely allied species of Aldrovada, like so many water-plants, have a wide range from Central Europe to Bengal and Australia. Drosophyllum includes only one species, limited to Portugal and Morocco. Roridula and Byblis each have two species; the former confined to the western parts of the Cape of Good Hope, and the latter to Australia. It is a strange fact that Dionaea, which is one of the most beautifully adapted plants in the vegetable kingdom, should apparently be on the high road to extinction. This is all the more strange as the organs of Dionaea are more highly differentiated than those of Drosera; its filaments serve exclusively as organs of touch, thc lobes for capturing insects, and the glands, when excited, for secretion as well as for absorption; whereas with Drosera the glands serve all these purposes, and secrete without being excited.

By comparing the structure of the leaves, their degree of complication, and their rudimentary parts in the six genera, we are led to infer that their common parent form partook of the characters of Drosophyllum, Roridula, and Byblis. The leaves of this ancient form were almost certainly linear perhaps divided, and bore on their upper and lower surfaces glands which had the power of secreting and absorbing. Some of these glands were mounted on pedicels, and others were almost sessile; the latter secreting only when stimulated by the absorption of nitrogenous matter. In Byblis the glands consist of single layer of cells, supported on a unicellular pedicel; in Roridula they have a more complex structure, and are supported on pedicels formed of several rows of cells; in Drosophyllum they further include spiral cells, and the pedicels include a bundle of spiral vessels. But in these three genera these organs do not possess any power of movement, and there is no reason to doubt that they are of the nature of hairs or trichomes. Although in innumerable instances foliar organs move when excited, no case is known of a trichome having such power. We are thus led to inquire how the so-called tentacles of Drosera, which are manifestly of the same general nature as the glandular hairs of the above three genera, could have acquired the power of moving. Many botanists maintain that these tentacles consist of prolongation's of the leaf, because they include vascular tissue, but this can no longer reconsidered as trustworthy distinction. The possession of the power of movement on excitement would have been safer evidence. But when we consider the vast number of the tentacles on both surfaces of the leaves of Drosophyllum, and on the upper surface of the leaves of Drosera, it seems scarcely possible that each tentacle could have aboriginally existed as a prolongation of the leaf. Roridula, perhaps, shows us how we may reconcile these difficulties with respect to the homological nature of the tentacles. The lateral division of the leaves of this plant terminate in long tentacles; and these include spiral vessels which extend for only a short distance up them, with no line of demarcation between what is plainly the prolongation of the leaf and the pedicel of a glandular hair. Therefore there would be nothing anomalous or unusual in the basal parts of these tentacles, which correspond with the marginal ones of Drosera, acquiring the power of movement; and we know that in Drosera it is only the lower part, which becomes inflected. But in order to understand how in this latter genus not only the marginal but all the inner tentacles have become capable of movement, we must further assume, either that through the principle of correlated development this power was transferred to the basal parts of the hairs, or that the surface of the leaf has been prolonged upwards at numerous points, so as to unite with the hairs, thus forming the bases of the inner tentacles.

The above-named three genera, namely Drosophyllum, Roridula, and Byblis, which appear to have retained a primordial condition, still bear glandular hairs on both surfaces of their leaves; but these on the lower surface have since disappeared in the more highly developed genera, with the partial exception of one species, *Drosera binata*. The small sessile glands have also disappeared in some of the genera, being replaced in Roridula by hairs and in most species of Drosera by absorbent papillae. *Drosera binata*, with its linear and bifurcating leaves, is in an intermediate condition. It still bears some sessile glands on both surfaces of the leaves and on the lower surface a few irregularly placed tentacles, which are incapable of movement. A further slight change would convert the linear leaves of this latter species into the along leaves of *Drosera anglica*, and these might easily pass into orbicular ones with footstalks like those of *Drosera rotundifolia*. The footstalks of this latter species bear multicellular hairs, which we have good reason to believe represent aborted tentacles.

The parent form of Dionaea and Aldrovanda seems to have been closely allied to Drosera, and to have had rounded leaves, supported on distinct footstalks, and furnished with tentacles all round the circumference, with other tentacles and sessile glands on the upper surface. We think so because the marginal spikes of Dionaea apparently represent the extreme marginal tentacles of Drosera, the six (sometimes eight) sensitive filaments on the upper surface, as well as the more numerous ones in aldrovanda, representing the central tentacles of Drosera, with their glands aborted, but their sensitiveness retained. Under this point of view we should bear in mind that the summits of the tentacles of Drosera, close beneath the glands, are sensitive.

The three most remarkable characters possessed by the several members of the Droseraceae consist in the leaves of some having the power of movement when excited, in their glands secreting a fluid which digests animal matter, and in their absorption of the digested matter. Can any light be thrown on the steps by which these remarkable powers were gradually acquired?

As the walls of the cells are necessarily permeable to fluids, in order to allow the glands to secrete, it is not surprising that they should readily allow fluids to pass inwards; and this inward passage would deserve to be called an act of absorption, if the fluids combined with the contents of the glands. Judging from the evidence above given, the secreting glands of many other plants can absorb salts of ammonia, of which they must receive small quantities from the rain. This is the case with two species of Saxifraga and the glands of one of them apparently absorb matter from captured insects, and certainly from an infusion of raw meat. There is, therefore, nothing anomalous in the Droseraceae having acquired the power of absorption in a much more highly developed degree.

It is a far more remarkable problem how the members of this family, and Pinguicula, and Nepenthes, could all have acquired the power of secreting a fluid which dissolves or digests animal matter. The six genera of the Droseraceae have probably inherited this power from a common progenitor, but this cannot apply to Pinguicula or Nepenthes, for these plants are not at all closely related to the Droseraceae. But the difficulty is not nearly so great as it at first appears. Firstly, the juices of many plants contain an acid, and, apparently, and acid serves for digestion. Secondly, as Dr. Hooker has remarked in relation to the present subject and as Sachs repeatedly insist, the embryos of some plants secrete a fluid which dissolved albuminous substances out of endosperm; although the endosperm is not actually united with, only in contact with, the embryo. All plants, moreover, have the power of dissolving albuminous or proteid substances, such as protoplasm, chlorophyll, gluten, aleurone, and of carrying them from one part to other parts of their tissues. This must be effected by a solvent, probably consisting of ferment together with an acid. Now, in the case of plants which are able to absorb already soluble matter from captured insects, though not capable of true digestion, the solvent just referred to, which must be occasionally present in the glands, would be apt to exude from the glands together with the viscid secretion, in as much as endosmose is accompanied by exosmose. If such exudation did ever occur, the solvent would act on the animal matter contained within the capture insects, and this would be an act of true digestion. As it cannot be doubted that this process would be of high service to plants growing in very poor soil, it would tend to be perfected through natural selection. Therefore, any ordinary plant having viscid glands, which occasionally caught insects, might thus be converted under favourable circumstances into a species capable of true digestion. It ceases, therefore, to be any great mystery how several genera of plants, in no way closely related together, have independently acquired this same power.

As there exist several plants the glands of which cannot, as far as it is known, digest animal matter, yet can absorb salts of ammonia and animal fluids, it is probable that this latter power forms the first stage towards that of digestion. It might, however, happen, under certain condition, that a plant, after having acquired the power of digestion, should degenerate into one capable only of absorbing animal matter in solution, or in a state of decay, or the final products of decay, namely the salts of ammonia. It would appear that this has actually occurred to a partial extent with the leaves of Aldrovanda;

the outer parts of which possess absorbent organs, but no glands fitted for the secretion of any digestive fluid, these being confined to the inner parts.

Little light can be thrown on the gradual acquirement of the third remarkable character possessed by the more highly developed genera of the Droseraceae, namely the power of movement when excited. It should, however, be borne in mind that leaves and their homologues as well as flower-peduncles, have gained this power, in innumerable instances, independently of inheritance from any common parent form; for instance, in tendril-bearers and leaf-climbers (*i.e.*, plants with their leaves, petioles and flower-peduncles, & c, modified for prehension) belonging to large number of the most widely distinct orders—in the leaves of many plants which go to sleep at night, or move when shaken—and in irritable stamens and pistils of not a few species. We may therefore infer that the power of movement can be by some means readily acquired. Such movement imply irritability or sensitiveness, but, as Cohn has remarked, the tissues of the plants thus endowed do not differ in any uniform manner from those of ordinary plants; it is therefore probable that all leaves are to slight degree irritable. Even if an insect alights on a leaf, a slight molecular change is probably transmitted to some distance across its tissue, with the sole difference that no perceptible effect is produced. We have some evidence in favour of this belief, for we know that a single touch on the glands of Drosera does not excite inflection; yet it must produce some effect, for if the glands have been immersed in a solution of camphor, inflection follows within a shorter time than would have followed from the effects of camphor alone. So again with Dionaea, the blades in their ordinary state may be roughly touched without their closing yet some effect must be thus caused and transmitted across the whole leaf, for if the glands have recently absorbed animal matter, even a delicate touch causes them to close instantly. On the whole we may conclude that the acquirement of high degree of sensitiveness and of the power of movement by certain genera of the Droseraceae presents no greater difficulty than that presented by the similar but feebler powers of a multitude of other plants.

The specialized nature of the sensitiveness possessed by Drosera and Dionaea, and by certain other plants, well deserves attention. A gland of Drosera may be forcibly hit once, twice, or even thrice, without any effect being produced, whilst the continued pressure of an extremely minute particle excited movement. On the other hand, a particle many times heavier may be gently laid on one of the filaments of Dionaea with no effect; but it touched only once by the slow movement of a delicate hair, the lobes close; and this difference in the nature of the sensitiveness of these two plants stands in manifest adaptation their manner of capturing insects. So does the fact, that when the central glands of Drosera absorb nitrogenous matter, they transmit a motor impulse to the exterior tentacles much more quickly than when they are mechanically irritated; whilst with Dionaea the absorption of nitrogenous matter causes the lobes to press together with extreme slowness, whilst a touch excites rapid movement. Somewhat analogous cases may be observed, with the tendrils of various plants; some being most excited by contact with fine fibres, other by contact with bristles, other with a flat or a serviced surface. The sensitive organs of Drosera and Dionaea are also specialised, so as not to be uselessly affected by the weight impact of drops of rain, or by blasts of air. This may be accounted for by supposing that these plants and their progenitors have grown accustomed to the repeated action of rain and wind, so that no molecular change is thus induced; whilst they have been rendered more sensitive by means of natural selection to the rarer impact or pressure of solid bodies. Although the absorption by the glands of Drosera of various fluids excited movement, there is a great difference in the action of allied fluids; for instance, between certain vegetable acids, and between citrate and phosphate of ammonia. The specialized nature and perfection of the sensitiveness in these two plants is all the more astonishing as no one supposes that they possess nerves; and by testing Drosera with several substances which act powerfully on the nervous system of animals, it does not appear that they include any diffused matter analogous to nerve-tissue.

Although the cells of Drosera and Dionaea are quite as sensitive to certain stimulants as are the tissues which surround the terminations of the nerves in the higher animals, yet these plants are inferior even to animals low down in the scale, is not being affected except by stimulants in contact with their sensitive parts. They would, however, probably be affected by radiant heat; for warm water excites energetic movement. When a glands of Drosera, or one of the filaments of Dionaea, is excited, the motor impulse radiates in all direction, and in not, as in the case of animals, directed towards special points or organs. This holds good even in the case of Drosera when some exciting substance has been placed at two points on the disc, and when the tentacles all round are inflected with marvellous precision towards the two points. The rate at which the motor impulse is transmitted, though rapid in Dionaea, is much slower than in most of all animals This fact, as well as that of the motor impulse not being specially directed to certain points, are both no doubt due to the absence of nerves. Nevertheless we perhaps see the pre-figurement of the formation of nerves in animals in the transmission of the motor impulse being so much more rapid down the confined space within the tentacles of Drosera than elsewhere, and somewhat more rapid in a longitudinal than in a transverse direction across the disc. These plants exhibit still more plainly their inferiority to animals in the absence of any reflex action, except in so far as the glands of Drosera, when excited from a distance, send back some influence which causes the contents of the cells to become aggregated down to the bases of the tentacles. But the greatest inferiority of all is the absence of a central organ, able to receive impressions from all points, to transmit their effects in any definite direction, to store them up and reproduce them.

❑❑❑

16
Pinguicula

Pinguicula vulgaris—Structure of leaves—Number of insects and other objects caught—Movement of the margins of the leaves—Uses of this movement—Secretion, digestion, and absorption—Action of the secretion on various animal and vegetable substances—The effects of substances not containing soluble nitrogenous matter or the glands—Pinguicula grandiflora—Pinguicula lusitanica, catches insects—Movement of the leaves, secretion and digestion.

Pinguiculia Vulgaris

This plant grows in moist places, generally on mountains. it bears on an average eight, rather thick, along, light green leaves, having scarcely any footstalk. A full-sized leaf is about $1\frac{1}{2}$ inch in breadth. The young central leaves are deeply concave, and project upwards; the older ones towards the outside are flat or convex, and lie close to the ground, forming a rosette from 3 to 4 inches in diameter. The margins of the leaves are incurved. Their upper surfaces are thickly covered with two sets of glandular hairs, differing in the size of the glands and in the length of their pedicels. The larger glands have a circular outline as seen from above, and are of moderate thickness; they are divided by radiating partitions into sixteen cells, containing light-green, homogeneous fluid. They are supported on elongated, unicellular pedicels (containing a nucleus with a nucleolus) which rest on slight prominence. The small glands differ only in being formed of about half the number of cells, containing much paler fluid, and supported on much shorter pedicels. Near the midrib, towards the base of the leaf, the pedicels are multicellular, are longer then elsewhere, and bear smaller glands. All the glands secrete a colourless fluid, which is so viscid that we have seen a fine thread drawn out to length of 18 inches; but in this case was secreted by glands, which had been excited. The edge of the leaf is translucent, and does not bear any glands; and here the spiral vessels, proceeding from the midrib, terminate in cells marked by a spiral line, somewhat like those within the glands of Drosera.

The roots are short. Three plants were dug up in North Wales on June 20, and carefully washed; each bore five or six unbranched roots, the longest of which was only 1-2 of an inch. Two rather young plants were examined on September 28; these had a greater number of roots, namely eight and eighteen, all under 1 inch in length, and every little branched.

Of these leaves, thirty-two had caught 142 insects or on an average 4.4 per leaf, minute fragments of insects not being included. Besides the insects, small leaves belonging to four different kinds of plants, those *Erica tetralix* being much the commonest, and three minute seedling plants, blown by wind, adhered to nineteen of the leaves. One had caught as many as ten leaves of the Erica. Seeds or fruits, commonly of Carex and one of Juncus, besides bits of moss and other rubbish, likewise adhered to six of the thirty-nine leaves. The same friend, on June 27, collected nine plants bearing seventy-four leaves, and all of these, with the exception of three young leaves, had caught insects; thirty insects were counted on one leaf, eighteen on a second, and sixteen on a third. Another friend examined on August 22 some plants in Donegal, Ireland, and found insects on 70 out of 157 leaves; fifteen of these leaves were sent me, each having caught on an average 2.4 insects. There was a seed on three leaves on the same plant. The sixteen seeds belonged to nine different kinds, which could not be recognised, excepting one of Ranunculus, and several belonging to three of four distinct species of Carex. It appears that fewer insects are caught late in the year than earlier; thus in Cumberland from twenty to twenty-four insects were observed in the middle of July on several leaves, whereas in the beginning of September the average number was only 2.27. Most of the insects, in all the foregoing cases, were Diptera, but with many minute Hymenoptera, including some ants, a few small Coleoptera, Larvae, spiders, and even small moths.

We thus see that numerous insects and other objects are caught by the viscid leaves; but we have no right to infer from this fact that the habit is beneficial to the plant, any more than in the before – given case of the Mirabilis, or of the horse-chestnut. But it will presently be seen that dead insects and other nitrogenous bodies excite the glands to increased secretion; and that the secretion then becomes acid and has the power of digesting animal substances, such as albumen, fibrin, &c. Moreover, the dissolved nitrogenous matter is absorbed by the gland, as shown by their limpid contents being aggregated into slowly moving granular masses of protoplasm. The same results follows when insects are naturally captured, and as the plant lives in poor soil and has small roots, there can be no doubt that it profits by its power of digesting and absorbing matter from the prey which it habitually captures in such large numbers. It will, however, be convenient first to describe the movement of the leaves.

Movement of the Leaves—That such thick, large leaves as those of *Pinguicula vulgaris* should have the power of curving inwards when excited has never even been suspected. It is necessary to select for experiment leaves with their glands secreting freely, and which have been prevented from capturing many insects; as old leaves, at least those growing in a state of nature, have their margins already curled so much inwards that they exhibit little power of movement, or move very slowly. We will first give in detail the more important experiments which tried, and then make some concluding remarks.

Experiment 1: A young and almost upright leaf was selected, with its two lateral edges equally and very slightly incurved. A row of small flies was placed along one margin. When looked at next day, after 15 hrs., this margin, but not the other, was found folded inwards, like the helix of the human ear, to the breadth of $\frac{1}{10}$ of an inch, so as to lie partly over the row of flies (fig.15). The glands on which the flies rested, as well as those on the over-lapping margin which had been brought into contact with the flies, were all secreting copiously.

Fig. 15
(Pinguicula vulgaris)
Outline of leaf with left margin inflected over a row of small flies.

Experiment 2: A row of flies was placed on one margin of a rather old leaf, which lay flat on the ground; and in this case the margin, after the same interval as before, namely 15 hrs., had only just begun to curl inwards; but so much secretion had been poured forth that the spoon-shaped tip of the leaf was filled with it.

Experiment 3: Fragments of a large fly were placed close to the apex of a vigorous leaf, as well as along half one margin. After 4 hrs. 20 m. there was decided incurvation, which increased a little during the afternoon, but was in the same state on the following morning. Near the apex both margins were inwardly curved. We have never seen a case of the apex itself being in the least curved towards the base of the leaf. After 48 hrs. (always reckoning from the time when the flies were placed on the leaf) the margin had everywhere begun to unfold.

Experiment 4: A large fragment of a fly was placed on a leaf, in a medial line, a little beneath the apex. Both lateral margins were perceptibly incurved in 3 hrs., and after 4 hrs. 20m. to such a degree that the fragment was clasped by both margins. After 24 hrs. the two infolded edges near the apex (for the lower part of the leaf was not at all affected) were measured and found to be .11 of an inch (2.795 mm.) apart. The fly was now removed, and a stream of water poured over the leaf so as to wash the surface; and after 24 hrs. the margins were .25 of an inch (6.349 mm.) apart, so that they were largely unfolded. After an additional 24 hrs. they were completely unfolded. Another fly was now put on the same spot to see whether this leaf, on which the first fly had been left 24 hrs., would move again; after 10 hrs. there was a trace of incurvation, but this did not increase during the next 24 hrs. A bit of meat was also placed on the margin of leaf, which four days previously had become strongly incurved over a fragment of a fly and had afterwards re-expanded; but the meat did not cause even a trace of incurvation. On the contrary, the margin became somewhat reflexed, as if injured, and so remained for the three following days, as long as it was observed.

Experiment 5: A large fragment of a fly was placed halfway between the apex and base of a leaf and halfway between the midrib and one margin. A short space of this margin, opposite the fly, showed a trace of incurvation after 3 hrs., and this became strongly pronounced in 7 hrs. After 24 hrs. the infolded edge was only 16 of an inch (4.064 mm.) from the midrib. The margin now began to unfold though the fly was left on the leaf; so that by the next morning (*i.e.*, 48 hrs. from the time when the fly was first put on) the infolded edge had almost completely recovered its original position, being now 3 of an inch (7.62 mm.), instead of .6 of an inch, from the midrib. A trace of flexure was, however, still visible.

Experiment 6: A young and concave leaf was selected with its margins slightly and naturally incurved. Two rather large, oblong, rectangular pieces of roast meat placed with their ends touching the infolded edge, and .46 of an inch (11.68 mm.) apart from one another. After 24 hrs. the margin was greatly and equally incurved (see fig.16) throughout this space, and for a length of .12 or .13 of an inch (3.048 or 3.302 mm.) above and below each bit; so that the margin had been affected over a greater length betweren the two bits, owing to their conjoint action, than beyond them. The bits of meat were too large to be clasped by the margin, but they were tilted up, one of them so as to stand almost vertically. After 48 hrs. the margin was almost unfolded, and the bits had sunk down. When again examined after two days, the margin was quite unfolded, with the exception of the naturally inflected edge; and one of the bits of meat, the end of which had at first touched the edge, was now .067 of an inch (1.70 mm.) distant from it; so that this bit had been pushed thus far across the blade of the leaf.

Fig. 16
(Pinguicula vulgaris)
Outline of leaf with right margin inflected against two square bits of meat.

Experiment 7: A bit of meat was placed close to the incurved edge of a rather young leaf, and after it had re-expanded, the bit was left lying .11 of an inch (2.795 mm.) from the edge. The distance from the edge to the midrib of the fully expanded leaf was .35 of an inch (8.89 mm.); so that the bit had been pushed inwards and across nearly one-third of its semi-diameter.

Experiment 8: Cubes of sponge, soaked in a strong infusion of raw meat, were placed in close contact with the incurved edges of two leaves – an older and younger one. The distance from the edges to the midribs was carefully measured. After 1 hr. 17m. there appeared to be a trace of incurvation. After 2 hrs. 17m. both leaves were plainly inflected; the distance betweren the edges and midribs being now only half what it was at first. The incurvation increased slightly during the next 4½ hrs., but remained nearly the same for the next 17 hrs. 30 m. In 35 hrs. from the time when the sponges were placed on the leaves, the margins were a little unfolded – to a greater degree in the young than in the older leaf. The latter was not quite unfolded until the third day, and now both of sponge were left at the distance of 1 of an inch (2.54 mm.)

from the edges; or about a quarter of the distance betweren the edge and midrib. A third bit of sponge adhered to the edge, and, as the margin unfolded, was dragged backwards, into its original position.

Experiment 9: A chain of fibres of roast meat, a thin as bristles and moistened with saliva, were placed down one whole side, close to the narrow, naturally incurved edge of a leaf. In 3 hrs. this side was greatly incurved along its whole length, and after 8 hrs. formed a cylinder, about $\frac{1}{20}$ of an inch (1.27 mm.) in diameter, quite concealing the meat. This cylinder remained closed for 32 hrs., but after 48 hrs. was half unfolded, and in 72 hrs. was as open as the opposite margin where no meat had been placed. As the thin fibres of meat were completely overlapped by the margin, they were not pushed at all inwards, across the blade.

Experiment 10: Six cabbage seeds, soaked for a night in water, were placed in a row close to the narrow incurved edge of a leaf. We shall hereafter see that these seeds yield soluble matter to the gland. In 2 hrs. 25 m. the margin was decidedly inflected; in 4 hrs. it extended over the seeds for about half their breadth, and in 7 hrs. over three-fourths of their breadth, forming a cylinder not quite closed along the inner side. After 24 hrs. the inflection had not increased, perhaps had decreased. The glands which had been brought into contact with the upper surfaces of the seeds were now secreting freely. In 36 hrs. from the time when the seeds were put on the leaf the margin had greatly, and after 48 hrs. had completely, re-expanded. As the seeds were no longer held by the inflected margin, and as the secretion was beginning to fail, they rolled some way down the marginal channel.

Experiment 11: Fragment of glass were placed on the margins of two fine young leaves. After 2 hrs. 30 m. the margin of one certainly became slightly incurved; but the inflection never increase, and disappeared in 16 hrs. 30 m. from the time when the fragments were first applied. With second leaf there was a trace of incurvation in 2 hrs. 15 m., which became decided in 4 hrs. 30 m., and still more strongly pronounced in 7 hrs., but after 19 hrs. 30 m. had plainly decreased. The fragments excited at most a slight and doubtful increase of the secretion; and in two other trials, no increase could be perceived. Bits of coal-cinders, placed on a leaf, produced no effect, either owing to their lightness or the leaf being torpid.

Experiment 12: We will now turn to fluids. A row of drops of a strong infusion of raw meat were placed along the margins of two leaves; squares of sponge soaked in the same infusion being placed on the opposite margins. My object was to ascertain whether a fluid would act as energetically as a substance yielding the same soluble matter to the glands. No distinct difference was perceptible; certainly none in the degree of incurvation; but the incurvation round the bits of sponge lasted rather longer, as might perhaps have been expected from the sponge remaining damp and supplying nitrogenous matter for a longer time. The margins, with the drops, became plainly incurved in 2 hrs. 17 m. The incurvation subsequently increased somewhat, but greatly decreased.

Experiment 13: Drops of the same strong infusion of raw meat were placed along the midrib of a young and rather deeply concave leaf. The distance across the broadest part of the leaf, betweren the naturally incurved edges, was .55 of an inch (13.97 mm.). In 3 hrs. 27 m. this distance was a trace less; in 6 hrs. 27 m. it was exactly .45 of an inch (11.43 mm.), and had therefore decreased by .1 of an inch (2.54 mm.). After only 10 hrs. 37 m. the margin began to re-expand, for the distance from edge to edge was now a trace wider, and after 24 hrs. 20 m. was as great, within a hair's breadth, as when the drops were first placed on the leaf. From this experiment we learn that the motor impulse can be transmitted to a distance of .22 of an inch (5.590 mm.) in a transverse direction from the midrib to both margins; but it would be safer to say .2 of an inch (5.08 mm.) as the drops spread a little beyond the midrib. The incurvation thus caused lasted for an unusually short time.

Experiment 14: Three drops of a solution of one part of carbonate of ammonia to 218 of water (2 grs. to 1 oz.) were placed on the margin of a leaf. These excited so much secretion that in 1 hr. 22 m. all three drops ran together; but although the leaf was observed for 24 hrs., there was no trace of inflection. We know that a rather strong solution of this salt, though it does not injure the leaves of Drosera, paralyses their power of movement, and we have no doubt, from (this and) the following case, that this holds good with Pinguicula.

Experiment 15: A row of drops of a solution of one part of carbonate of ammonia to 875 of water (1 gr. to 2 oz.) was placed on the margin of leaf. In 1 hr. there was apparently some slight incurvation, and this was well marked in 3 hrs. 30 m. After 24 hrs. the margin was almost re-expanded.

Experiment 16: A row of large drops of a solution of one part of phosphate of ammonia to 4375 of water (1 gr. to 10 oz.) was placed along the margin of a leaf. No effect was produced, and after 8 hrs. fresh drops were added along the same margin without the least effect. We know that a solution of this strength acts powerfully on Drosera, and it is just possible that the solution was too strong.

Experiment 17: As the pressure from bits of glass causes incurvation, we scratched the margins of two leaves for some minutes with a blunt needle, but no effect was produced. The surface of a leaf beneath a drop of a strong infusion of raw meat was also rubbed for 10 m. with the end of a bristle, so as to imitate the struggles of a captured insect; but this part of the margin did not bend sooner than the other parts with undisturbed drops of the infusion.

We learn from the foregoing experiments that the margins of the leaves curl inwards when excited by the mere pressure of objects not yielding any soluble matter, by objects yielding such matter, and by some fluids—namely, an infusion of raw meat and a weak solution of carbonate of ammonia. A stronger solution of two grains of this salt to an ounce of water, though exciting copious secretion, paralyses the leaf. Drops of water and of a solution of sugar or gum did not cause any movement. Scratching the surface of the leaf for some minutes produced no effect. Therefore, as far as we at present know, only two causes—namely, slight continued pressure and the absorption of nitrogenous matter-excite movement. It is only the margins of the leaf which bend, for the apex never curves towards the base. The pedicels of the glandular hairs have no power of movement. We observed on several occasions that the surface of the leaf became slightly concave where bits of meat or large flies had long lain, but this may have been due to injury from over-stimulation.

The shortest time in which plainly marked movement was observed was 2 hrs. 17 m., and this occurred when either nitrogenous substances or fluids were placed on the leaves; but in some cases there was a trace of movement in 1 hr. or 1 hr. 30 m. The pressure from fragments of glass excites movement almost as quickly as the absorption of nitrogenous matter, but the degree of incurvation thus caused is much less. After a leaf has become well incurved and has again expanded, it will not soon answer to a fresh stimulus. The margin was affected longitudinally, upwards or downwards, for a distance of .13 of an inch (3.302 mm.) from an excited point, but for a distance of .46 of an inch betweren two excited points, and transversely for a distance of .2 of an inch (5.08 mm.). The motor impulse is not accompanied, as in the case of Drosera, by any influence causing increased secretion; for when a single gland was strongly stimulated and secreted copiously, the surrounding glands were not in the least affected. The incurvation of the margin is independent of increased secretion, for fragments of glass cause little or no secretion, and yet excite movement; whereas a strong solution of carbonate of ammonia excites secretion, but no movement.

One of the most curious facts with respect to the movement of the leaves is the short time during which they remain incurved, although the exciting object is left on them. In the majority of cases there was well-marked re-expansion within 24 hrs. from the time when even large pieces of meat, &c. were placed on the leaves, and in all cases within 48 hrs. In one instance the margin of leaf remained for 32 hrs. closely inflected round thin fibres of meat; in another instance, when a bit of sponge, soaked in a strong infusion of raw meat, had been applied to a leaf, the margin began to unfold in 35 hrs. Fragments of glass keep the margin incurved for a shorter time than do nitrogenous bodies; for in the former case there was complete re-expansion in 16 hrs. 30 m. Nitrogenous fluids act for a shorter time than nitrogenous substances; thus, when drops of an infusion of raw meat were placed on the midrib of a leaf, the incurved margins began to unfold in only 10 hrs. 37 m., and this was the quickest act of re-expansion observed by me; but it may have been partly due to the distance of the margins from the midrib where the drops lay.

We are naturally led to inquire what is the use of this movement, which lasts for so short a time? If very small objects, such as fibres of meat, or moderately small objects, such as little flies or cabbage-seeds, are placed close to the margin, they are either completely or partially embraced by it. The glands of the overlapping margin are thus brought into contact with such objects and pour forth their secretion, afterwards absorbing the digested matter. But as the incurvations lasts for so short a time any such benefit can be of only slight importance, yet perhaps greater than at first appears. The plant lives in humid districts, and the insects which adhere to all parts of the leaf are washed by every heavy shower of rain into the narrow channel formed by the naturally incurved edges. For instance, we placed several insects on some leaves, and two days afterwards (there having been heavy rain in the interval) found some of them quite washed away, and many others safely tucked under the now closely inflected margins, the glands of which all round the insects were no doubt secreting. We can thus, also, understand how it is that so many insects, and fragments of insects, are generally found lying within the incurved margins of the leaves.

The incurvation of the margin, due to the presence of an exciting object, must be serviceable in another and probably more important way. We have seen that when large bits of meat, or of sponge soaked in the juice of meat, were placed on a leaf, the margin was not able to embrace them, but, as it became incurved, pushed them very slowly towards the middle of the leaf, to a distance from the outside of fully .1 of an inch (2.54 mm.), that is, across betweren one-third and one-fourth of the space between the edge and midrib. Any object, such as moderately sized insect, would thus be brought slowly into contact with a far larger number of glands, inducing much more secretion and absorption, than would otherwise have been the case. That this would be highly serviceable to the plant, we may infer from the fact that Drosera has acquired highly developed powers of movement, merely for the sake of bringing all its glands into contact with captured insects. So again, after a leaf of Dionaea has caught an insect, the slow pressing together of the two lobes serves merely to bring the glands on both sides into contact with it, causing also the secretion charged with animal matter to spread by capillary attraction over the whole surface. In the case of pinguicula, as soon as an insect has been pushed for some little distance towards the midrib, immediate re-expansion would be beneficial, the margins could not capture fresh prey until they were unfolded. The service rendered by this pushing action, as well as that from the marginal glands being brought into contact for a short time with the upper surfaces of minute captured insects, may perhaps account for the peculiar move-ments of the leaves: otherwise, we must look at these move-ments as a remnant of a more highly developed power formerly possessed by the progenitors of the genus.

In the four British species, and in most or all these species of the genus, the edges of the leaves are in some degree naturally and permanently incurved. This incurvation serves, as already shown, to prevent insects from being washed away by the rain; but it likewise served for another end. When a number of glands have been powerfully excited by bits of meat, insects, or any other stimulus, the secretion often tricks down the meat, insects, or any other stimulus, the secretion often trickles down the leaf, and is caught by the incurved edges, instead of rolling off and being lost. As it runs down the channel, fresh glands are able to absorb the animal matter held in solution. Moreover, the secretion often collects in little pools within the channel, or in the spoon-like tips of the leaves; and we ascertained that bits of albumen, fibrin, and gluten are here dissolved more quickly and completely than on the surface of the leaf, where the secretion cannot accumulate; and so it would be with naturally caught insects. The secretion was repeatedly seen thus to collect on the leaves of plants protected from the rain; and with exposed plants there would be still greater need of some provision to prevent, as far as possible, the secretion, with its dissolved animal matter, being wholly lost.

It has already been remarked that plants growing in a state of nature have the margins of their leaves much more strongly incurved than those grown in pots and prevented from catching many insects. We have seen that insects washed down by the rain from all parts of the leaf often lodge within the margins; which are thus excited to curl farther inwards; and we may suspect that this action, many times repeated during the life of the plant, leads to their permanent and well-marked incurvation.

It may here be added, though not immediately bearing on our subject, that when a plant is pulled up, the leaves immediately curl downwards so as to almost conceal the roots—a fact which has been noticed by many persons. We suppose that this is due to the same tendency, which causes the outer and older leaves to lie flat on the ground. It further appears that the flower-stalks are to a certain extent irritable, for Dr. Johnson stated that they "bend backwards if rudely handled".

Secretion, Absorption, and Digestion—We will first give my observations and experiments, and then a summary of the results.

The Effects of Objects Containing Soluble Nitrogenous Matter—

1. *Flies* were placed on many leaves, and excited the glands to secrete

copiously; the secretion always becoming acid, though not so before. After a time these insects rendered so tender that their limbs and bodies could be separated by a mere touch, owing no doubt to the digestion and disintegration of their muscles. The glands in contact with a small fly continued to secrete for four days, and then became almost dry. A narrow strip of this leaf was cut off, and the glands of the longer and shorter hairs, which had lain in contact for the four days with the fly, and those which had not touched it, were compared under the microscope, and presented a wonderful contrast. Those which had been in contact were filled with brownish granular matter, the others with homogeneous fluid. There could therefore be no doubt that the former had absorbed matter from the fly.

2. Small bits of *roast meat*, placed on a leaf, always caused much acid secretion in the course of a few hours – in on case within 40 m. when thin fibres of meat were laid along the margin of a leaf which stood almost upright, the secretion ran down to the ground. Angular bits of meat, placed in little pools of the secretion near the margin, were in the course of two or three days much reduced in size, rounded, rendered more or less colourless and transparent, and so much softened that they fell to pieces on the slightest touch. In only one instance was a very minute particle completely dissolved, and this occurred within 48 hrs. When only a small amount of secretion was excited, this was generally absorbed in from 24 hrs. to 48 hrs.; the glands being left dry. But when the supply of secretion was copious, round either a single rather large bit of meat, or round several small bits, the glands did not become dry until six or seven days had elapsed. The most rapid case of absorption observed by me was when a small drop of an infusion of raw meat was placed on a leaf, for the glands here became almost dry in 3 hrs. 20 m. Glands excited by small particles of meat, and which have quickly absorbed their own secretion, begin to secrete again in the course of seven or eight days from the time when the meat was given them.

3. Three minute cubes of tough *cartilage* from the leg-bone of a sheep were laid on a leaf. After 10 hrs. 30 m. some acid secretion was excited, but the cartilage appeared little or not at all affected. After 24 hrs. the cubes were rounded and much reduced in size; after 32 hrs. they were softened to the

centre, and one was quite liquefied; after 35 hrs. mere traces of solid cartilage were left; and after 48 hrs. a trace could still be seen through a lens in only one of the three. After 82 hrs. not only were all three cubes completely liquefied, but all the secretion was absorbed and the glands left dry.

4. Small cubes of *albumen* were placed on a leaf; in 8 hrs. Feebly acid secretion extended to a distance of nearly $\frac{1}{10}$ of an inch round them, and the angles of one cube were rounded. After 24 hrs. the angles of all the cubes were rounded, and they were rendered throughout very tender; after 30 hrs. the secretion began to decrease, and after 48 hrs. the glands were left dry; but very minute bits of albumen were still left undissolved.
5. Smaller cubes of *albumen* (about $\frac{1}{50}$ or $\frac{1}{60}$ of an inch, .508 or .423 mm.) were placed on four glands; after 18 hrs. one cube was completely dissolved, the other being much reduced in size, softened and transparent. After 24 hrs. two of the cubes were completely dissolved, and already the secretion on these glands was almost wholly absorbed. After 42 hrs. the two other cubes were completely dissolved. These four glands began to secrete again after eight or nine days.
6. Two large cubes of *albumen* (fully $\frac{1}{20}$ of an inch, 1.27 mm.) were placed, one near the midrib and the other near the margin of a leaf; in 6 hrs. there was much secretion, which after 48 hrs. accumulated in a little pool round the cube near the margin. This cube was much more dissolved than that on the blade of the leaf; so that after three days it was greatly reduced in size, with all the angles rounded, but it was too large to be wholly dissolved. The secretion was partially absorbed after four days. The cube on the blade was much less reduced, and the glands on which it rested began to dry after only two days.
7. *Fibrin* excites less secretion than does meat or *albumen*. Several trials were made, but we will give only three of them. Two minute shreds were placed on some glands, and in 3 hrs. 45 m. their secretion was plainly increased. The smaller shred of the two was completely liquefied in 6 hrs. 15 m., and other in 24 hrs.; but even after 48 hrs. a few granules of fibrin could still be seen through a lens floating in both drops of secretion. After 56 hrs. 30 m. these granules were completely dissolved. A third shred was placed in a

little pool of secretion, within the margin of leaf where a seed had been lying, and this was completely dissolved in the course of 15 hrs. 30 m.

8. Five very small bits of *gluten* were placed on a leaf, and they excited so much secretion that one of the bits glided down into the marginal furrow. After a day all five bits seemed much reduced in size, but none were wholly dissolved. On the third day we pushed two of them which had begun to dry, on to fresh glands. On the fourth day undissolved traces of three out of the five bits could still be detected, the other two having quite disappeared. Two fresh bits were now placed, one near the middle and the other near the margin of another leaf; both excited an extraordinary amount of secretion; that near the margin had little pool formed round it, and was much more reduced in size than that on the blade, but after four days was not completely dissolved. Gluten, therefore, excites the glands greatly, but is dissolved with much difficulty, exactly as in the case of Drosera.

9. A small square thin piece of pure *gelatine*, moistened with water, was placed on a leaf, and excited very little secretion in 5 hrs. 30 m., but later in the day a greater amount. After 24 hrs. the whole square was completely liquefied; and this would not have occurred had it been left in water. The liquid was acid.

10. Small particles of chemically prepared *casein* excited acid secretion, but were not quite dissolved after two days; and the glands then began do dry. Nor could their complete dissolution have been expected from what we have seen with Drosera.

11. Minute drops of skimmed *Milk* were placed on a leaf, and these caused the glands to secrete freely. After 3 hrs. the milk was found curdled, and after 23 hrs. the curds were dissolved. On placing the now clear drops under the microscope, nothing could be detected except some oil-globules. The secretion, therefore, dissolves fresh casein.

12. Two fragments of a leaf were immersed for 17 hrs., each in a drachm of a solution of *carbonate of ammonia*, of two strengths, namely of one part to 437 and 218 of water. The glands of the longer and shorter hairs were then examined, ad their contents found aggregated into granular matter of a brownish-green colour. These granular masses were seen by my son

slowly to change their forms, and no doubt consisted of protoplasm. The aggregation was more strongly pronounced, and the movements of the protoplasm more rapid, wthin the glands subjected to the stronger solution than in the other. The experiment was repeated with the same result; and on this occcasion we observed that the protoplasm had shrunk a little from the walls of the single elongated cells forming the pedicels. In order to observe the process of aggregation, a narrow strip of leaf was laid edgeways under the microscope, and the glands were seen on to be quite transparent; a little of the stronger solution (*viz.*, one part to 218 of water) was now added under the covering glass; after an hour or two the glands contained very fine granular matter, which slowly became coarsely granular and slightly opaque; but even after 5 hrs. not as yet of a brownish tint. By this time a few rather large, transparent, globular masses appeared within the upper ends of the pedicels, and the protoplasm lining their walls had shrunk a little. It is thus evident that the glands of Pinguicula absorb carbonate of ammonia; but they do not absorb it, or are not acted on by it, nearly so quickly as those of Drosera.

13. Little masses of the orange-coloured *pollen* of the common pea, placed on several leaves, excited the glands to secrete freely. Even a very few grains which accidentally fell on a single gland caused the drop surrounding it to increase so much in size, in 23 hrs., as to be manifestly larger than the drops on the adjoining glands. Grains subjected to the secretion for 48 hrs. did not emit their tubes; they were quite discoloured, and seemed to contain less matter then before; that which was left being of a dirty colour, including globules of oil. They thus differed in appearance from other grains kept in water for the same length of time. The glands in contact with the pollen-grains had evidently absorbed matter from them; for they had lost their natural pale-green tint, and contained aggregated globular masses of protoplasm.

14. Square bits of the leaves of spinach, cabbage, and a saxifrage, and the entire leaves of *Erica tetralix*, all excited the glands to increase secretion. The spinach was the most effective, for it caused the secretion evidently to increase in 1 hr. 40 m, and ultimately to run some way down the leaf; but the glands soon began to dry, *viz.*, after 35 hrs. The leaves of *Erica tetralix* began to act in 7 hrs. 30 m., but never caused much secretion; nor

did the bits of leaf of the Saxifrage, though in this case the glands continued to secrete for seven days. Some leaves of Pinguicula were adhered with leaves of *Erica tetralix* and of an unknown plant; and the glands in contact with them had their contents plainly aggregated, as if they had been in contact with insects; whilst the other glands on the same leaves contained only clear homogeneous fluid.

15. A considerable number of seeds or fruits selected by hazard, some fresh and some a year old, some soaked for a short time in water and some not soaked, were tried. The ten following kinds, namely, cabbage, radish, *Anemone nemorosa, Rumex acetosa, Carex sylvatica*, mustard, turnip, cress, *Ranunculus acris*, and *Avena pubescens*, all excited much secretion, which was in several cases tested and found always acid. The five first-named seeds excited the glands more than the others. The secretion was seldom copious until about 24 hrs. had elapsed, no doubt owing to the coats of the seeds not being easily permeable. Nevertheless, cabbage seeds excite some secretion in 4 hrs. 30 m.; and this increased so much in 18 hrs. as to run down the leaves. The seeds, or properly the fruits, of Carex are much oftener found adhering to leaves in a state of nature than those of any other genus; and the fruits of *Carex sylvatica* excited so much secretion that in 15 hrs. it ran into the incurved edges; but the gland ceased to secrete after 40 hrs. On the other hand, the glands on which the seeds of the Rumex and Avena rested continued to secrete for nine days.

The nine following kinds of seeds excited only a slight amount of secretion, namely, celery, parsnip, caraway, *Linum grandiflorum, Cassia, Trifolium pannonicum*, Plantago, onion, and Bromus. Most of these seeds did not excite any secretion until 48 hrs. had elapsed, and in the case of the Trifolium only one seed acted, and this not until the third day. Although the seeds of the Plantago excited very little secretion, the glands continued to secrete for six days. Lastly, the five following kinds excited no secretion, though left on the leaves for two or three days, namely, lettuce, *Erica tetralix, Atriplex hortensis, Phalaris canariensis*, and wheat. Nevertheless, when the seeds of the lettuce, wheat, and Atriplex were split open and applied to leaves, secretion was excited in considerable quantity in 10 hrs., and we believe that some was excited in six hours. In the case of the Atriplex the secretion ran down to the margin, and after 24 hrs. we concluded it "as immense in quantity, and acid". The split seeds also of the Trifolium and celery acted powerfully and quickly, though the whole seeds caused, as we have seen, very little secretion, and only after a long interval of time. A slice of the common pea, which however was not tried whole, caused secretion in 2 hrs. From these facts we may conclude that the great difference in the degree and rate at which various kinds of seeds excite secretion, is chiefly or wholly due to the different permeability of their coats.

Some thin slices of the common pea, which had been previously soaked for 1 hr. in water, were placed on a leaf, and quickly excited much acid secretion. After 24 hrs. these slices were compared under a high power with others left in water for the same time; the latter contained so many fine granules of legume in that the slide was rendered muddy; whereas the slices which had been subjected to the secretion were much cleaner and more transparent, the granules of legume apparently having been dissolved. A cabbage seed which had lain for two days on a leaf and had excited much acid secretion, was cut into slices, and these were compared with those of seed which had been left for the same time in water. Those subjected to the secretion were of a paler colour; their coats presenting the greatest differences, for they were of a pale dirty tint instead of chestnut-brown. The glands on which the cabbage seeds had rested, as well as those bathed by the surrounding secretion, differed greatly in appearance from the other glands on the same leaf, for they all contained brownish granular matter, proving that they had absorbed matter from the seeds.

That the secretion acts on the seeds was also shown by some of them being killed or by the seedlings being injured. Fourteen cabbage seeds were left for three days on leaves and excited much secretion; they were then placed on damp sand under conditions known to be favourable for germination. Three never germinated, and this was a far larger proportion of deaths than occurred with seeds of the same lot, which had not been subjected to the secretion, but were otherwise treated in the same manner. Of the eleven seedlings raised, three had the edges of their cotyledons slightly browned, as if scorched; and the cotyledons of one grew into a curious indented shape. Two mustard seeds germinated; but their cotyledons were marked with brown patches and their radicles deformed. Of two radish seeds, neither germinated; whereas of many seeds of the same lot not subjected to the secretion, all, excepting one, germinated. Of the two Rumex Seeds, one died and the other germinated; but its radicle was brown and soon withered. Both seeds of the Avena germinated, one grew well, the other had its radicle brown and withered. Of six seeds of the Erica none germinated, and when cut open after having been left for five months on damp sand, one alone seemed alive. Twenty-two seeds of various kinds were found adhering to the leaves of plants growing in a state of nature; and of these, though kept for five months on damp send, none germinated, some being then evidently dead.

The Effects of Objects not Containing Soluble Nitrogenous Matter—

16. It has already been shown that bits of glass, placed on leaves, excite little or no secretion. The small amount which lay beneath the fragments was tested and found not acid. A bit of wood excited no secretion; nor did the several kinds of seeds of which the coats are not permeable to the secretion, and which, therefore, acted like inorganic bodies. Cubes of fat, left two days on a leaf, produced no effect.

17. A particle of white *sugar*, placed on a leaf, formed in 1 hr. 10 m. a large drop of fluid, which in the course of 2 additional hours ran down into the naturally inflected margin. This fluid was not in least acid, and began to dry up, or more probably was absorbed in 5 hrs. 30 m. The experiment was repeated; particles being placed on a leaf, and other of the same size on a slip of glass in a moistened state; both being covered by a bell-glass. This was done to see whether the increased amount of fluid on the leaves could be due to mere deliquescence; but this was proved not to be the case. The particle on the leaf caused so much secretion that in the course of 4 hrs. it ran down across two-thirds of the leaf. After 8 hrs. the leaf, which was concave, was actually filled with very viscid fluid; and it particularly deserves notice that this, as on the former occasion, was not in the least acid. This great amount of secretion may be attributed to exosmose. The glands which had been covered for 24 hrs. by this fluid did not differ, when examined under the microscope, from other on the same leaf, which had not come into contact with it. This is an interesting

fact in contrast with the invariably aggregated condition of glands which have been bathed by the secretion, when holding animal matter in solution.

18. Two particles of *gum Arabic* were placed on a leaf, and they certainly caused in 1 hr. 20 m. a slight increase of secretion. This continued to increase for the next 5 hrs., that is for as long a time as the leaf was observed.

19. Six small particles of dry *starch* of commerce were placed on a leaf, and one of these caused some secretion in 1 hr. 15 m., and the others in from 8 hrs. to 9 hrs. The glands which had thus been excited to secrete soon became dry, and did not begin to secrete again until the sixth day. A larger bit of starch was then placed on a leaf, and no secretion was excited in 5 hrs. 30 m.; but after 8 hrs. there was a considerable supply, which increased so much in 24 hrs. as to run down the leaf to the distance of ¾ of an inch. This secretion, though so abundant, was not in the least acid. As it was so copiously excited, and as seeds not rarely adhere to the leaves of naturally growing plants, it occurred that the glands might perhaps have the power of secreting a ferment, like ptyaline, capable of dissolving starch; so we carefully observed the above six small particles during several days, but they did not seem in the least reduced in bulk. A particle was also left for two days in a little pool of secretion, which had run down from a piece of spinach leaf; but although the particle was so minute no diminution was perceptible. We may therefore conclude that the secretion cannot dissolve starch. The increase caused by this substance may be attributed to exosmose. But we were surprised that starch acted so quickly and powerfully as it did, though in a less degree than sugar. Colloids are known to possess some slight power of dialysis; and on placing the leaves of a primula in water, and others in syrup and diffused starch, those in the starch became flaccid, but to less degree and at a much slower rate than the leaves in syrup; those in water remaining all the time crisp.

From the foregoing experiments and observations we see that objects not containing soluble matter have little or no power of exciting the glands to secrete. Non-nitrogenous fluids, if dense, cause the glands to pour forth a large supply of viscid fluid, but this is not in the least acid. On the other hand, the secretion from glands excited by contact with nitrogenous solids or liquids is invariably acid, and is so copious that it often runs down the leaves and collects within the naturally incurved margins. The secretion in this state has the power of quickly dissolving, that is of digesting, the muscles of insects, meat, cartilage, albumen, fibrin, gelatine, and casein as it excited by chemically prepared casein and gluten; but these substances (the latter not having been soaked in weak hydrochloric acid) are only partially dissolved, as was likewise the case with Drosera. The secretion, when containing animal matter in solution, whether derived from solids or from liquids, such as an infusion of raw meat, milk, or a weak solution of carbonate of ammonia, is quickly absorbed; and the glands, which were before limpid and of a greenish colour, become brownish and contain masses of aggregated granular matter. This matter, from its spontaneous movements, no doubt consists of protoplasm. No such effect is produced by the action of non-nitrogenous fluids. After the glands have been excited to secrete freely, they cease for time to secrete, but begin again in the course of a few days.

Glands in contact with pollen, the leaves of other plants, and various kinds of seeds, pour forth much acid secretion, and afterwards absorb matter probably of an albuminous nature from them. Nor can the benefit thus derived be insignificant, for a considerable amount of pollen must be blown from the many wind-fertilized carices, grasses, &c., growing where Pinguicula lives, on to the leaves thickly covered with viscid glands and forming large rosettes. Even a few grains of pollen on a single gland cause it to secrete copiously. We have also seen how frequently the small leaves of *Erica tetralix* and of other plants, as well as various kinds of seeds and fruits, especially of Carex, adhere to the leaves. One leaf of the Pinguicula had caught ten of the little leaves of the Erica; and three leaves on the same plant had each caught a seed. Seeds subjected to the action of the secretion are sometimes killed, or the seedlings injured. We may therefore conclude that *Pinguicula vulgaris*, with its small roots, is not only supported to a large extent by the extraordinary number of insects which it habitually captures, but likewise draws some nourishment from the pollen, leaves, and seeds of other plants which often adhere to its leaves. It is therefore partly a vegetable as well as an animal feeder.

Pinguicula Grandiflora

This species is so closely allied to the last that it is ranked by Dr. Hooker as sub-species. It differs chiefly in the larger size of its leaves, and in the glandular hairs near the basal part of the mid-rib being longer. But it likewise differs in constitution; it grows in rather different sites; and Dr. Moore, informed that it is much more manageable under culture, growing freely and flowering annually; whilst *Pinguicula vulgaris* had to be renewed every year. Mr. Ralfs found numerous insects and fragments of insects adhering to almost all the leaves. These consisted chiefly of Diptera, with some Hymenoptera, Homoptera, Coleoptera and a moth; on one leaf there were nine dead insects, besides a few still alive. He also observed a few fruits of *Carex pulicaris*, as well as the seeds of this same Pinguicula, adhering to the leaves. We tried only two experiments with this species; firstly, a fly was placed near the margin of a leaf, and after 16 hrs. this was found well inflected. Secondly, several small flies were placed in a row along one margin of another leaf, and by the next morning this whole margin was curled inwards, exactly as in the case of *Pinguicula vulgaris*.

Pinguicula Lusitanica

This species is very distinct from the two foregoing ones. The leaves are rather smaller, much more transparent and are marked with purple branching veins. The margins of the leaves are much more involved; those of the older ones extending over a third of the space betweren the midrib and the outside. As in the two other species, the glandular hairs consist of longer and shorter ones, and have the same structure; but the glands differ in being purple, and are often containing granular matter before they have been excited. In the lower part of the leaf, almost half the space on each side between the midrib and margin is destitute of glands; these being replaced by long, rather stiff, multi-cellular hairs, which intercross over the midrib. These hairs perhaps serve to prevent insects from setting on this part of the leaf, where there are no viscid glands by which they could be caught; but it is hardly probable that they were developed for this purpose. The spiral vessels proceeding from the midrib terminate at the extreme margin of the leaf in spiral cells; but these are not so well developed as in the two preceding species. The flower—peduncles, sepals, and petals are studded with glandular hairs, like those on the leaves.

The leaves catch many small insects, which are found, chiefly beneath the involuted margins, probably washed there by the rain. The colour of the glands on which insects have long lain is changed, bearing either brownish or pale purple, with their contents coarsely granular; so that they evidently absorb matter from their prey. Leaves of the *Erica tetralix* flowers of a *Galium*, scales of grasses, &c., likewise adhered to some of the leaves. Several of the experiments which were tried on *Pinguicula vulgaris* were repeated on *Pinguicula lusitanica*, and these will now be given.

1. A moderately sized and angular bit of *albumen* was placed on one side of leaf, halfway betweren the midrib and the naturally involuted margin. In 2 hrs. 15 m. the glands poured forth much secretion, and this side became more enfolded than the opposite one. The inflection increased, and in 3 hrs. 30 m. extended up almost to the apex. After 24 hrs. the margin was rolled into cylinder, the outer surface of which touched the blade of the leaf and reached within the 1/20 of an inch of the midrib. After 48 hrs. it began to unfold, and in 72 hrs. was completely unfolded. The cube was rounded and greatly reduced in size; the remainder being in a semi-liquefied state.
2. A moderately sized bit of *albumen* was placed near the apex of a leaf, under the naturally incurved margin. In 2 hrs. 30 m. much secretion was excited, and next morning the margin on this side was more incurved than the opposite one, but not to so great a degree as in the last case. The margin unfolded at the same rate as before. A larger proportion of the albumen was dissolved, a remnant being still left.
3. Large bits of *albumen* were laid in a row on the midribs of two leaves, but produced in the course of 24 hrs. no effect; nor could this have been expected, for even had glands existed here, the long bristles would have prevented the albumen from coming in contact with them. On both leaves the bits were now pushed close to one margin, and in 3 hrs. 30 m. this became so greatly inflected that the outer surface touched the blade; the opposite margin not being in the least affected. After three days the margins of both leaves with the albumen were still as much inflected as ever, and the glands were still secreting copiously. With *Pinguicula vulgaris* we have never seen inflection lasting so long.
4. Two *cabbage seeds*, after being soaked for an hour in water, were placed

near the margin of leaf, and caused in 3 hrs. 20 m. increased secretion and incurvation. After 24 hrs. the leaf was partially unfolded, but the glands were still secreting freely. These began to dry in 48 hrs., and after 72 hrs. were almost dry. The two seeds were then placed on damp sand under favourable condition for growth; but they never germinated, and after a time were found rotten. They had no doubt been killed by the secretion.

5. Small bits of a *spinach leaf* caused in 1 hr. 20 m. increased secretion; and after 3 hrs. 20 m. plain incurvation of the margin. The margin was well inflected after 9 hrs. 15 m., but after 24 hrs. was almost fully re-expanded. The glands in contact with the spinach became dry in 72 hrs. Bits of albumen had been placed the day before on the opposite margin of this same leaf, as well as on that of a leaf with cabbage seeds, and these margins remained closely inflected for 72 hrs., showing how much more enduring is the effect of albumen than of spinach leaves or cabbage seeds.

6. A row of small *fragments of glass* was laid along one margin of a leaf; no effect was produced in 2 hrs. 10 m., but after 3 hrs. 25 m. there seemed to be a trace of inflection, and this was distinct, though nor strongly marked, after 6 hrs. The glands in contact with the fragments now secreted more freely than before; so that they appear to be more easily excited by the pressure of inorganic object's than are the glands of *Pinguicula vulgaris*. The above slight inflection of the margin had not increased after 24 hrs., and the glands were now beginning to dry. The surface of leaf, near the midrib and towards the base, was rubbed and scratched for some time, but no movement ensued. The long hairs which are situated here were treated in the same manner, with no effect. This latter trial was made because we thought that the hairs might perhaps be sensitive to a touch, like the filaments of Dionaea.

7. The flower-peduncles, sepals and petals bear glands in general appearance like those on the leave. A piece of a flower-peduncle was therefore left for 1 hr. in a solution of one part of carbonate of ammonia to 437 of water, and this caused the glands to change from bright pink to a dull purple colour; but their contents exhibited no distinct aggregation. After 8 hrs. 30 m. they became colourless. Two minute cubes of albumen were placed on the glands of a flower-peduncle, and another cube on the glands of a

sepal; but they were not excited to increase secretion, and the albumen after two days was not in the least softened. Hence these glands apparently differ greatly in function from those on the leaves.

From the foregoing observations on *Pinguicula lusitanica* we see that the naturally much incurved margins of the leaves are excited to curve still farther inwards by contact with organic and inorganic bodies; that albumen, cabbage seeds, bits of spinach leaves, and fragments of glass, cause the glands to secrete more freely; that albumen is dissolved by the secretion, and cabbage seeds killed by it; and lastly that matter is absorbed by the glands from the insects which are caught in large numbers by the viscid secretion. The glands on the flower-peduncles seem to have no such power. This species differs from *Pinguicula vulgaris and grandiflora* in the margins of the leaves, when excited by organic bodies, being inflected to a greater degree, and in the inflection lasting for a longer time. The glands, also, seem to be more easily excited to increased secretion by bodies no yielding soluble nitrogenous matter. In other respects, as far as our observations serve, all their species agree in their functional powers.

❑❑❑

17

Utricularia

Utricularia neglecta – Structure of the bladder – The uses of the several parts – Number of imprisoned animal – Manner of capture – The bladders cannot digest animal matter, but absorb the products of its decay – Experiments on the absorption of certain fluids by the quadrifid processes – Absorption by the glands – Summary of the observations on absorption – Development of the bladders – *Utricularia vulgaris* – *Utricularia minor* – *Utricularia clandestina.*

Utricularia Neglecta

Structure of Bladder – The general appearance of a branch (about twice enlarged), with the pinnatifid leaves bearing bladders, is represented in the following sketch (fig. 17). The leaves continually bifurcate, so that a full-grown one terminates in from twenty to thirty points. Each point is tipped by a short, straight bristle; and slight notches on the sides of the leaves bear similar bristles. On both surfaces there are many small papillae, crowned with two hemispherical cells in close contact. The plants float near the surfaces of the water, and are quite destitute of roots, even during the earliest period of growth. They commonly in habit, as more than one observer has remarked to me, remarkably foul ditches.

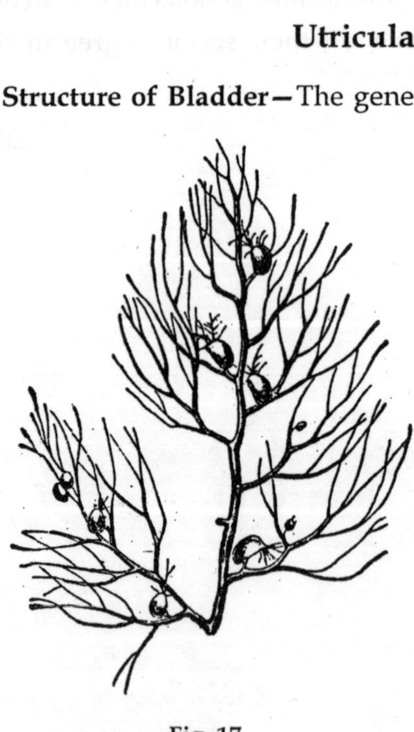

Fig. 17
(*Utricularia neglecta*)
Branch with divided leaves bearing bladders; about twice enlarged.

The bladders offer the chief point of interest. There are often two or three on the same divided leaf, generally near the base. They are supported on short footstalks. When fully grown, they are nearly $\frac{1}{10}$ of an inch (2. 54 mm) in length. They are translucent, of a green colour, and the walls are formed of two layers of cells. The exterior cells are polygonal and rather large; but at many of the points where the angles meet, there are smaller rounded cell. These latter supported short conical projections, surmounted by two hemispherical cells in such close apposition that they appear united; but they often separate a little when immersed in certain fluids. The papillae thus formed are exactly like those on the surfaces of the leaves. Those on the same bladder vary much in size; and there are a few, especially on very young bladders, which have an elliptical instead of a circular outline. The two terminal cells are transparent, but must hold much matter in solution, judging from the quantity coagulated by prolonged immersion in alcohol or ether.

The bladders are filled with water. They generally, but by no means always, contain bubbles of air. According to the quantity of the contained water and air, they vary much in thickness, but are always somewhat compressed. At an early stage of growth, the flat or ventral surface faces the axis or stem; but the foot stalks must have some power of movement; for in plants kept in greenhouse the ventral surface was generally turned either straight or obliquely downwards. The Rev. H.M. Wilkinson examined plants in a state of nature, and found this commonly to be the case, but the younger bladders often had their valves turned upwards.

The general appearance of a bladder viewed laterally, with the appendages on the near side alone represented, is shown on the preceding page (fig. 18). The lower side, where the footstalks arise, is nearly straight, and we have called it the ventral surfaces. The other or dorsal surface is convex, and terminates in two prolongations, formed of several rows of cells, containing cholorophyll, and bearing, chiefly on the outside, six or seven long, pointed, multicellular bristles. These prolongations of the bladder

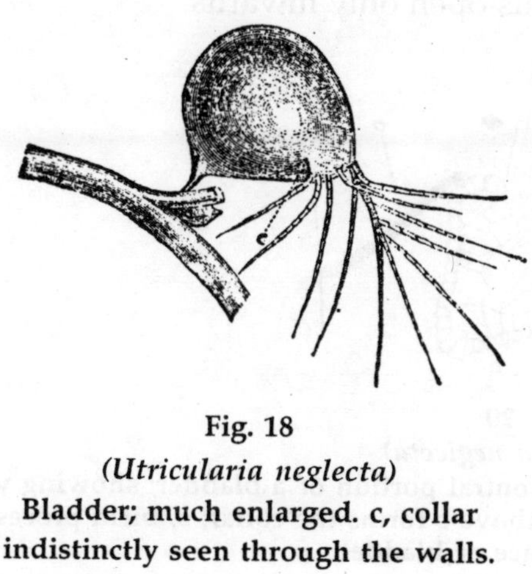

Fig. 18
(Utricularia neglecta)
Bladder; much enlarged. c, collar indistinctly seen through the walls.

may be conveniently called the antennae, for the whole bladder (see fig. 17) curiously resembles an entomos-tracan crustacean, the short footstalk representing the tail. In fig. 18, the near antenna alone are shown. Beneath the two antennae the end of the bladder is slightly truncated, and here is situated the most important part of the whole structure, namely the entrance and valve. On each side of the entrance from three to rarely seven long, multicellualr bristles project outwards; but only those (four in number) on the near side are shown in the drawing. These bristles, together with those borne by the antennae, form a sort of hollow cone surrounding the entrance.

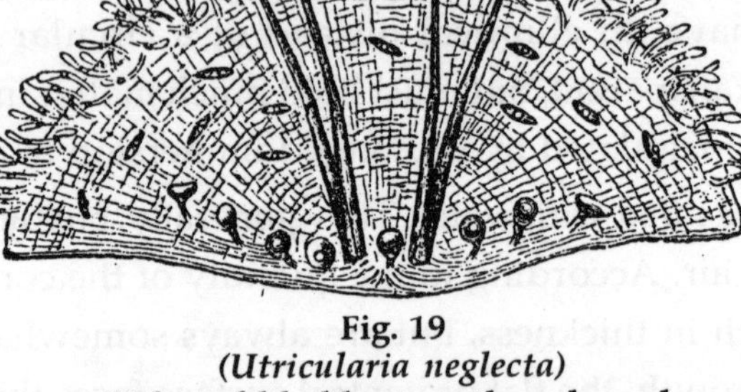

Fig. 19
(Utricularia neglecta)
Valve of bladder; greatly enlarged.

The valve slopes into the cavity of the bladder, or upwards in fig. 18. It is attached on all sides to the bladder, excepting by its posterior margin, or the lower one in fig. 19, which is free, and forms one side of the slit-like orifice leading into the bladder. This margin is sharp, thin, and smooth, and rests on the edge of a rim or collar, which dips deeply into the bladder, as shown in the longitudinal section (fig. 20) of the collar and valve; it is also shown at c, in fig. 18. The edge of the valve can thus open only inwards.

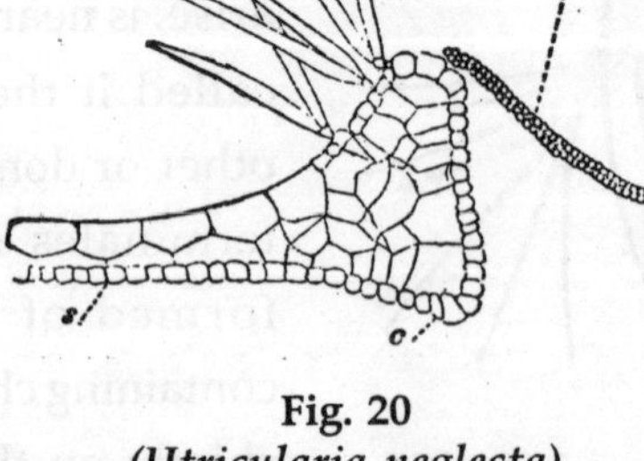

Fig. 20
(Utricularia neglecta)
Longitudinal vertical section through the ventral portion of a bladder; showing valve and collar. *v*, valve; the whole projection above *c* forms the collar; *b*, bifid processes; *s*, ventral surface of bladder.

The valve is colourless, highly transparent, flexible and elastic. It is convex in a transverse direction, but has been drawn (fig. 19) in a flattened state, by which its apparent breadth is increased. It is formed, according to Cohn, of two layers of small cells, which are continuous with the two layers of larger cells forming the walls of the bladder, of which it is evidently a prolongation. Two pairs of transparent pointed bristles, about as long as the valve itself, arise from near the free posterior margin (fig.19), and point obliquely outwards in the direction of the antennae. There are also on the surface of the valve numerous glands. They consist of three kinds, which to a certain extent graduate into one another. Those situated round the anterior margin of the valve (upper margin in fig.19) are very numerous and crowded together; they consist of oblong head on a long pedicel. The pedicel itself is formed of an elongated cell, surmounted by a short one. The glands towards the free posterior margin are much larger, few in number, and almost spherical, having short footstalks; the head is formed by the confluence of two cells, the lower one answering to the short upper cell of the pedicel of the oblong glands. The glands of the third kind have transversely elongated heads, and are scated on very short footstalks; so that they stand paralleled close to the surface of the valve; they maybe called the two-armed glands. The cells forming all these glands contain a nucleus, and are lined by a thin layer of more or less granular protoplasm, the primordial utricle of Mold. They are filled with fluid, which must hold much matter in solution, judging from the quantity coagulated after they have been long immersed in alcohol or ether. The depression in which the valve lies is also lined with innumerable glands; those at the sides having oblong heads and elongated pedicels, exactly like the glands on the adjoining parts of the valve.

The collar (called the peristome by Cohn) is evidently formed, like the valve, by inward projection of the walls of the bladder. The cells composing the outer surface, or that facing the valve, have rather thick walls, are of a brownish colour, minute, very numerous, and elongated; the lower ones being divided into two by vertical partitions. The whole presents a complex and elegant appearance. The cells forming the inner surface are continuos with those over the whole inner surface of the bladder. The space between the inner and outer surface consists of coarse cellular tissue (fig. 20). The inner side is thickly covered with delicate bifid processes, hereafter to be described.

The collar is thus made thick; and it is rigid, so that it retains the same outline whether the bladder contains little or much air and water. This is of great importance, as otherwise the thin and flexible valve would be liable to be distorted, and in this case would not act properly.

Altogether the entrance into the bladder, formed by the transparent valve, with its four obliquely projecting bristles, its numerous diversely shaped glands, surrounded by the collar, bearing glands on the insides and bristles on the outside, together with the bristles borne by the antennae, presents an extraordinary complex appearance when viewed under the microscope.

We will now consider the internal structure of the bladder. The whole inner surface, with the exception of the valve, is seen under a moderately high power to be covered with a serried mass of processes (fig. 21). Each of these consists of four divergent arms; whence their name of quadrifid processes. They arise from small angular cells, at the junctions of the angles of the larger cells, which form the interior of the bladder. The middle part of the upper surface of these small cells projects a little, and then contracts into a very short and narrow footstalk which bears the four arms (fig. 22). Of these, two are long, but often of not quite equal length, and project obliquely inwards and towards the posterior end of the bladder. The two others are much shorter, and project at a smaller angle, that is, are more nearly horizontal, and are towards the anterior end of the bladder. These arms are only moderately sharp; they are composed of extremely thin transparent membrane, so that they can be bend or doubled in any direction without being broken. They are lined with a delicate layer of protoplasm, as is likewise the short conical projection from which they arise. Each arm generally (but not invariably) contains a minute faintly brown particle either rounded or more commonly elongated, which exhibits incessant Brownian movements. These particles slowly change their positions, and travel from one end to the other of the arms, but are commonly found near their bases. They are present in the quadrifids of young bladders, when only about a third of their full size. They do not resemble ordinary nuclei, but we believe that they are nuclei in a modified condition, for when absent, we could occasionally just distinguish in their places a delicate halo of matter, including a darker spot. Moreover, the quadrifids of *Utricularia montana* contain rather larger and much more regularly spherical, but otherwise similar, particles, which closely resemble the nuclei in the cells forming the wells of the bladders. In the present cause there were sometimes two, there, or even more, nearly similar particles within a single arm; but, as we shall hereafter see, the presence of more than one seemed always to be connected with the absorption of decayed matter.

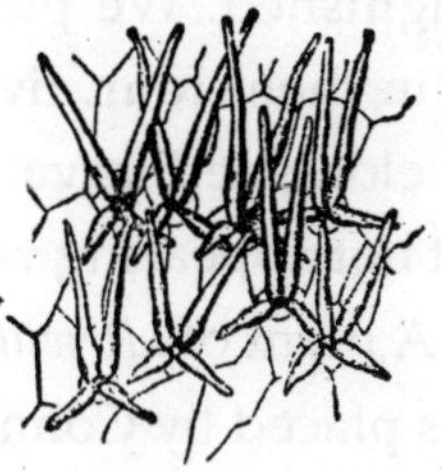

Fig. 21
(Utricularia neglecta)
Small portion of inside of bladder, much enlarged, showing quadrifid processes.

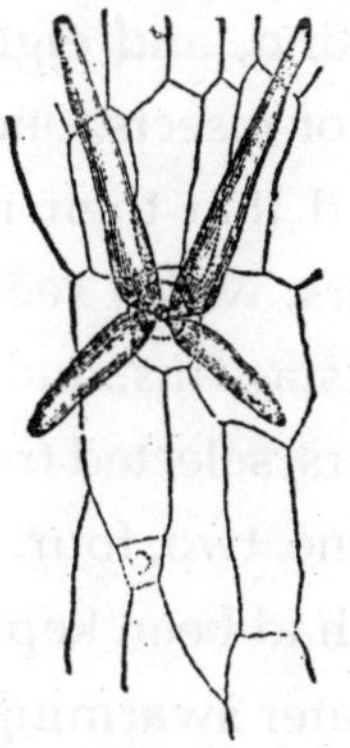

Fig. 22
(Utricularia neglecta)
One of the quadrifid processes greatly enlarged.

The inner side of the collar (see the previous fig. 20) is covered with several crowded rows of processes, differing in no important respect from the quadrifids, except in bearing only two arms instead of four; they are, however, rather narrower and more delicate. We shall call them the bifids. They project into the bladder, and are directed towards its posterior end. The quadrifid and bifid processes no doubt are homologous with the papillae on the outside of the bladder and of the leaves; and we shall see that they are developed from closely similar papillae.

The Uses of the Several Parts—After the above long but necessary description of the parts, we will turn to their uses. The bladders have been supposed by some authors to serve as floats; but branches which bore no bladders, and other from which they had been removed, floated perfectly, owing to the air in the intercellular spaces. Bladders containing dead and captured animals usually include bubbles of air, but these cannot have been generated solely by the process of decay, as we have often seen air in young, clean and empty bladders; and some old bladders with much decaying matter had no bubbles.

The real use of the bladders is to capture small aquatic animals, and this they do on a large scale. In the first lot of plants a large proportion of the fully grown bladders contained prey; in a second lot most of the bladders were empty, but plants had been selected which had grown in unusually pure water. In the first lot, we examined seventeen bladders, including prey of some kind, and eight of these contained entomostracan crustaceans, three larvae of insects, one being still alive, and six remnants of animals so much decayed that their nature could not be distinguished. We picked out five bladders, which seemed very full, and found in them four, five, eight, and ten crustaceans, and in the fifth a single much elongated larva. In five other bladders, selected from containing remains, but not appearing very full, there were one, two, four, two, and five crustaceans. A plant of *Utricularia vulgarus*, which had been kept in almost pure water, was placed by Cohn one evening into water swarming with crustaceans, and by the next morning most of the bladders contained these animals entrapped and swimming round and round their prisons. They remained alive for several days; but at last perished, asphyxiated by the oxygen in the water having been all consumed. Freshwater worms were also found by Cohn in some bladders. In all cases the bladders with decayed remains swarmed with living algae of many kinds, Infusoria, and other low organisms, which evidently lived as intruders.

Animals enter the bladders by bending inwards the posterior free edge of the valve, which from being highly elastic shuts again instantly. As the edge is extremely thin, and fits closely against the edge of the collar, both projecting into the bladder (see section, fig. 20), it would evidently be very difficult for any animal to get out when once imprisoned, and apparently they never do escape. To show how closely the edge fits, I may mention that my son found a Daphnia, which had inserted one of its antennae into the slit, and it was thus held fast during a whole day. On three or four occasions I have seen long narrow larvae, both deaf and alive, wedged between the corner of the valve and collar, with half their bodies within the bladder and half out.

As we felt much difficulty in understanding how such minute and week animals, as are often captured, could force their way into the bladders, we tried many experiments to ascertain how this was affected. The free margin of the valve bends so easily that no resistance is felt when a needle or thin bristle is inserted. A thin human hair fixed to a handle, and cut off so as to project barely 14 of an inch, entered with some difficulty; a longer piece yielded instead of entering. On three occasions minute particles of blue glass (so as to be easily distinguished) were placed on valves whilst under water; and on trying gently to move them with a needle, they disappeared so suddenly that, not seeing what had happened, we thought that we had flirted them off; but on examining the bladders, they were found safely enclosed. The same thing occurred to my son, who placed little cubes of green box-wood (about $\frac{1}{60}$ of an inch,. 432 mm.) on some valves; and thrice in the act of placing them on, or whilst gently moving them to another spot, the valve suddenly opened and they were engulfed. He then placed similar bits of wood on other valves, and moved them about for some time, but they did not enter. Again, particles of blue glass on three valves and extremely minute shavings of lead on two other valves; after 1 or 2 hrs. none had entered, but in from 2 to 5 hrs. all five were enclosed. One of the particles of glass was a long splinter, of which one end rested obliquely on the valve, and after a few hours it was found fixed, half within the bladder and half projecting out, with the edge of the valve fitting closely all round, except at one angle, where a small one space was left. It was so firmly fixed, like the above-mentioned larvae, that the bladder was torn from the branch and shaken, and yet the splinter did not fall out. We also placed little cubes (about 1/65 of an inch, .391 mm.) of green box-wood, which were just heavy enough to sink in water, on three valves. These were examined after 19 hrs. 30 m., and were still lying on the valves; but after 22 hrs. 30 m. one was found enclosed. We may here mention that we found in a bladder on a naturally growing plant a grain of sand, and in another bladder three grains; these must have fallen by some accident on the valves, and then entered like the particles of glass.

The slow bending of the valve from the weight of particles of glass and even of box-wood, though largely supported by the water, is analogous to the slow bending of colloid substances. For instances, particles of glass were placed on various points of narrow strips of moistened gelatine and these yielded and became bend with extreme slowness. It is much more difficult to understand how gently moving particles from one part of a valve to another causes it suddenly to open. To ascertain whether the valves were endowed with irritability, the surfaces of several were scratched with a needle or brushed with a fine camel-hair brush, so as to imitate the crawling movement of small crustaceans, but the valve not open. Some bladders, before being brushed, were left for a time in water at temperatures between 80° and 130° F (26.6–54.4° C) as, judging from a wide-spread analogy, this would by itself have excited movement but no effect was produced. We may therefore conclude that animals enter merely by forcing their way through the slit-like orifice; their heads serving as a wedge. but we were surprised that such small and weak creatures as are often captured (for instance, the nauplius of a crustacean, and a tardigrade) should be strong enough to act in this manner, seeing that it was difficult to push in one end of bit of hair ¼ of an inch in length. Nevertheless, it is certain that weak and small creatures do enter, and has often been witnessed in the case of *Utricularia clandestina* the whole process. We saw a tardigrade slowly walking round a bladder, as if reconnoitring; at last it crawled into the depression where the valve lies, and then easily entered. We also witnessed the entrapment of various minute crustaceans. Cypris "was quite wary, but nevertheless was often caught. Coming to the entrance of a bladder, it would sometimes pause a moment, and then dash away; at other times it would come close up, and even venture part of the way into the entrance and back out as if afraid. Another, more heedless, would open the door and walk in; but it was no sooner in than it manifested alarm, drew in its feet and antenneae, and closed its shell". Larvae, apparently of gnats, when "feeding near the entrance, are pretty certain to run their heads into the net, whence there is no retreat. A large larva is sometimes three or four hours in being swallowed, the process bringing to mind what we have witnessed when a small snake makes a large frog its victim". But as the valve does not appear to be in the least irritable, the slow swallowing process must be the effect of the onward movement of the larva.

It is difficult to conjecture what can attract so many creatures, animal- and vegetable-feeding crustaceans, worms, tardigrades, and various larvae just referred to are vegetable feeders, and seem to have special liking for the long bristles round the valve, but this taste will not account for the entrance of animal-feeding crustaceans. Perhaps small aquatic animal habitually try to enter every small crevice, like that between the valve is an accidental circumstance, and the spot of light thus formed may serve as a guide. The long bristles round the entrance apparently serve for the same purpose. We believe that this is the case, because the bladders of some epiphytic and marsh species of Utricularis which live embedded either in entangled vegetation or in mud, have no bristles round the entrance, and these under such conditions would be of no service as a guide. Nevertheless, with these epiphytic and marsh species, two pairs of bristles project from the surface of the valve, as in the aquatic species; and their use probably is to prevent too large animals from trying to force an entrance into the bladder, thus rupturing the orifice.

As under favourable circumstances most of the bladders succeed in securing prey, in one case as many as ten crustaceans;-as the valve is so well fitted to allow animals to enter and to prevent their escape;-and as the inside of the bladder presents so singular a structure, clothed with innumerable quadrifid and bifid recesses, it is impossible to doubt that the plant has been specially adapted for securing prey. From the analogy of Pinguicula, belonging to the same family, we naturally expected that the bladders would have digested their prey; but this is not the case, and there are no glands fitted for secreting the proper fluid. Nevertheless, in order to test their power of digestion, minute fragments of roast meat, three small cubes of albumen, and three of cartage were pushed through the orifice into the bladders of vigorous plants. They were left from one day to three days and a half within, and the bladders were then cut open: but none of the above substances exhibited the least signs of digestion or dissolution; the angles of the cubes being as sharp as ever. These observations were made subsequently to those on Drosera, Dionaea, Drosophyllum, and Pinguicula; so that I was familiar with the appearance of these substances when under going the early and final stages of digestion. We may therefore conclude that Utricularia cannot digest the animal, which it habitually captures.

In most of the bladders the captured animal are so much decayed that they form a pale brown, pulpy mass, with their chitionous coats so tender that they fall to pieces with the greatest ease. The black pigment of the eye-spots is preserved better than anything else. Limbs, jaws &c. are often found quite detached; and this is the result of the vain struggles of the later captured animals. We have sometimes felt surprised at the small proportion of imprisoned animals in a fresh state compared with those utterly decayed. Mrs. Treat stated with respect to the larvae above referred to, that "usually in less than two days after a large one was captured the fluid contents of the bladders began to assume a cloudy or muddy appearance, and often became so dense that the outline of the animal was lost to view". This statement raises the suspicion that the bladders secrete some ferment hastening the process of decay. There is no inherent improbability in this supposition, considering that meat soaked for ten minutes in water mingled with the milky juice of the papaw becomes quite tender and soon passes, as Browne remarked in his "Natural History of Jamaica", into a state putridity.

Whether or not the decay of the imprisoned animals is in any way hastened, it is certain that matters is absorbed from them by the quadrifid and bifid processes. The extremely delicate nature of the membrane of which these processes are formed, and the large surface which they expose, owing to their number crowned over the whole interior of the bladder, are circumstances all favouring the process of absorption. Many perfectly clean bladders which had never caught any prey were opened, and nothing could be distinguished with a No. 8 object—glass of Hartnack within the delicate, structureless protoplasmic lining of the arms, excepting to each a single yellowish particle or modified nucleus. Sometimes two or even three such particles were present; but in this case trace of decaying matter could generally be detected. On the other hand, in bladders containing either one large or several small delayed animals, the processes presented a widely different appearance. Six such bladder were carefully examined; one contained an elongated, coiled-up larva; another a single large entomostracan crustacean, and the others from two to five smaller ones, all in a decayed state. In these six bladders, a large number of the quadrifid processes contained transparent, often yellowish, more or less confluent, spherical or irregularly shaped, masses of matter. Some of the processes, however, contained only fine granular matter, the particles of which were so small that they could not be defined clearly with No. 8 Hartnack. The delicate layer of protoplasm lining their walls was in some cases a little shrunk. On three occasions the above small masses of matter were observed and sketched at short intervals of time; and they certainly changed their positions relatively to each other and to the walls of the arms. Separate masses sometimes became confluent, and then again divided. A single little mass would send out a projection, which after a time separated itself. Hence there could be no doubt that these masses consisted of protoplasm. Bearing in mind that many clean bladders were examined with equal care, and that these presented no such appearance, we may confidently believe that the protoplasm in the above cases had been generated by the absorption of nitrogenous matter from the decaying animals. In two or three other bladders, which at first appeared quite clean, on careful search a few processes were found, with their outsides clogged with a little brown matter, showing that some minute animal had been captured and had decayed, and the arm here included a very few more or less spherical and aggregated masses; the processes in other parts of the bladders being empty and transparent. On the other hand, it must be stated that in three ladders containing dead crustaceans, the processes were likewise empty. This fact may be accounted for by the animals not having been sufficiently decayed, or by time enough not having been allowed for the generation of protoplasm or by its subsequent absorption and transference to other parts of the plant. It will hereafter be seen that in three or four other species of Utricularia the quadrifid processes in contact with decaying animals likewise contained aggregated masses of protoplasm.

The Absorption of Certain Fluids by the Quadrifid and Bifid Processes— These experiments were tried to ascertain whether certain fluids, which seemed adapted for the purpose would produce the same effects on the processes as the absorption of decayed animal matter. Such experiments are, however, troublesome; for it is not sufficient merely to place a branch in the fluid, as the valve shuts so closely that the fluid apparently does not enter soon, if at all. Even when bristles were pushed into the orifices, they were in several cases wrapped so closely round by the tin flexible edge of the valve that the fluids was apparently excluded; so that the experiments tried in this manner are doubtful and not worth giving. The best plan would have been to puncture the bladders. In all such trials, however, it cannot be ascertained positively that the bladder, though translucence, does not contain some minute animal in the last stage of decay. Therefore most of our experiments were made by cutting bladders longitudinally into two; the quadrifids were examined with No. 8 of Hartnack, then irrigated, whilst under the covering glass, with a few drops of the fluids under trial, kept in a damp chamber, and re-examined after stated intervals of time with the same power as before.

Four bladders were first tried as a control experiment, in the manner just described, in a solution of one part of gum Arabic to 218 of water, and two bladders in a solution of one part of sugar to 437 of water; and in neither case was any change perceptible in the quadrifids or bifids after 21 hrs. Four bladders were then treated in the same manner with a solution of one part of nitrate of ammonia to 437 of water, and re-examined after 21 hrs. In two of these the quadrifids now appeared full of very finely granular matter, and their protoplasmic lining or primordial utricle was a little shrunk. In the third bladder, the quadrifids included distinctly visible granules, and the primordial utricle was a little shrunk after only 8 hrs. In the fourth bladder the primordial utricle in most of the processes was here and there thickened into little irregular yellowish specks; and from the gradations which could be trace in this and other cases, these specks appear to give rise to the larger free granules contained within some of the process. Other bladders, which, as far as could be judged, had never caught any prey, were punctured and left in the same solution for 17 hrs.; and their quadrifids now contained very fine granular matter.

A bladder was dissected, examined, and irrigated with a solution of one part of carbonate of ammonia to 437 of water. After 8 hrs. 30 m. the quadrifids contained a good many granules, and the primordial utricle was somewhat shrunk; after 23 hrs. the quadrifids and bifids contained many spheres of hyaline manner, and in one arm twenty-four such spheres of moderate size were counted. Two bisected bladders which had been previously left for 21 hrs. in the solution of gum (one part to 218 of water) without being affected, were irrigated with the solution of carbonate of ammonia; and both had their quadrifids modified in nearly the same manner as just described—one after only 9 hrs., and the other after 24 hrs. Two bladders which appeared never to have caught any prey were punctured and placed in the solution; the quadrifids of one were examined after 17 hrs., and found slightly opaque; the quadrifids of the other, examined after 45 hrs., had their primordial utricles more or less shrunk with thickened yellowish specks like those due to the action of nitrate of ammonia. Several uninjured bladders were left in the same solution, as well as in a weaker solution of one part of 1750 of water, or 1 gr. to 4 oz.; and after two days the quadrifids were more or less opaque, with their contents finely granular.

Two bisected bladders were irrigated with a solution of one part of urea to 218 of water; but when this solution was employed, we forgot that it had been kept for some days in a warm room, and had there fore probably generated ammonia; anyhow, the quadrifids were affected after 21 hrs. as if a solution of carbonate of ammonia had been used; for the primordial utricle was thickened in specks, which seemed to graduate into separate granules. Three bisected bladders were also irrigated with a fresh solution of urea of the same strength; their quadrifids after 21 hrs. were much less affected than in the former case; nevertheless, the primordial utricle in some of the arms was a little shrunk, and in others was divided into two almost symmetrical sacks.

Three bisected bladders, after being examined, were irrigated with a putrid and very offensive infusion of raw meat. After 23 hrs. the quadrifids and bifids in all three specimens abounded with minute, hyaline, spherical masses; and some of their primordial utricles were a little shrunk. Three bisected bladders were also irrigated with a fresh infusion of raw meat; to my surprise the quadridfids in one of them appeared, after 23 hrs., finely granular, with their primordial utricles somewhat shrunk and marked with thickened yellowish specks; so that they had been acted on in the same manner as by the putrid infusion or by the salts of ammonia. In the second bladder some of the quadrifid were similarly acted on, though to a very slight degree; whilst the third bladder was not at all affected.

From these experiments it is clear that the quadrifid and bifid processes have the power of absorbing carbonate and nitrate of ammonia, and matter of some kind from a putrid infusion of meat. Salts of ammonia were selected for trial, as they are known to be rapidly generated by the decay of animal matter in the presence of air and water, and would therefore be generated within the bladders containing captured prey. The effect produced on the processes by these salts and by a putrid infusion of raw meat differs from that produced by the decay of the naturally captured animals only in the aggregated masses of protoplasm being in the latter case of larger size; but it is probable that the fine granules and small hyaline spheres produced by the solution would coalesce into larger masses, with time enough allowed. We have seen with Drosera that the first effect of a weak solution of carbonate of ammonia on the cell-contents is the production of the finest granules, which afterwards aggregated into larger, more or less rounded, masses; and that the granules in the layer of protoplasm which flows round the walls ultimately coalesce with these masses. Changes of this nature are, however, far more rapid in Drosera than in Utricularia. Since the bladders have no power of digesting albumen, cartilage, or roast meat, we were surprised that matter was absorbed, at least in one case, from afresh infusion of raw meat. We were also surprised, from what we shall presently see with respect to the glands round the orifice, that a fresh solution of urea produced only a moderate effect on the quadrifids.

As the quadrifids are developed from papillae which at first closely resemble those on the outside of the bladders and on the surfaces of the leaves, two hemispherical cells with which these latter papillae are rowed, and which in their natural state are perfectly transparent, likewise absorb carbonate and nitrate of ammonia; for, after an immersion of 23 hrs. in solution of one part of both these salts to 437 of water, their primordial utricles were a little shrunk and of a pale brown tint, and sometimes finely granular. The same result followed from the immersion of a whole branch for nearly three days in a solution of one part of the carbonate in 1750 of water, their primordial utricles were a little shrunk and of a pale brown tint, and sometimes finely granular. The same result followed from the immersion of a whole branch for nearly three days in a solution of one part of the carbonate to 1750 of water. The grains of cholorophyll, also, in the cells of the leaves on this branch became in many placed aggregated into little green masses, which were often connected together by the finest threads.

The Absorption of Certain Fluids by the Glands on the Valve and Collar—The glands round the orifices of bladders which are still young, or which have been long kept in moderately pure water, are colourless; and their primordial utricles are only slightly or hardly at all granular. But in the greater number of plants in a state of nature-and we must remember that they generally grown in very foul water—and with plants kept in an aquarium in foul water, most of the glands were of a pale brownish tint; their primordial utricles were more or less shrunk, sometimes ruptured, with their contents often coarsely granular or aggregated into little masses. That this state of the glands is due to their having absorbed matter from the surrounding water, we cannot doubt; for, as shall immediately see, nearly the same result follow from their immersion for a few hours in various solution. Nor is it probable that this absorption is useless, seeing that it is almost universal with plants growing in a state of nature, excepting when the water is remarkably pure.

The pedicels of the glands which are situated close to the slit-like orifice, both those on the valve and on the collar, are short; whereas the pedicels of the more distant glands are much elongated and project inwards. The glands are thus well placed so as to be washed by any fluid coming out of the bladder through the orifice. The valve fits so closely, judging from the result of immersing uninjured bladders in various solutions, that it is doubtful whether any putrid fluid habitually passes outwards. But we must remember that a bladder generally captured several animals; and that each time a fresh animal enters, a puff of foul water must pass out and bathe the glands. Moreover, we have repeatedly found that, by gently pressing bladders, which contained air, minute bubbles were driven out through the orifice; and if a bladder is laid on blotting paper and gently pressed, water oozes out. In this latter case, as soon as the pressure is relaxed, air is drawn in, and the bladder recovers its proper form. If it is now placed under water and again gently pressed, minute bubbles issue from the orifice and nowhere else, showing that the walls of the bladder have not been ruptured. We mention this because Cohn quotes a statement by Treviranus, that air cannot be forced out of a bladder without rupturing it. We may therefore conclude that whenever air is secreted within a bladder already full of water, some water will be slowly driven out through the orifice. Hence we can hardly doubt that the numerous glands crowned round the orifice are adapted to absorb matter from the putrid water, which will occasionally escape from bladders including decayed animals.

In order to test this conclusion, we experimented with various solution on the glands. As in the case of the quadrifids, salts of ammonia were tried, since these are generated by the final decay of animal matter under water. Unfortunately the glands cannot be carefully examined whilst attached to the bladders in their entire state. Their summits, therefore, including the valve, collar, and antennae, were sliced off and the condition of the glands observed; they were then irrigated, whilst beneath a covering glass, with the solution, and after a time re-examined with the same power as before, namely No.8 of Hartnack. The following experiments were thus made.

As a control experiment solutions of one part of white sugar and of one part of gum to 218 of water first used, to see whether these produced any change in the glands. It was also necessary to observe whether the glands were affected by the summits of the bladders having been cut off. The summits of four were thus tried; one being examined after 2 hrs. 30 m., and the other three after 23 hrs.; but there was no marked change in the glands of any of them.

Two summits bearing quite colourless glands were irrigated with a solution of carbonate of ammonia of the same strength (*viz.*, one part to 218 of water), and in 5 m. the primordial utricles of most of the glands were somewhat contracted; they were also thickened in specks or patches, and had assumed a pale brown tint. When looked at again after 1 hr. 30., most of them presented a somewhat different appearance. A third specimen was treated with a weaker solution of one part of the carbonate to 437 of water, and after 1 hr. the glands were pale brown and contained numerous granules.

Four summits were irrigated with a solution of one part of nitrate of ammonia to 437 of water. One was examined after 15 m., and the glands seemed affected; after 1 hr. 10 m. there was a greater change, and the primordial utricles in most of them were somewhat shrunk, and included many granules. In the second specimen, the primordial utricles were considerably shrunk and brownish after 2 hrs. Similar effects were observed in the two other specimens, but these were not examined until 21 hrs. had elapsed. The nuclei of many of the glands apparently had increased in size. Five bladders on a branch, which had been kept for a long time in moderately pure water, were cut off and examined, and their glands found very little modified. The remainder of this branch was placed in the solution of the nitrate, and after 21 hrs. two bladders were examined, and all their glands were brownish, with their primordial utricles somewhat shrunk and finely granular.

The summit of another bladder, the glands of which were in a beautifully clear condition, was irrigated with a few drops of a mixed solution of nitrate and phosphate of ammonia, each of one part to 437 of water. After 2 hrs. some few of the glands were brownish. After 8 hrs. almost all the oblong glands were brown and much more opaque than they were before; their primordial utricles were somewhat shrunk and contained a little aggregated granular matter. The spherical glands were still white, but their utricles were broken up into three or four small hyaline spheres, with an irregularly contracted mass in the middle of the basal part. These smaller spheres changed their forms in the course of a few hours, and some of them disappeared. By the next morning, after 23 hrs. 30 m., they had all disappeared, and the glands were brown; their utricles now formed a globular shrunken mass in the middle. The utricles of the oblong glands had shrunk very little, but their contents were somewhat aggregated. Lastly, the summit of a bladder which had been previously irrigated for 21 hrs. with a solution of one part of sugar to 218 of water without being affected, was treated with the above mixed solution; and after 8 hrs. 30 m. all the glands became brown, with their primordial utricles slightly shrunk.

Four summits were irrigated with a putrid infusion of raw meat. No change in the glands was observable for some hours, but after 24 hrs. most of them had become brownish, and more opaque and granular than they were before. In these specimens, as in these irrigated with the salts of ammonia, the nuclei seemed to have increased both in size and solidity, but they were not measured. Five summits were also irrigated with a fresh infusion of raw meat; three of these were not at all affected in 21 hrs. but the glands of the other two had perhaps become more granular. One of the specimens which was not affected was then irrigated with the mixed solution of the nitrate and phosphate of ammonia, and after only 25 m. the glands contained from four or five to a dozen granules. After six additional hours their primordial utricles were greatly shrunk.

The summit of bladder was examined, and all the glands found colourless, with their primordial utricles not at all shrunk; yet many of the oblong glands contained granules just resolvable with No. 8 of Hartnack. It was then irrigated with a few drops of a solution of one part of urea to 218 of water. After 2 hrs. 25 m. the spherical glands were still colourless; whilst the oblong and two-armed ones were of a brownish tint, and their primordial utricles much shrunk, some containing distinctly visible granules. After 9 hrs. some of the spherical glands were brownish, and the oblong glands were still more changed, but they contained fewer separate granules; their nuclei, on the other hand, appeared larger, as if they had absorbed the granules. After 23 hrs. all the glands were brown, their primordial utricles greatly shrunk, and in many cases ruptured.

A bladder was now experimented on, which was already somewhat affected by the surrounding water; for the spherical glands, though colourless, had their primordial utricles slightly shrunk; and the oblong glands were brownish, with their utricles much, but irregularly, shrunk. The summit was treated with the solution of urea, but was little affected by it in 9 hrs.; nevertheless, after 23 hrs. the spherical glands were brown, with their utricles more shrunk; several of the other glands were still browner, with their utricles contracted into irregular little masses.

Two other summits, with their glands colourless and their utricles not shrunk, were treated with the same solution of urea. After 5 hrs. many of the glands presented a shade of brown, with their utricles slightly shrunk. After 20 hrs. 40 m. some few of them were quite brown, and contained irregularly aggregated masses; others were still colourless, though their utricles were shrunk; but the greater number were not much affected. This was a good instance of how unequally the glands on the same bladder are sometimes affected, as likewise often occurs with plants growing in foul water. Two other summits were treated with a solution which had been kept during several days in a warm room, and their glands were not at all affected when examined after 21 hours.

A weaker solution one part of urea to 437 of water was next tried on six summits, all carefully examined before being irrigated. The first was re-examined after 8 hrs. 30 m., and the glands, including the spherical ones, were brown; many of the oblong glands having their primordial utricles much shrunk and including granules. The second summit, before being irrigated, had been somewhat affected by the surrounding water, for the spherical glands were not quite uniform in appearance; and a few of the oblong ones were brown, with their utricles shrunk. Of the oblong glands, those, which were before colourless, became brown in 3 hrs. 12 m. after irrigation, with their utricles slightly shrunk. The spherical glands did not become brown, but their contents seemed changed in appearance, and after 23 hrs. still more changed and granular. Most of the oblong glands were now dark brown, but their utricles were not greatly shrunk. The four other specimens were examined after 3 hrs. 30 m., after 4 hrs. and 9 hrs.; a brief account of their condition will be sufficient. The spherical glands were not brown, but some of them were finely granular. Many of the oblong glands were brown; and these, as well as others, which still remained colourless, had their utricles more or less shrunk, some of them including small aggregated masses of matter.

Summary of the Observations on Absorption – From the facts now given there can be no doubt that the variously shaped glands on the valve and round the collar have the power of absorbing matter from weak solutions of certain salts of ammonia and urea, and from a putrid infusion of raw meat. Prof. Cohn believes that they secrete slimy matter; but we were not able to perceive any trace of such action, excepting that, after immersion in alcohol, extremely fine lines could sometimes be seen radiating from their surfaces. The glands are variously affected by absorption: they often become of a brown colour; sometimes they contain very fine granules, or moderately sized grain, or irregularly aggregated little masses; sometimes the nuclei appear to have increase in size; the primordial utricles are generally more or less shrunk and sometimes ruptured. Exactly the same changes may be observed in the glands of plants growing and flourishing in foul water. The spherical glands are generally affected rather differently from the oblong and two-armed ones. The former does not so commonly become brown, and are acted on more slowly. We may therefore infer that they differ sometimes in their natural functions.

It is remarkable how unequally the glands on the bladders on the same branch, and even the glands of the same kind on the same bladder, are affected by the foul water in which the plants have grown, and by the solution which were employed. In the former case we presume that this is due either to little currents bringing matter to some glands and not to others, or to unknown differences in their constitution. When the glands on the same bladder are differently affected by a solution, we may suspect that some of them had previously absorbed a small amount of matter from the water. However this may be, we have seen that the glands on the same leaf of Drosera are sometimes very unequally affected, more especially when exposed to certain vapours.

If glands, which have already become brown, with their primordial utricles shrunk, are irrigated with one of the effective solution, they are not acted on, or only slightly and slowly. If, however, a gland contains merely a few coarse granules, this does not prevent a solution from acting. We have never seen any appearance making it probable that glands which have been strongly affected by absorbing matter of any kind are capable of recovering their pristine, colourless, and homogeneous condition, and of regaining the power of absorbing.

From the nature of the solution which were tried, we presume that nitrogen is absorbed by the glands; but the modified, brownish, more of less shrunk, and aggregated contents of the oblong glands were never seen by me or by my son to undergo those spontaneous changes of form characteristic of protoplasm. On the other hand, the contents of the larger spherical glands often separated into small hyaline globules or irregularly shaped masses, which changed their forms very slowly and ultimately coalesced, forming a central shrunken mass. Whatever may be the nature of the contents of several kinds of glands, after they have been acted on by foul water or by one of the nitrogenous solutions, it is probable that the matter thus generated is of service to the plant, and is ultimately transferred to other parts.

The glands apparently absorb more quickly than do the quadrifid and bifid processes; and on the view above maintained, namely that they absorb matter from putrid water occasionally emitted from the bladders, they ought to act more quickly than the processes; as these latter remain in permanent contact with captured and decaying animals.

Finally, the conclusion to which we are led by the foregoing experiments and observations that the bladders have no power of digesting animal matter, though it appears that the quadrifids are somewhat affected by a fresh infusion of raw meat. It is certain that the process within the bladders, and the glands outside, absorbed matter from salts of ammonia, from a putrid infusion of raw meat, and from urea. The glands apparently are acted on more strongly by a solution of urea, and less strongly by an infusion of raw meat, than are the processes. The case of urea is particularly interesting, because we have seen that it produces no effect on Drosera, the leaves of which are adapted to digest fresh animal matter. But the most important fact of all is, that in the present and following species the quadrifid and bifid processes of bladders containing decayed animals generally include little masses of spontaneously moving protoplasm; whilst such masses are never seen in perfectly clean bladders.

Development of the Bladders—Our observations apply to the present species and to *Utricularia vulgaris*, but were made chiefly on the latter, as the bladders are twice as large as those of *Utricularia neglecta*. In the early part of autumn the stems terminate in large buds, which fall off and lie dormant during the winter at the bottom. The young leaves forming these buds bear bladders in various stages of early development. When the bladders of *Utricularia vulgaris* are about $\frac{1}{100}$ inch (.254 mm.) in diameter (or $\frac{1}{200}$ in the case of *Utricularia neglecta*), they are circular in outline, with a narrow, almost closed, transverse orifice, leading into a hollow filled with water; but the bladders are hollow when much under $\frac{1}{100}$ of an inch in diameter. The orifices face inwards or towards the axis of the plant. At this early age the bladders are flattened in the plant in which the orifice lies, and therefore at right angles to that of the mature bladders. They are covered exteriorly with papillae of different sized, many of which have an elliptical outline. A bundle of vessels, formed of simple elongated cells, runs up the short footstalk, and divided at the base of the bladder. One branch extends up the middle of the dorsal surface, and the other up the middle of the ventral surface. In full grown bladders the ventral bundle divides close beneath the collar, and the two branches run on each side to near where the corners of the valve unite with the collar; but these branches could not be seen in very young bladders.

The accompanying figure (fig. 23) shows a section, which happened to be strictly medial, though the footstalk and between the nascent antennae of a bladder of a *Utricularia vulgaris*, $\frac{1}{100}$ inch in diameter. The specimen was soft, and the young valve became separated from the collar to a greater degree than is natural, and is thus represented. We here clearly see that the valve and collar are in folded prolongations of the wall of the bladder. Even at this early age, glands could be detected on the valve. The state of the quadrifid processes will presently be described. The antennae at this period consist of minute cellular projections (not shown in the above figure, as they do not lie in the medial plane), which soon bear incipient bristles. In five instances the

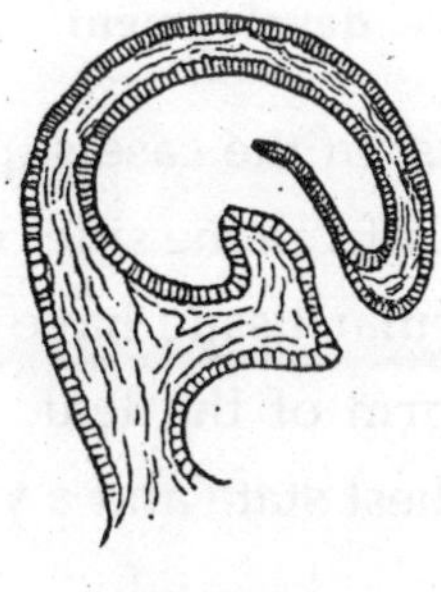

Fig. 23
(Utricularia vulgaris)
Longitudinal section through a young bladder, $\frac{1}{100}$ of an inch in length, with the orifice too widely open.

young antennae were not of quite equal length; and this fact is intelligible if we are right in believing that they represent two divisions of the leaf, rising from the end of the bladder; for, with the true leaves, whilst very young, the divisions are never strictly opposite; they must therefore be developed one after the other, and so it would be with the two antennae.

At a much earlier age, when the half formed bladders are only $\frac{1}{300}$ inch (.0846 mm.) in diameter or a little more, they present a totally different appearance. One is represented on the left side of the accompanying drawing (fig. 24). The young leaves at this age have broad flattened segments, with their future divisions represented by prominence, one of which is shown on the right side. Now, in a large number of specimens examined by my son, the young bladders appeared as if formed by oblique folding over of the apex and of one margin with a prominence, against the opposite margin. The circular hollow between the infolded apex and infolded prominence apparently contracts into the narrow orifice, wherein the valve and collar will be developed; the bladder itself being formed by the confluence of the opposed margins of the rest of the leaf. But strong objections may be urged against the view, for we must in the case suppose that the valve and collar are developed as symmetrically from the sides of the apex and prominence. Moreover, the bundles of vascular tissue have to be formed in lines quite irrespective of the original form of the leaf. Until gradations can be shown to exist between this the earliest state and a young yet preface bladder, the case must be left doubtful.

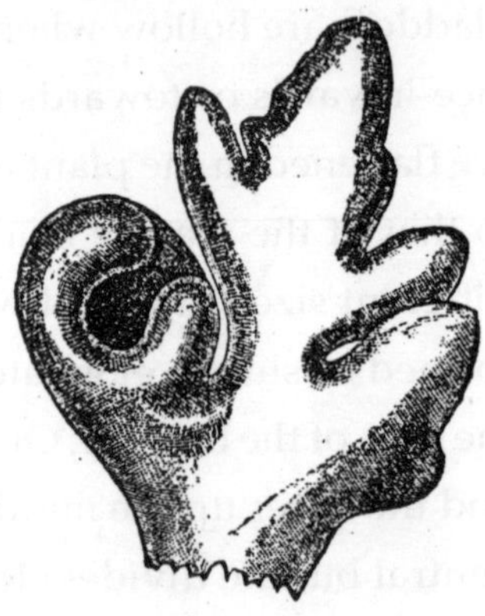

Fig. 24
(Utricularia vulgaris)
Young leaf from a winter bud, showing on the left side a bladder in its earliest stage of development.

As the quadrifid and bifid processes offer one of the greatest peculiarities in the genus, we carefully observed their development in *Utricularia neglecta.* In bladders about 1/100 in an inch in diameter the inner surface is studded with papillae, rising from small cells at the junctions of the larger ones. These papillae consist of a delicate conical protuberance, which narrows into a very short footstalk, surmounted by two minute cells. They thus occupy the same relative position, and closely resemble, except in being smaller and rather more prominent, the papillae on the outside of the bladders, and on the surfaces of the leaves. The two terminal cells of the papillae first become much elongated in a line parallel to the inner surface of the bladder. Next, each is divided by a longitudinal partition. Soon the two half-cells thus formed separate from one another; and we now have four cells or an incipient quadrifid process. As there is not space for the two new cells to increase in breadth in their original plane, the one slides partly under the other. Their manner of growth now changes, and their outer sides, instead of their apices, continue to grow. The two lower cells, which have slid partly beneath the two upper ones, form the longer and more upright pair of processes: whilst the two upper cells form the shorter and more horizontal pair; the four together forming a perfect quadrifid. A trace of the primary division between the two cells on the summits of the papillae can still be seen between the bases of the longer processes. The development of the quadrifids is very liable to be arrested. We have seen a bladder $\frac{1}{50}$ of an inch in length including only primordial papillae; and another bladder, about half its full size, with the quadrifids in an early stage development.

As the quadrifid and bifid processes offer one of the greatest peculiarities in the genus, we carefully observed their development in *Utricularia neglecta*. In bladders about $\frac{1}{100}$ papillae, rising from small cells at the junctions of the larger ones. These papillae consist of a delicate conical protuberance, which narrows into a very short footstalk, surmounted buy two minute cells. They thus occupy the same relative position, and closely resemble, except in being smaller and rather more prominent, the papillae on the outside of the bladders, and on the surfaces of the leaves. The two terminal cells of the papillae first become much elongated in a line parallel to the inner surface of the bladder. Next, each is divided by a longitudinal partition. Soon the two half-cells thus formed separate from one another; and we now have four cells or an incipient quadrifid process. As there is not space for the two new cells to increase in breadth in their original plane, the one slides partly under the other. Their manner growth now changes, and their outer sides, instead of their apices, continue to grow. The two lower cells, which have slid partly beneath the two upper ones, form the longer and more upright pair of processes; whilst the two upper cells form the shorter and more horizontal pair; the four together forming a perfect quadrifid. A trace of the primary division between the two cells on the summits of the papillae can still be seen between the bases of the longer processes. The development of the quadrifids is very liable to be arrested. We have seen bladder $\frac{1}{50}$ of an inch in length including only primordial papillae; and another bladder, about half its full size, with the quadrifids in an early stage of development.

The bifid processes are developed in the same manner as the quadrifids, excepting that the two primary terminal cells never become divided, and only increase in length. The glands on the valve and collar appear at so early an age that we could not trace their development; but we may reasonably suspect that they are developed from papillae like those on the outside of the bladder, but with their terminal cells not divided into two. The two segments forming the pedicels of the glands probably answer to the conical protuberance and short footstalk of the quadrifid and bifid processes. The glands are developed from papillae like those on the outside of the bladder, from the fact that in *Utricularia amethystine* the glands extent along the whole ventral surface of the bladder close to the footstalk.

Utricularia Vulgaris

This species differs from the last in the stems and leaves being thicker or coarser; their divisions form a more acute angle with tone another; the notches on the leaves bear three or four short bristles instead of one; and the bladders are twice as larger, or about $\frac{1}{5}$ of an inch (5.08 mm.) in diameter. In all essential respects the bladders resemble those of *Utricularia neglecta*, but the sides of the peristome are perhaps a little more prominent, and always bear, as far as we have seen, seven or eight long multicellualr bristles. There are eleven long bristles on each antenna, the terminal pairs being included. Five bladders, containing prey of some kind, were examined. The first included five Cypris, a large copepod and a Diaptomus; the second, four Cypris; the third, a single rather large crustacean; the fourth, six crustaceans; and the fifth, ten. We examined the quadrifid processes in a bladder containing the remains of two crustaceans, and found some of them full of spherical or irregularly shaped masses of matter, which were observed to move and to coalesce. These therefore consisted of protoplasm.

Utricularia Minor

The leaves and bladders are much smaller than those of *Utricularia neglecta*. The leaves bear fewer and shorter bristles, and the bladders are more globular. The antennae, instead of projecting in front of the bladders, are curled under the valve, and are armed with twelve or fourteen extremely long multicellualr bristles, generally arranged in pairs. These, with seven or eight long bristle on both sides of the peristome form a sort of net over the valve, which would tend to prevent all animals, excepting very small ones, entering the bladder. The valve and collar have the same essential structure as in the two previous species; but the glands are not quite so numerous; the oblong ones are rather more elongated, whilst the two armed ones are rather less elongated. The four bristles which project obliquely from the lower edge of the valve are short. Their shortness, compared with those on the valves of the foregoing species, is intelligible if my view is correct that they serve to prevent too large animals forcing an entrance through the valve, thus inuring it; for the valve is already protected to a certain extent by the incurved antennae, together with the lateral bristles. The bifid processes are like those in the previous species; but the quadrifids differ in the four arms (fig. 25) being directed to the same side; the two longer ones being central, and the two shorter ones on the outside.

Fig. 25
(Utricularia minor)
Quadrifid process; greatly enlarged.

The plants were collected in the middle of July; and the contents of five bladders, which from their opacity seemed full of prey were examined. The first contained no less than twenty-four minute fresh-water crustaceans, most of them consisting of empty shells, or including only a few drops of red oily matter; the second contained twenty; the third, fifteen; the fourth, ten, some of them being rather larger then usual; an the fifth, which seemed stuffed quite full, contained only seven, but five of these were of unusual large size. The prey, therefore, judging from these five bladders, consists exclusively of fresh-water crustaceans, most of which appeared to be distinct species from those found in the bladder of the two former species. In one bladder the quadrifids in contact with a decaying mass contained numerous spheres of granular matter, which slowly changed their forms and positions.

Utricularia Clandestina

This North American species, which is aquatic like the three foregoing ones, appears to be lined with quadrifid processes. A vast number of captured animals were found within the bladders; some being crustaceans, but the greater number delicate, elongated larvae, we suppose of Culicidae. On some stems, "fully nine out of every ten bladders contained these larvae or their remains". The larvae "showed signs of like from twenty-four to thirty-six hours being imprisoned", and then perished.

❑❑❑

18

Utricularia-II

Utricularia montana—Description of the bladders on the subterranean rhizomes—Prey captured by the bladders of plants under culture and in a state of nature—Absorption by the quadrifid processes and glands—Tubers serving as reservoirs for water—Various other species of Utricularia—Polypompholyx—Genlisea, different nature of the trap for capturing prey—[Sarracenia] Diversified methods by which plants are nourished.

Utricularia Montana

Description of Bladders—This species inhabits the tropical parts of south America, and is said to be epiphytic; but, judging from the state of the roots (rhizomes) of some dried specimens from the herbarium at Kew, it likewise lives in earth, probably in crevices of rocks. In English hot-houses it is grown in peaty soil. The leaves are entire, instead of being much divided, as in the foregoing aquatic species. They are elongated, about 1/½ inch in breadth, furnished with a distinct footstool. The plant produces numerous colourless rhizomes, as thin as thread, which bear minute bladders, and occasionally swell into tubers, as will hereafter be decribed. These rhizomes appear exactly like roots, but occasionally throw up green shoots. They penetrate the earth sometimes to the depth of more than 2 inches: but when the plant grows as an epiphyte, they must creep amidst the mosses, roots, and the trees of these countries are thickly covered.

As the bladders are attached to the rhizomes, they are necessarily subterranean. They are produced in extraordinary numbers. One of my plants, though young must have borne several hundreds; for a single branch out of an entangled mass had thirty-two, and another branch about 2 inches in length (but with its end and one side branch broken off), had seventy three bladders.

The bladders are compressed and rounded, with the ventral surface, or that between the summit of the long delicate footstalk and valve, extremely short (fig. 27). They are colourless and almost as transparent as glass, so that they appear smaller than they really are, the largest being under the $\frac{1}{20}$ of an inch (1.27 mm.) in its longer diameter. They are formed of rather large angular cells, at the junctions of which oblong papillae project, corresponding with those on the surfaces of the bladders of the previous species. Similar papillae abound on the rhizomes, and even on the entire leaves, but they are rather broader on the latter. Vessels, marked with parallel bars instead of by a spiral line, run up the footstalks, and just enter the bases of the bladders; but they do not bifurcate and extend up the dorsal and ventral surfaces, as in the previous species.

Fig. 26
(Utricularia Montana)
Rhizome swollen into a tuber; the branches bearing minute bladders; of natural size.

The antennae are of moderate length, and taper to a fine point; they differ conspicuously from those before described, in not being armed with bristles. Their bases are so adrupty curved that their tips generally rest one on each margin. Their curved bases thus form a roof over the cavity in which the valve lies; but there is always left on each side a little circular passage into the cavity, as may be seen in the drawing, as well as a narrow passage between the bases of the two antennae. As the bladders are subterranean, had it not been for the roof, the cavity in which the valve lies would have been liable to be blocked up with earth and rubbish; so that the curvature of the antennae is a serviceable character. There are no bristles on the outside of the collar or peristome, as in the foregoing species.

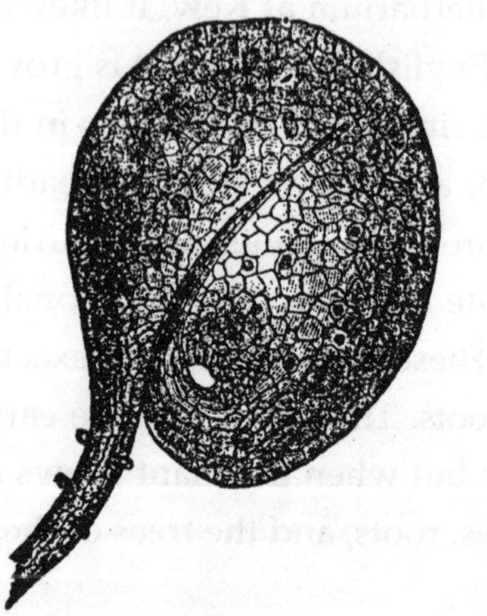

Fig. 27
(Utricularia Montana)
Bladder; about 27 times enlarged.

The valve is small and steeply inclined, with its free posterior edge abutting against a semicircular, deeply depending collar. It is moderately transparent, and bears two pairs of short stiff bristles, in same position as in the other species. The presence of these four bristles, in contrast with the absence of those on the antennae and collar, indicates that they are of functional importance, namely, to prevent too large animal forcing an entrance through the valve. The many glands of diverse shapes attached to the valve and round the collar in the previous species are here absent, with the exception of about a dozen of the two-armed or transversely elongated kind, which are seated near the borders of the valve, and are mounted on very short footstalks. These glands are only the $\frac{3}{4000}$ of an inch (.0191 mm.) in length; though so small, they act as absorbents. The collar is thick, stiff and almost semicircular; it is formed of the same peculiar brownish tissue as in the former species.

The bladders are filled with water, and sometimes include bubbles of air. They bear internally rather short, thick, quadrifid processes arranged in approximately concentric row. The two pairs of arms of which they are formed differ only a little in length, and stand in a peculiar position (fig. 28); the two longer ones forming one line, and the two shorter ones another parallel line. Each arm includes a small spherical mass of brownish matter, which, when crushed, breaks into angular pieces. We have no doubt that these spheres are nuclei, for closely similar ones are present in the cells forming the walls of the bladder. Bifid processes, having rather short oval arms, arise in the usual position of the inner side of the collar.

Fig.. 28
(Utricularia Montana)
One of the quadrifid processes; much enlarged.

These bladders, therefore, resemble in all essential respects the larger ones of the foregoing species. They differ chiefly in the absence of the numerous glands on the valve and round the collar, a few minutes ones of one kind alone being present on the valve. They differ more conspicuously in the absence of the long bristles on the antennae and on the outside of the collar. The presence of these bristles in the previously mentioned species probably relates to the capture of aquatic animals.

It seemed to me an interesting question whether the minute bladders of *Utricularia montana* served, as in the previous species, to capture animals living in the earth, or in the dense vegetation covering the trees on which this species is epiphytic; for in this case we should have a new sub-class of carnivorous plants, namely, subterranean feeders. Many bladders, therefore, were examined, with the following results:

1. A small bladder, less than $\frac{1}{3}$ of an inch (.847 mm.) in diameter contained a minute mass of brown, much decayed matter; and in this, a tarsus with four or five joints, terminating in a double hook, was clearly distinguished under the microscope. We suspect that it was a remnant of one of the Thysanoura. The quadrifids in contact with this decayed remnant contained either small mass of translucent, yellowish matter, generally more or less globular, or fine granules. In distant parts of the same bladders, the processes were transparent and quite empty, with the exception of their solid nuclei. We made at short intervals of time sketches of one of the above aggregated masses, and found that they continually and completely changed their forms; sometimes separating from one another and again coalescing. Evidently protoplasm had been generated by the absorption of some element from the decaying animal matter.
2. Another bladder included a still smaller speck of decayed brown matter, and the adjoining quadrifids contained aggregated matter, exactly as in the last case.
3. A third bladder included larger organism, which was so much decayed that we could only make out that it was spinose or hairy. The quadrifids in this case were not much affected, excepting that the nuclei in the several arms differed much in size; some of them containing two masses having a similar appearance.
4. A fourth bladder contained an articulate organism, for we distinctly saw the remnant of limb, terminating in hook. The quadrifids were not examined.
5. A fifth included much decayed matter apparently of some animal, but with no recognisable features. The quadrifids in contact contained numerous spheres of protoplasm.
6. Some few bladders on the plant were examined; and in one, there was worm-shaped animal very little decayed, with a distinct remnant of a similar one greatly decayed. Several of the arms of the processes in contact with these

remains contained two spherical masses, like the single solid nucleus, which is properly found in each arm. In another bladder there was a minute grain of quartz, reminding me of two similar cases with *Utricularia neglecta*.

As it appeared probable that this plant would capture a greater number of animals in its native country than under capture, we did not at first find out that it was advisable to soak the rhizomes for two or three days, and that it was necessary to open the bladders and spread out their contents on glass: as from their state of decay and from having been dried and pressed, their nature could not otherwise be well distinguished. Several bladders on a plant which had grown in black earth were first examined; and four of these included remnants of animals. The first contained a hairy Acarus, so much decayed that nothing was left except its transparent coat; also a yellow chitinous head of some animal with an internal fork, to which the oesophagus was suspended, but we could see no mandibles; also the double hook of the tarsus of some animal; also an elongated greatly decayed animal; and lastly, a curious flask-shaped organism, having the walls formed of rounded cells. Professor Claus has looked at this latter organism, and thinks that it is the shell of a rhizopod, probably one of the arcellidae. In this bladder, as well as in several others, there were some unicellular algae, and one multicellualr Alga, which no doubt had lived as intruders.

A second bladder contained an Acarus much less decayed than the former one, with its eight legs preserved; as well as remnants of several other articulate animals. A third bladder contained the end of the abdomen with the two hinder limbs of an Acarus. A fourth contained remnants of a distinctly articulated bristly animal, and of several other organisms, as well as much dark brown organic matter, the nature of which could not be made out.

Some bladders from a plant, which had lived as an epiphyte in Trinidad, in the West Indies, were next examined, but not so carefully as the other; nor had they been soaked long enough. Four of them contained much brown, translucent granular matter, apparently organic, with no distinguishable parts. The quadrifids in two were brownish, with their contents granular; and it was evident that they had absorbed matter. In a fifth bladder there was a flask-shaped organism, like that above mentioned. A sixth contained a very long, much decayed, worm-shaped animal. Lastly, a seventh bladder contained an organism, but of what nature could not be distinguished.

Only one experiment was tried on the quadrifid processes and glands with reference to their power of absorption. A bladder was punctured and left for 24 hrs. in a solution of one part of urea to 437 of water, and the quadrifid and bifid processes were found much affected. In some arms there was only a single symmetrical globular mass, larger than the proper nucleus, and consisting of yellowish matter, generally translucent but sometimes granular; in others there were two masses of different sizes, one large and the other small; and in others there were irregularly shaped globules; so that it appeared as if the limpid contents of the processes, owing to the absorption of matter from the solution, had become aggregated sometimes round the nucleus, and sometimes into separate masses; and that these then tended to coalesce. The primordial utricle or protoplasm lining the processes was also thickened here and there into irregular and variously shaped specks of yellowish translucent matter, as occurred in the case of *Utricularia neglecta* similar treatment. These specks apparently did not change their forms.

The minute two-armed glands on the valve were also affected by the solution; for they now contained several, sometimes as many as six or eight, almost spherical masses of translucent matter, tinged with yellow, which slowly changed their forms and position. Such masses were never observed in these glands in their ordinary state. We may therefore infer that they serve for absorption. Whenever a little water is expelled from a bladder containing animal remains (by the means formerly specified, more especially by the generation of bubbles of air), it will fill the cavity in which the valve lies; and thus the glands will be able to utilize decayed matter which otherwise would have been wasted.

Finally, as numerous minute animals are captured by this plant in its native country and when cultivated, there can be no doubt that the bladders, though so small, are far from being in a rudimentary condition; on the contrary, they are highly efficient traps. Nor can there be any doubt that matter is absorbed from the decayed prey by the quadrifid and bifid processes, and that protoplasm is thus generated. What tempts animals of such diverse kinds to enter the cavity beneath the bowed antennae, and then force their way through the little slit-like orifice between the valve and collar into the bladders filled with water.

Tubers—These organs, one of which is represented in a previous figure (fig. 26) of the natural size, deserve a few remarks. Twenty were found on the rhizomes of single plant, but they cannot be strictly counted; for, besides the twenty, there were all possible gradations between a short length of a rhizome just perceptibly swollen and one so much swollen that it might be doubtfully called a tuber. When well developed, they are oval and symmetrical, more so than appears in the figure. The largest which we saw was 1 inch (25.4 mm.) in length and .45 inch (11.43 mm.) in breadth. They commonly lie near the surface, but some are buried at the depth of 2 inches. The buried ones are dirty white, but those partly exposed to the light become greenish from the development of chlorophyll in their superficial cells. They terminate in a rhizome, but this sometimes decays and drops off. They do not contain any air, and they sink in water; their surfaces are covered with the usual papillae. The bundle of vessels which runs up each rhizome, as soon as it enters the tuber, separates into three distinct bundles, which reunite at the opposite end. A rather thick slice of tuber is almost as transparent as glass, and is seen to consist of large angular cells, full of water and not containing starch or any other solid matter. Some slices were left in alcohol for several days, but only a few extremely minute granules of matter were precipitated on the walls of the cells; and these were much smaller and fewer then those precipitated on the cell-walls of the rhizomes and bladders. We may therefore conclude that the tubers do not serve as reservoirs for food, but for water during the dry season to which the plant is probably exposed. The many little bladders filled with water would aid towards the same end.

To test the correctness of this view, a small plant, growing in light peaty earth in a pot (only 4½ by 4½ inches outside measure) was copiously watered, and then kept without a drop of water in the hot-house. Two of the upper tubers were beforehand uncovered and measured, and then loosely covered up again. In a fortnight's time the earth in the pot appeared extremely dry; but not until the thirty-fifth day were the leaves in the least affected; they then became slightly reflexed, though still soft and green. This plant, which bore only ten tubers, would no doubt have resisted the drought for even a longer time, had we not previously removed three of the tubers and cut off several long rhizomes. When, on the thirty-fifth day, the earth in the pot was turned out, it appeared as dry as the dust on the road. All the tubers had their surfaces much wrinkled, instead of being smooth and tense. They had all shrunk, but we cannot say accurately how much; for as they were at first symmetrically oval, we measured only their length and thickness; but they contracted in a transverse line much more in one direction than in another, so as to become greatly flattened. One of the two tubers which had been measured was now three-fourths of its original length, and two-thirds of its original thickness in the direction in which it had been measured, but in another direction only one-third shorter, one-eighth less thick in the direction in which it had been measured, and only half as thick in another direction.

A slice was cut from one of these shrivelled tubers and examined. The cells still contained much water and no air, but they were more rounded or less angular than before, and their walls not nearly so straight; it was therefore clear that the cells had contracted. The tubers, as long as they remain alive, have a strong attraction for water; the shrivelled one, from which a slice had been cut, was left in water for 22 hrs. 30 m., and its surfaces became a smooth and tense as it originally was. On the other hand, a shrivelled tuber, which by some accident had been separated from its rhizome, and which appeared dead, did not swell in the least, though left for several days in water.

With many kinds of plants, tubers, bulbs, &c., no doubt serve in part as reservoirs for water. Prof. Oliver informed that two or three other species of Utricularia are provided with these appendages; and the group containing them has in consequence received the name of *orchidioides*. All the other species of Utricularia, as well as of certain closely related genera, are either aquatic or marsh plants generally having a similar constitution, a never-failing supply of water would probably be of great importance to our present species. We can thus understand the meaning of the development of its tubers, and of their number on the same plant, amounting in one instance to at least twenty.

Utricularia Nelumbifolia, AmetHystina, Griffithii, Caerulea, Orbiculata, Multicaulis [Cornuta]

Utricularia nelumbifolia – The habitat of this species is remarkable. According to its discoverer, Mr. Gardner, it is aquatic, but "is only to be found growing in the water which collects in the bottom of the leaves of a large Tillandsia, that inhabits abundantly an arid rocky part of the mountain, at an elevation of about 5000 feet above the level of the sea. Besides the ordinary method by seed, it propagates itself by runners, which it throws cut from the base of the flower-stem; this runner is always found directing itself towards the nearest Tillandsia, when it inserts its point into the water and gives origin to a new plant, which in its turn sends out another shoot. In this manner we have seen not less than six plants united." The bladders resemble those of *Utricularia monotana* in all essential respect, even to the presence of a few minute two-armed glands on the valve. Within one bladder there was the remnant of the abdomen of some larva or crustacean of large size, having a brush of long sharp bristles at the apex. Other bladders included fragments of articulate animals, and many of them contained broken pieces of a curious organism, the nature of which was not recognised by any one of whom it was shown.

Utricularia amethystina — This species had small entire leaves, and is apparently a marsh plant; but it must grow in places where crustaceans exist, for there were two small species within one of the bladders. The bladders are nearly of the same shape as those of *Utricularia montana,* and are covered outside with the usual papillae but they differ remarkably in the antennae being reduced to two short points, united by a membrane hollowed out in the middle. This membrane is covered with innumerable oblong glands supported on long footstalks; most of which are arranged in two rows covering towards the valve. Some, however, are scatted on the margins of the membrane; and the short ventral surface of the bladder, between the petiole and valve, is thickly covered with glands. Most of the heads had fallen off, and the footstalks alone remained; so that the ventral surface and the orifice, when viewed under a weak power, appeared as if clothed with fine bristles. The valve is narrow, and bears a few almost sessile glands. The collar against which the edge shuts is yellowish, and presents the usual structure. From the large number of glands on the ventral surface and round the orifice, it is probable that this species lives in very foul water, from which it absorbs matter, as well as from its captured and decaying prey.

Utricularia griffithii — The bladders are transparent and minute; one which was measured being only $\frac{28}{1000}$ of an inch (.711 mm.) in diameter. The antennae are of moderate length, and project straight forward; they are united for a short space at their bases by a membrane; and they bear a moderate number of bristles or hairs, not simple as heretofore, but surmounted by glands. The bladders also differ remarkably from those of the previous species, as within there are no quadrifid, only bifid processes. In one bladder there was a minute aquatic larva; in another the remains of some articulate animal; and in most of them grains of sand.

Utricularia caerulea — The bladders resemble those of the last species, both in the general character of the antennae and in the processes within being exclusively bifid. They contained remnants of entomostracan crustaceans.

Utricularia orbiculata — The orbicular leaves and the stems bearing the bladders apparently float in water. The bladders do not differ much from those of the two last species. The antennae, which are united for a short distance at their bases, bear on their outer surfaces and summits numerous, long, multicellular hairs, surmounted by glands. The processes within the bladders are quadrifid, with the four diverging arms of equal length. The prey which they had captured consisted of entomostracan crustaceans.

Utricularia multicaulis — The bladders, attached to rhizomes, are remarkable from the structure of the antennae. These are broad, flattened, and of large size; they bear on their margins multicellular hairs, surmounted by glands. Their bases are united into a single, rather narrow pedicel, and they thus appear like a great digitate expansion at one end of the bladder. Internally the quadrifid processes have divergent arms of equal length. The bladders contained remnants of articulate animals.

***Utricularia cornuta*, Michx.** — This species has been studied by A. Schimper in America, and is the subject of a short paper in the 'Botanische Zeitung.' It grows in swampy ground, and presents a remarkable appearance; the aerial part of the plant seems at first sight to consist of nothing but almost naked flower-stems a foot in height, bearing from two to five large yellow flowers. U. *Cornuta* has no roots, its underground stem or rhizome is much branched and bears numerous minute bladders. The breaches of the rhizome throw up here and there grass-like leaves, which cover the ground without having any apparent connection with the flower-stem. The structure of the bladders is not in any way remarkable, resembling in its general features that of the European species. The bladders generally contain organic remains; out of 114 only 11 contained no debris. The contents include diatoms and small animals—worms, rotifers, small crustaceans; and the hairs lining the inside of the bladders give evidence of having absorbed matter from the decaying mass.

Polypompholyx

This genus, which is confined to Western Australia, is characterized by having a "quadripartite calyx." In other respects, as Prof. Oliver remarks, "it is quite a Utricularia."

Polypompholyx multifida—The bladders are attached in whorls round the summits of stiff stalks. The two antennae are represented by a minute membranous fork, the basal part of which forms a sort of hood over the orifice. This hood expands into two wings on each side of the bladder. A third wing or crest appears to be formed by the extension of the dorsal surface of the petiole; but the structure of these three wings could not be clearly made out, owing to the state of the specimens. The inner surface of the hood is lined with long simple hairs, containing aggregated matter, like that within the quadrifid processes or the previously described species when in contact with decayed animals. These hairs appear therefore to serve as absorbents. A valve was seen, but its structure could not be determined. On the collar round the valve there are in the place of glands numerous one-celled papillae, having very short footstalks. The quadrifid processes have divergent arms of equal length. Remains of entomostracan crustaceans were found within the bladders.

Polypompholyx tenella—The bladders are smaller than those of the last species, but have the same general structure. They were full of debris, apparently organic, but no remains of articulate animals could be distinguished.

Genlisea

This remarkable genus is technically distinguished from Utricularia, as it has a five partite calyx. Species are found in several parts of the world, and are said to be "herbae annuae paludosae."

Genlisea ornata—This species has been described and figured by Dr. Warming, who states that it bears two kinds of leaves, called by him spathulate and utriculiferous. The latter include cavities; and as these differ much from the bladders of the foregoing species, it will be convenient to speak of them as utricles. The accompanying figure (fig. 29) of one of the utriculiferous leaves, about thrice enlarged, will illustrate the following description by my son, which agrees in all essential points with that given by Dr. Warming. The Utricle *(b)* is formed by a slight enlargement of the narrow blade of the leaf. A hollow neck *(n)*, no less than fifteen times as long as the utricle itself, forms a passage from the transverse slit-like orifice *(o)* into the cavity of the utricle. A utricle which measured $\frac{1}{36}$ of an inch (.705mm.) in its longer diameter had a neck $\frac{15}{36}$ of an inch (10.538 mm.) in length, and $\frac{1}{100}$ of an inch (.254 mm.) in breadth. On each side of the orifice there is a long spiral arm or tube *(a)*; the structure of which will be best understood by the following illustration. Take a narrow ribbon and wind it spirally round a thin cylinder, so that the edges come into contact along its whole length; then pinch up the two edges so as to form a little crest, which will of course wind spirally round the cylinder like a thread round a screw. If the cylinder is now removed, we shall have a tube like one of the spiral arms. The two projecting edges are not actually united, and a needle can be pushed in easily between them. They are indeed in many placed little separated, forming narrow entrances into the tube; but this may be the result of the drying of the specimens. The lamina of which the tube is formed seems to be a lateral prolongation of the lip of the orifice; and the spiral line between the two projecting edges is continuous with the corner of the orifice. If a fine bristle is pushed down one of the arms, it passes

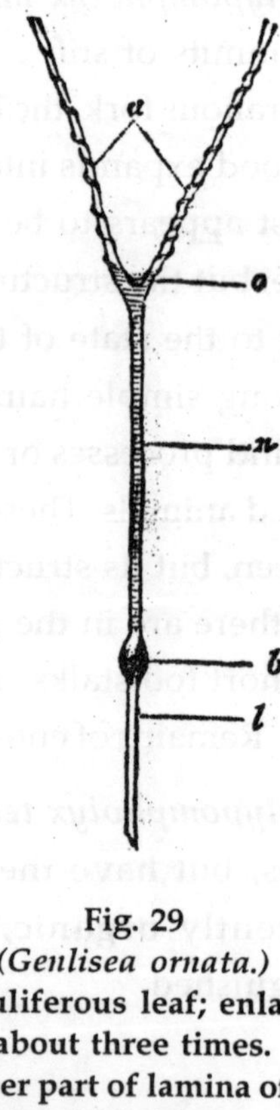

Fig. 29
(Genlisea ornata.)
Utriculiferous leaf; enlarged about three times.

- *l* **Upper part of lamina of leaf.**
- *b* **Utricle of bladder.**
- *n* **Neck of utricle.**
- *o* **Orifice.**
- *a* **Spirally wound arms, with their ends broken off.**

into the top of the hollow neck. Whether the arms are open or closed at their extremities could not be determined, as all the specimens were broken; nor does it appear that Dr. Warming ascertained this point.

So much for the external structure. Internally the lower part of the utricle is covered with spherical papillae, formed of four cells (sometimes eight according to Dr. Warming), which evidently answer to the quadrifid processes within the bladders of Utricularia. These papillae extend a little way up the dorsal and ventral surfaces of the utricle; and a few, according to Warming, may be found in the upper part. This upper region is covered by many transverse rows, one above the other, of short, closely approximate hairs, pointing downwards. These hairs have broad bases, and their tips are formed by a separate cell. They are absent in the lower part of the utricle where the papillae abound. The neck is likewise lined throughout its whole length with transverse rows of long, thin, transparent hairs, having broad bulbous (fig. 30) bases, with similarly constructed sharp points. They arise from little projecting ridges, formed of rectangular epidemic cells. The hairs very a little in length, but their points generally extent down to the row next below; so that if the neck is split open and laid flat, the inner surface resembles a paper of pins—the hairs representing the pins, and the little transverse ridges representing the folds of paper through which the pins are thrust. These rows of hairs are indicated in the previous figure (29) by numerous transverse lines crossing the neck. The inside of the neck is also studded with pappillae; those in the lower part are spherical and formed of four cells, as in the lower part of the utricle; those in the upper part are formed of two cells, which are much elongated downwards beneath their points of attachment.

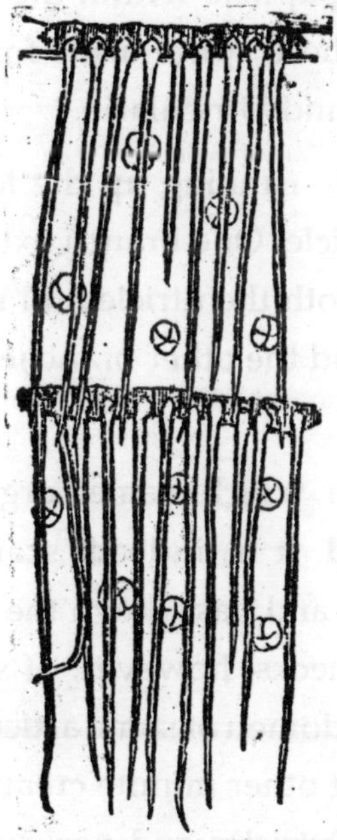

Fig. 30
(Genlisea ornata.)
Portion of inside of neck leading into the utricle, greatly enlarged, showing the downward pointed bristles, and small quadrifid cells or processes.

These two-celled papillae apparently correspond with the bifid process in the upper part of the bladders of Utricularia. The narrow transverse orifice (*o*, fig. 29) is situated between the bases of the two spiral arms. No valve could be detected here, nor was any such structure seen by Dr. warming. The lips of the orifice are armed with many short, thick, sharply pointed somewhat incurved hairs or teeth.

The two projecting edges of the spirally wound lamina, forming the arms, are provided with short incurved hairs or teeth, exactly like those on the lips. These project inwards at right angles to the spiral line of junction between the two edges. The inner surface of the lamina supports two celled, elongated papillae, resembling those in the upper part of the neck, but differing slightly form them, according to warming, in their footstalks being formed by prolongations of large epidemic cells; whereas the papillae within the neck rest on small cells sunk amidst the larger ones. These spiral arms form a conspicuous difference between the present genus and Utricularia.

Lastly, there is a bundle of spiral vessels which, running up the lower part of the linear leaf, divides close beneath the utricle. One branch extends up the dorsal and the other up the ventral side of both the utricle and neck. Of these two branches, one enters one spiral arm, and the other branches the other arm.

The utricles contained much debris or dirty matter, which seemed organic, though no distinct organisms could be recognised. It is, indeed, scarcely possible that any object could enter the small orifice and pass down the long narrow neck, except a living creature. Within the necks, however, of some specimens, a worm with retracted horny jaws, the abdomen of some articulate animal, and specks of dirt, probably the remnants of other minute creatures, were found. Many of the papillae within both the utricles and necks were discoloured, as if they had absorbed matter.

From this description it is sufficiently obvious how Genlisea secures its prey. Small animals entering the narrow orifice – but what induces them to enter is not known any more than in the case of Utricularia – would find their egress rendered difficult by the sharp incurved hairs on the lips, and as soon as they passed some way down the neck, it would be scarcely possible for them to return, owing to the many transverse rows of long, straight, downward pointing hairs, together with the ridges from which these project. Such creatures would, therefore, perish either within the neck or utricle; and the quadrifid and bifid papillae would absorb matter from their decayed remain. The transverse rows of hairs are so numerous that they seem superfluous merely for the sake of preventing the escape of prey, and as they are thin and delicate, they probably serve as additional absorbents, in the same manner as the flexible bristles on the infolded margins of the leaves of Aldrovanda. The spiral arms no doubt act as accessory traps. Until fresh leaves are examined, it cannot be told whether the line of junction of the spirally wound lamina is a little open along its whole course, or only in parts, but a small creature which forced its way into the tube at any point, would be prevented from escaping by the incurved hairs, and would find an open path down the tube into the neck, and so into the utricle. If the creature perished within the spiral arms, its decaying remains would be absorbed and utilized by the bifid papillae. We thus see that animals are captured Genlisea, not by means of an elastic valve, as with the foregoing species, but by a contrivance resembling an eel-trap, though more complex.

Genlisea africana — Fragments of the utriculiferuos leaves of this species exhibited the same structure as those of *Genlisea ornata*. A nearly perfect Acarus was found within the utricle or neck of one leaf, but in which of the two was not recorded.

Genlisea aurea — A fragment of the neck of a utricle was lined with transverse rows of hairs, and was furnished with elongated papillae, exactly like those within the neck of *Genlisea ornata*. It is probable, therefore, that the whole utricle is similarly constructed.

Genlisea filiformis — Many leaves were examined and none were found provided with utricles, whereas such leaves were found without difficulty in the three previous species. On the other hand, the rhizomes bear bladders resembling in essential character those on the rhizomes of Utricularia. These bladders are transparent, and very small, *viz.*, only $\frac{1}{100}$ of an inch (.254 mm.) in length. The antennae are not united at their bases, and apparently bear some long hairs. On the outside of the bladders there are only a few papillae, and internally very few quadrifid processes. These latter, however, are of unusually large size, relatively to the bladder, with the four divergent arms of equal length. No prey could be seen within these minute bladders. As the rhizomes of this species were furnished with bladders, those of *Genlisea africana, ornata,* and *aurea* were carefully examined, but none could be found. What are we to infer from these facts? Did the three species just named, like their close allies, the several species of Utricularia, aboriginally possess bladders on their rhizomes, which they afterwards lost, acquiring in their place utriculiferous leaves? In support of this view it may be urged that the bladders of *Genlisea filiformis* appear from their small size and from the fewness of their quadrifid processes to be tending towards abortion; but why has not this species acquired utriculiferous leaves, like its congeners?

Conclusion

It has now been that many species of Utricularia and of two closely allied Genera, inhabiting the most distant parts of the world – Europe, Africa, India, the Malay Archipelago, Australia, North and South America – are admirably adapted for capturing by two methods small aquatic or terrestrial animals, and that they absorb the products of their decay.

Ordinary plants of the higher classes procure the requisite inorganic elements from the soil by means of their roots, and absorb carbonic acid from the atmosphere by means of their leaves and stems. but we have seen in a previous part of this work that there is class of plants which digest and afterwards absorb animal mater, namely, all the Droseraceae, Pinguicula, and as discovered by Dr. Hooker, Nephethes, and to this class other species will almost certainly soon be added. These plants can dissolve matter out of certain vegetable substances, such as pollen, seeds, and bits of leaves. No doubt their glands likewise absorb the salts of ammonia brought to them by the rain. It has also been shown that some other plants can absorb ammonia of their glandular hairs; and these will profit by that brought to them by the rain. There is a second class of plants which, as we have just seen, cannot digest, but absorb the products of the decay of the animals which they capture, namely, Utricularia and its close allies; and from the excellent observations of Dr. Mellichamp and Dr. Canby, there can scarcely be a doubt that Sarracenia and Darlingtonia may be added to this class.

A. Schimper, in an interesting paper, give evidence that the products of decay are absorbed by the pitchers of *Sarracenia purpurea*. In the epidemic cells at the base of the pitcher the changes produced by presence of decaying animal matter are strikingly evident, and bear a strong resemble to the process of aggregation as seen in Drosera. The cell-sap is rich in tannin (as in Drosera), and when aggregation takes place the single vacuole containing the cell-sap is replaced by several highly refractive drops. The process resembles in fact of division and concentration of the vacuole as described by Dr. Vries. Schimper supposes that the cell-sap gives up to the protoplasm part of its water, and he describes the concentrated, tannin-containing drops which are thus formed, as lying the swollen watery protoplasm which now takes up more space than in the unstimulated condition. Schimper's paper also contains a good general description of the pitchers of Sarracenia.

There is a third class of plants which feed, as is now generally admitted, on the products of the decay of vegetable matter, such as the bird's-nest orchis (Neottia), &c. Lastly, there is the well-known class of parasites (such as the mistletoe), which are nourished by the juices of living plants. Most, however, of the plants belonging to these four classes obtain part of their carbon like ordinary species, from the atmosphere. Such are the diversified means, by which higher plants gain their subsistence.

❑❑❑

Index

□□□